ADVANCES IN MATERIALS RESEARCH 2

Physics and Astronomy ONLINE LIBRARY

http://www.springer.de/phys/

Springer-Verlag Berlin Heidelberg GmbH

ADVANCES IN MATERIALS RESEARCH

The series Advances in Materials Research reports in a systematic and comprehensive way on the latest progress in basic materials sciences. It contains both theoretically and experimentally oriented texts written by leading experts in the field. Advances in Materials Research is a continuation of the series Research Institute of Tohoku University (RITU).

1 **Mesoscopic Dynamics of Fracture**
Computational Materials Design
Editors: H. Kitagawa, T. Aihara, Jr., and Y. Kawazoe

2 **Advances in Scanning Probe Microscopy**
Editors: T. Sakurai and Y. Watanabe

3 **Amorphous and Nanocrystalline Materials**
Preparation, Properties, and Applications
Editors: A. Inoue and K. Hashimoto

Series homepage – http://www.springer.de/phys/books/amr/

T. Sakurai Y. Watanabe (Eds.)

Advances in Scanning Probe Microscopy

With 241 Figures

Springer

Professor Toshio Sakurai
Professor Yousuke Watanabe
Institute of Materials Research, Tohoku University
2-1-1 Katahira, Aoba-ku, Sendai 980-8577, Japan

Series Editor-in-Chief:

Professor Yoshiyuki Kawazoe
Institute of Materials Research, Tohoku University
2-1-1 Katahira, Aoba-ku, Sendai 980-8577, Japan

Series Editors:

Professor Masayuki Hasegawa
Professor Akihisa Inoue
Professor Norio Kobayashi
Professor Toshio Sakurai
Institute of Materials Research, Tohoku University
2-1-1 Katahira, Aoba-ku, Sendai 980-8577, Japan

Professor Luc Wille
Department of Physics, Florida Atlantic University
777 Glades Road, Boca Raton, FL 33431, USA

Library of Congress Cataloging-in-Publication Data: Advances in scanning probe microscopy/[edited by] T. Sakurai, Y. Watanabe. p. cm. – (Advances in materials research; 2) Includes bibliographical references. ISBN 3540667180 (alk. paper) 1. Scanning probe microscopy. I. Sakurai, T. (Toshio), 1945– II. Watanabe, Y. (Yousuke), 1938– III. Series. QH212.S33 A38 2000 621.3815'2–dc21 99-088035

ISSN 1435-1889
ISBN 978-3-642-63084-2 ISBN 978-3-642-56949-4 (eBook)

DOI 10.1007/978-3-642-56949-4

Originally published by Springer-Verlag Berlin Heidelberg New York in 2000
Softcover reprint of the hardcover 1st edition 2000

Typesetting: Data conversion by EDV-Beratung Frank Herweg, Hirschberg
Cover concept: eStudio Calamar Steinen
Cover design: *design & production*, Heidelberg

SPIN: 10747311 57/3144/ba - 5 4 3 2 1 0

Preface of Series Editor-in-Chief

The present book is the second volume of the Springer Series 'Advances in Materials Research', which is edited by scientists form the Institute of Materials Research (IMR) of Tohoku University. The series began with the book 'Mesoscopic Dynamics of Fracture' and further volumes are planned to appear at regular intervals. Together these books will provide high-quality reviews of topical subjects from the very active field of materials science. The aims and scope of this second volume 'Advances in Scanning Probe Microscopy' are described in the preface of the volume editor.

As series editor, I would like to express my sincere thanks to Dr. Claus Ascheron of Springer Verlag for his kind encouragement and continuing support.

Sendai, November 1999 *Yoshiyuki Kawazoe*

Preface

There have been many books published on scanning tunneling microscopy (STM), atomic force microscopy (AFM) and related subjects since Dr. Gerd Binnig and Dr. Heinrich Rohrer invented STM in 1982 and AFM in 1986 at IBM Research Center in Zurich, Switzerland. These two techniques, STM and AFM, now form the core of what has come to be called the 'scanning probe microscopy (SPM)' family. SPM is not just the most powerful microscope for scientists to image atoms on surfaces, but is also becoming an indispensable tool for manipulating atoms and molecules to construct man-made materials and devices. Its impact has been felt in various fields, from surface physics and chemistry to nano-mechanics, nano-electronics and medical science. Its influence will surely extend further as the years go by, beyond the reach of our present imagination, and new research applications will continue to emerge.

This book, therefore, is not intended to be a comprehensive review or textbook on SPM. Its aim is to cover only a selected part of the active research fields of SPM and related topics in which I have been directly involved over the years. These include the basic principles of STM and AFM, and their applications to fullerene film growth, SiC surface reconstructions, MBE (molecular beam epitaxy) growth of GaAs, atomic scale manipulation of Si surfaces and mesoscopic work function. I sincerely hope that readers will find this a stimulating and interesting compilation of representative work in the field, and one which may point the way towards further new applications of scanning probe microscopy.

During the preparation of this book and, indeed, throughout the research undertaken over the last ten years, I have benefited from the help of many associates and colleagues. I would like to especially acknowledge the great assistance and support provided by Dr. T. Hashizume, Dr. A. Sakai, Dr. L. Li, Dr. Q-K. Xue, and Dr. Y. Hasegawa. I also would like to thank Dr. Y. Watanabe for his help in compiling this book in its present form.

I consider myself immensely fortunate to have had the opportunity to work in this exciting field and to be part of its continuing rapid development, a circumstance made possible by the ingenuity of Heinrich Rohrer and Gerd Binnig.

Finally, I would like to express my deep gratitude to Dr. Yoshiyuki Kawazoe, editor of this book series, for his original suggestion and encouragement to prepare this book.

Tohoku, November 1999 *Toshio Sakurai*

Contents

List of Contributors

Y. Hasegawa
Institute for Materials Research
Tohoku University 2–1–1
Katahira
Aoba-ku
Sendai 980-8577
Japan

T. Hashizume
Advanced Research Laboratory
Hitachi Ltd.
Hatoyama
Saitama 350-0395
Japan

S. Heike
Advanced Research Laboratory
Hitachi Ltd.
Hatoyama
Saitama 350-0395
Japan

T. Hitosugi
School of Engineering
University of Tokyo
7–3–1 Hongo
Bunkyo-ku
Tokyo 113-8656
Japan

J.F. Jia
Institute for Materials Research
Tohoku University 2–1–1
Katahira
Aoba-ku
Sendai 980-8577
Japan

Y. Kawazoe
Institute for Materials Research
Tohoku University 2–1–1
Katahira
Aoba-ku
Sendai 980-8577
Japan

K. Kitazawa
School of Engineering
University of Tokyo
7–3–1 Hongo
Bunkyo-ku
Tokyo 113-8656
Japan

L. Li
Chemical Engineering Department
University of California
Los Angeles CA 90095
USA

Z.Q. Li
Institute for Materials Research
Tohoku University 2–1–1
Katahira
Aoba-ku
Sendai 980-8577
Japan

K. Ohno
Institute for Materials Research
Tohoku University
Sendai 980-8577
Japan

T. Ohno
National Research Institute
for Metals
Tsukuba 305-0047
Japan

A. Sakai
Mesoscopic Materials Research
Center
Faculty of Engineering
Kyoto University
Sakyo-ku
Kyoto 606–8501
Japan

T. Sakurai
Institute for Materials Research
Tohoku University 2–1–1
Katahira
Aoba-ku
Sendai 980-8577
Japan

N. Sasaki
Department of Physics
Graduate School of Science
University of Tokyo
7–3–1 Hongo
Bunkyo-ku
Tokyo 113-0033
Japan

M. Tsukada
Department of Physics
Graduate School of Science
University of Tokyo
7–3–1 Hongo
Bunkyo-ku
Tokyo 113-0033
Japan

Q.-K. Xue
Institute for Materials Research
Tohoku University 2–1–1
Katahira
Aoba-ku
Sendai 980-8577
Japan
and
Institute of Physics
Academia Sinica
Beijing 100080
P.R. China

1 Theory of Scanning Probe Microscopy

Naruo Sasaki and Masaru Tsukada

Summary. The first-principles simulation method of STM images is introduced and applied to various surface systems. It is clarified how atomic resolution can be achieved by STM, and tip effects are discussed. For the case of scanning force microscopy, we have been able to reproduce FFM images of graphite fairly well by use of a simple tip model involving only a single atom. Cantilever dynamics for dynamic-mode AFM are carefully studied using numerical simulations and analysis of the equation of motion. The mechanisms of dynamic force microscopy are discussed in the light of these results.

1.1 Introduction

Recent remarkable progress in Scanning Probe Microscopy (SPM) [1] has introduced an evolutional epoch in surface science. A novel research area, namely the science of nanoscale materials, has been brought about in materials science through the development of SPM. In spite of the rapid progress in SPM as an experimental tool, its firm theoretical basis has not yet been established in general. Only with regard to the fundamental aspects of Scanning Tunneling Microscopy (STM) does present theoretical understanding of the mechanisms seem to be satisfactory [2]. Indeed, there still remain various puzzles concerning certain tunneling processes involved in STM. In contrast, our theoretical understanding of Scanning Force Microscopy (SFM) [3] lies far behind. In particular, although the recent rapid development of Non-Contact mode Atomic Force Microscopy (NC-AFM) [4,5] has achieved real atomic resolution, the mechanism ensuring this is unclear. It is not yet known what kind of quantity is measured in order to form NC-AFM images.

Therefore, in the present article, we will describe the theoretical backgrounds of STM, Frictional Force Microscopy (FFM) [6], and dynamic-mode AFM (including NC-AFM) with several case studies based on theoretical simulations. By comparing various kinds of scanning probe microscopies, we will be able to investigate the characteristic features of each kind of spectroscopy. Roughly speaking, the real atomic resolution of STM is made possible by a remarkable concentration of the tunneling current onto a single apex atom of the mini-tip [2]. We might expect that the force contributing NC-AFM images also concentrates on the tip apex atom, but there have so far been no theories supporting this idea. Theories are also needed to investigate how the tip–surface interaction force can affect the dynamics of the cantilever in NC-AFM and how the subtle information from the atomic force reflects itself through images.

First we introduce the theory of STM based on first-principles electronic state theory [2]. By the use of Bardeen's perturbation theory [7], we can calculate the tunnel current from first principles for an atomistic model of the tip at any of its positions relative to the sample surface. The theoretically simulated image is often similar to the Local Density of States LDOS map of the surface for relatively small bias voltages, but for certain cases it shows remarkable deviations from the LDOS map. From theoretical simulations made using various models of the tip, we see the effects of atomic structure and tip shape on STM/STS. Furthermore, various strange features of STM such as the negative difference tunnel conductance and abnormal images are clarified only via simulations with atomistic tip models. The mechanism achieving atomic resolution by the tip can only be understood by such theoretical analysis. We might conclude that the mechanism for atomic resolution by STM is the predominant concentration of the tunneling current on the single apex atom of the tip. In Sect. 1.2.1, we will describe the theoretical framework of the simulation, and in Sects. 1.2.2 and 1.2.3, some remarkable features of STM/STS will be presented based on the numerical calculations by the theory.

On the other hand, the surface information from SFM is quite different from that observed by STM/STS. This can easily be understood by considering that the tip–surface interaction force is produced by the total electron charge and its response to the external field rather than the LDOS of the electrons near the Fermi level. The interaction force may have a long attractive tail, which might spread over the wide range of the tip and the sample surface and dominate the total tip–surface force. However, this part of the force does not possess atomic sensitivity, in particular to the lateral direction. Then the true atomic resolution of SFM images might be attained by the short range attractive or repulsive force which is a tiny portion of the net force and felt only by a single or a few atoms on the tip apex [8].

How such atomic sensitive force can be separated and detected by various modes of SFM is not a trivial problem. Recently, NC-AFM has been demonstrated to possess a real atomic-scale resolution of Si(111) [9–13], InP(110) [14], and NaCl(001) [15]. However the mechanism is still puzzling and a serious question is raised as regards which quantities actually form the NC-AFM images. To establish a firm basis for such problems, the first step is to investigate the relation between cantilever dynamics and tip–surface interaction force [16–24]. Next the nature of the tip–surface interaction force and its distribution over the tip and surface should be theoretically investigated for realistic models.

Thus in Sect. 1.3 we discuss the results of simulations for the case of FFM, where theory reproduces experimental images fairly well. We will investigate how Tomlinson's stick–slip model manifests itself in FFM images. Then we consider the fundamental problems of dynamic-mode AFM. Features of the non-linear dynamics of the cantilever will be carefully studied in Sect. 1.4. Various remarkable types of behavior will be discussed, such as the dynamic touching mode and hysteresis due to bimodality. The way the interaction potential influences the frequency shift curve of the large amplitude NC-AFM mode is discussed in Sect. 1.5.

1.2 Scanning Tunneling Microscopy

1.2.1 Theory of STM Tunneling Current

When the distance between the tip and the sample surface is not very narrow and the applied bias voltage is not large, a perturbation treatment of tunneling is valid. This approach allows us a detailed analysis of tunneling features and the mechanism for achieving atomic resolution by STM. Thus, starting from Bardeen's perturbation theory [7], the tunneling current in STM is obtained as follows:

$$I = \frac{2\pi e}{\hbar} \int \mathrm{d}E \Big[f(E) - f(E + eV) \Big] A(\boldsymbol{R}, E, E + eV) \ , \tag{1.1}$$

$$A(\boldsymbol{R}, E, E') = \int \mathrm{d}\boldsymbol{r} \int \mathrm{d}\boldsymbol{r}' V_{\mathrm{T}}(\boldsymbol{r}) V_{\mathrm{T}}(\boldsymbol{r}') G^{\mathrm{T}}(\boldsymbol{r}', \boldsymbol{r}; E') G^{\mathrm{S}}(\boldsymbol{r} + \boldsymbol{R}, \boldsymbol{r}' + \boldsymbol{R}; E) \ , \tag{1.2}$$

where A is the energy density of the tunnel current, f is the Fermi distribution function, $V_{\mathrm{T}}(\boldsymbol{r})$ is the potential in the tip and V is the tip bias voltage. G^{S} and G^{T} are the imaginary parts of the Green's function of the surface and the tip, respectively. Thus the integral of (1.1) ranges over the Fermi level offsets between the two electrodes. If there is a very small protrusion close to the sample surface, the tunnel current is almost totally concentrated on this region. This can be confirmed by expanding the smooth part of the surface Green's function in (1.2) around this protrusion. This leads to a moment expansion of the tunnel current, in which each term corresponds to the respective order of the derivative of the surface Green's function multiplied by the geometrical moment of the weight function, $w(\boldsymbol{r}, \boldsymbol{r}'; E)$,

$$\mu_{nm}(E) = \int \overbrace{\boldsymbol{r} \cdots \boldsymbol{r}}^{n} \overbrace{\boldsymbol{r}' \cdots \boldsymbol{r}'}^{m} w(\boldsymbol{r}, \boldsymbol{r}'; E) \mathrm{d}\boldsymbol{r} \mathrm{d}\boldsymbol{r}' \ , \tag{1.3}$$

$$w(\boldsymbol{r}, \boldsymbol{r}'; E) = V_{\mathrm{T}}(\boldsymbol{r}) V_{\mathrm{T}}(\boldsymbol{r}') \sum_{\nu} \psi_{\nu}(\boldsymbol{r}) \psi_{\nu}(\boldsymbol{r}') \gamma(z; E) \gamma(z'; E) \delta(E - E_{\nu}) . \tag{1.4}$$

In the above ψ_{ν} is the tip eigenfunction, and the decay function $\gamma(z; E)$ is defined by $\exp(-z\sqrt{2m|E|}/\hbar)$. If we take only the first term of the expansion, the naive relation [25]

$$\frac{\mathrm{d}I}{\mathrm{d}V} \propto \rho\left(\boldsymbol{R}, E_{\mathrm{F}} - eV\right) \tag{1.5}$$

results, where ρ is the LDOS (Local Density of States) of the surface at the protrusion. Indications of significant deviations from the naive relation have often been noticed and they are mostly related to microscopic states of the tip. For example, much finer structures of STM images are often observed than would be expected merely as a result of the surface LDOS. This could only be explained by the contribution from higher order derivatives of the surface wavefunction, which are contributed by higher order terms in the moment expansion. In deriving (1.5), the LDOS of the tip is assumed to be smooth compared with that of the surface. But this is not necessarily the case, depending on the tip. Typical examples are seen when the negative differential conductance is observed in STS. In such cases, complete theoretical simulations based on (1.1) and (1.2) are effective [27].

1.2.2 STM Images of Si(111) $\sqrt{3} \times \sqrt{3}$-Ag and -Sb Surfaces

The STM image for negative tip bias by van Loenen et al. [27] revealed a distribution of bright spots arranged in a honeycomb lattice. On the other hand the HCT model (Honeycomb-Chained Trimer model) (see Fig. 1.1a) proposed by Takahashi using the X-ray standing wave method [28] and by Katayama et al. using Coaxial Impact Collision Ion Scattering Spectroscopy (CAICISS) [29] turned out to be consistent with many experimental findings, but not with STM.

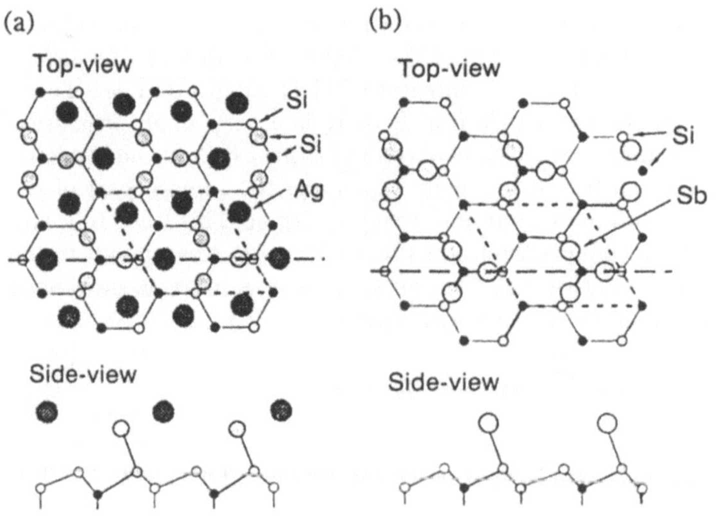

Fig. 1.1. (**a**) The HCT model of the Si(111) $\sqrt{3}\times\sqrt{3}$-Ag surface. (**b**) The milkstool model of the $\sqrt{3}\times\sqrt{3}$-Sb surface

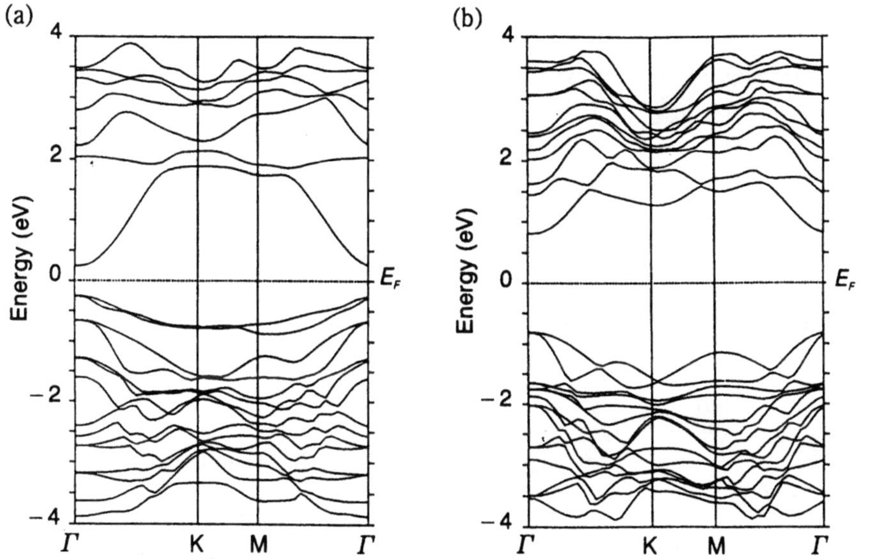

Fig. 1.2. (**a**) The energy band structure of the Si(111)$\sqrt{3} \times \sqrt{3}$-Ag surface (HCT model) by the thin layer model. (**b**) The energy band structure of the Si(111)$\sqrt{3} \times \sqrt{3}$-Sb surface (milkstool model) by the thin layer model

Watanabe et al. [30] calculated the energy band structure of the HCT model and found a semiconducting character as shown in Fig. 1.2a.

Moreover, the overall density of states DOS is consistent with the UPS experiment [31]. So how can the HCT model be reconciled with the observed STM honeycomb pattern? Watanabe, Aono and Tsukada [31] performed a theoretical simulation of the STM image of this surface with a tip model W_{10} [111] consisting of ten W atoms. The theoretical STM image showed good agreement with the observed image [27]. The reason why the HCT model gives the honeycomb pattern can be understood from Fig. 1.3. The reason is that bright spots do not correspond to any of the surface atoms, but are located at the inner region of the Ag trimer. This is an interesting example in which the naive interpretation of STM images sometimes fails. The Sb chemisorbed Si(111) $\sqrt{3} \times \sqrt{3}$ surface takes the milkstool structure as shown in Fig. 1.1b. This structure is related to the HCT model of the Si(111) $\sqrt{3} \times \sqrt{3}$-Ag surface (Fig. 1.1a). The milkstool structure is obtained from the HCT structure by removing the top Ag layers and replacing the Si atoms in the next layer by Sb atoms. We performed a simulation of the STM images [32] for this surface. Fig. 1.4 shows the theoretical STM images for the surface bias voltages, $V_s = -2.0$ V, $+2.0$ V, and $+1.0$ V, respectively. For the case of $V_s = -2.0$ V, each Sb atom in the milkstool can be observed as a distinct bright spot forming a trimer. For the case of $V_s = 2.0$ V, the three atoms of the trimer cannot be seen separately, but form a single large bright area centered at the milkstool.

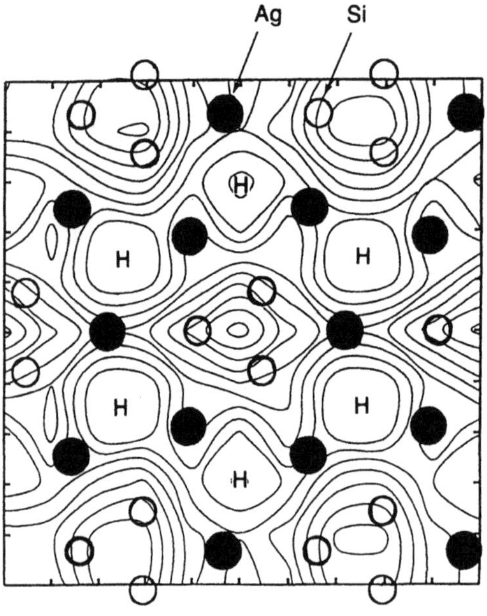

Fig. 1.3. Theoretical STM image of the Si(111) $\sqrt{3} \times \sqrt{3}$-Ag surface (HCT model) by the $W_{10}[111]$ tip. The tip–surface distance is 3.7 Å and the surface bias voltage is 2.1 V

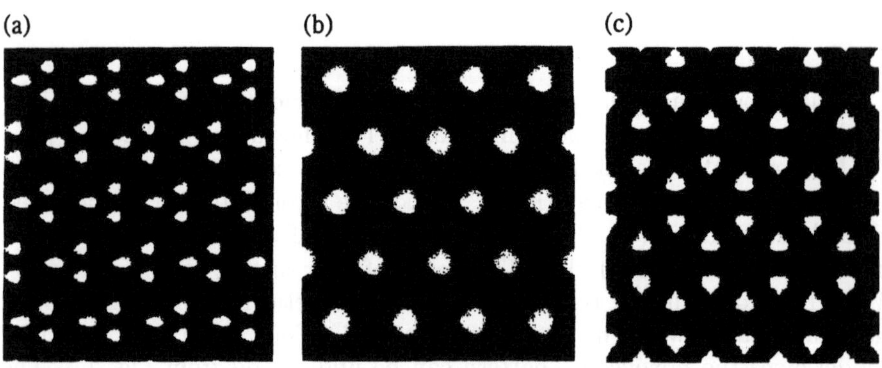

Fig. 1.4. Theoretical STM images of the Si(111) $\sqrt{3} \times \sqrt{3}$-Sb surface (milkstool model) by the $W_{10}[111]$ tip. The tip–surface distance is 3.7 Å and surface bias voltages are (**a**) −2.0 V, (**b**) +2.0 V, and (**c**) +1.0 V, respectively

Interestingly enough, almost the same honeycomb pattern appears for the case of $V_s = 1.0$ V as is found for the HCT model of the Ag chemisorbed surface. The spatial position at which the bright spot is located in the honeycomb of the Sb chemisorbed surface is outside the milkstool and it is thus in exactly the same location as in the case of the Ag HCT surface. Figures 1.2a and b show the energy band structure of Si(111) $\sqrt{3} \times \sqrt{3}$-Ag (HCT) and that of the Si(111) $\sqrt{3} \times \sqrt{3}$-Sb (milkstool) surface, respectively [32]. The surface energy band which contributes to honeycomb spots in the STM image is found to be the lowest unoccupied band for both the surfaces. We can see from Fig. 1.2 that the lowest surface state band in each surface has a quite similar dispersion although that in the Si(111) $\sqrt{3} \times \sqrt{3}$-Sb (milkstool) is shifted to a somewhat higher energy position and merged into the higher conduction bands. As indicated by the STM images, the wavefunction of the lowest unoccupied surface state relative to the substrate is almost the same for both surfaces.

1.2.3 Effects of Microscopic Tip Structure

A number of theoretical simulations proved that normal STM images can be reproduced by a tip with a single atom on its apex [1,2]. However, if there is more than one atom on the top plane of the tip, i.e., on the same level in the outermost atomic layer parallel to the sample surface, abnormal images result. This is due to the distribution of the tunnel current over those atoms on the tip.

With the tip models of $W_{10}[111]$ or $W_{14}[110]$ (Fig. 1.5), the normal images of the unoccupied state, which are observed by experiments, are reproduced quite well. If we remove the apex atom from these tips, the calculated image is changed drastically. These abnormal images are formed because the tunnel current is distributed over four atoms for the case of the W_{13} [110] tip, or three atoms for the case of the W_9 [111] tip (Fig. 1.5). The abnormal images change drastically with rotation of the tip around its symmetry axis, whereas

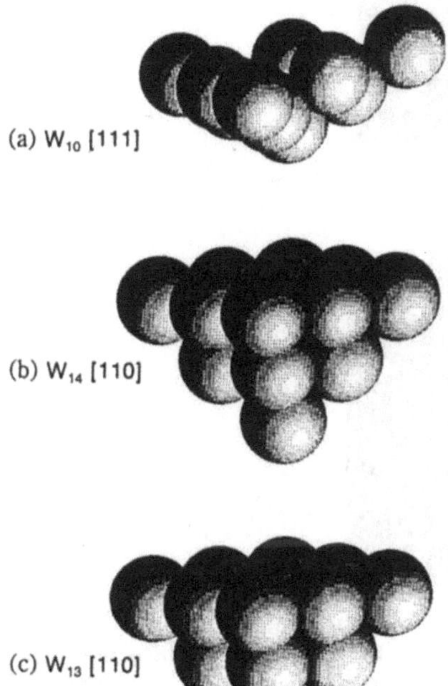

Fig. 1.5. Tip models used in the theoretical calculation: (**a**) $W_{10}[111]$, (**b**) $W_{14}[110]$, and (**c**) $W_{13}[110]$

normal STM images by the W_{10}[111] tip or by the W_{14}[110] tip hardly change with such rotation.

If the distribution of the tunnel current over several atoms is due to the presence of these atoms on the same top layer of the tip, a slight tilting of the tip axis from the normal direction of the sample surface will recover the correct STM images. To confirm this idea we attempted to make a simulation of the STM image with the tilted tips [33]. Figures 1.6a–d show the simulated results of the STM images by the tip W_{13} [110] for tilting angles $\theta = 0$, 10, 20 and 30 degrees from the surface normal, respectively. Notice that the simulated images gradually tend back to the normal image (Fig. 1.3) with increased tilting angle θ. For $\theta = 20°$, images are almost the same as the normal one. For the W_9[111] tip, similar results are obtained for reverting abnormal images; in this case only a small tilting angle $\theta = 10°$ is required to revert the STM image to the normal one. Features obtained by the simulation confirm the assumption that the tunnel current concentrates almost totally on the single apex atom when normal images are obtained.

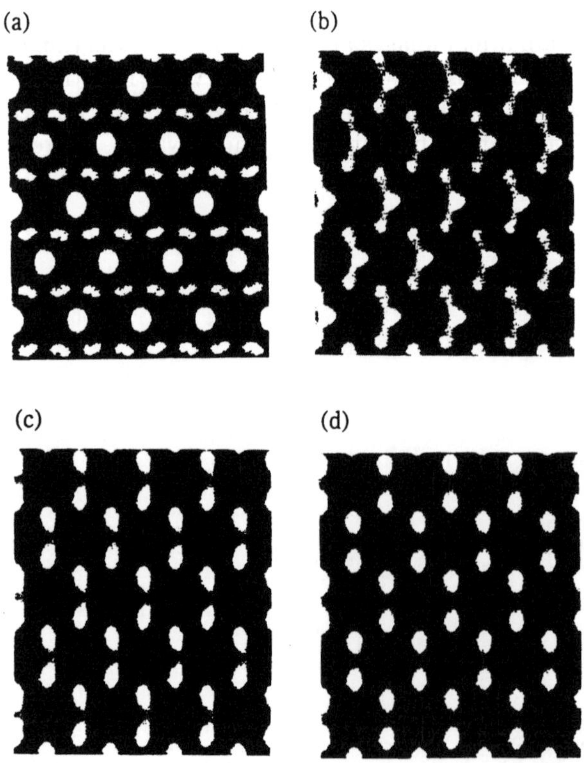

Fig. 1.6. Theoretical STM image of the Si(111)$\sqrt{3} \times \sqrt{3}$-Ag surface (HCT model) by the W_{13}[110] tip. The tip–surface distance is 3.7 Å and the surface bias voltage is 1.0 V. (**a**), (**b**), (**c**), and (**d**) correspond to tilting angles 0°, 10°, 20°, 30°, respectively

1.3 Frictional Force Microscopy

1.3.1 FFM in Micro-Tribology

The research area of microscopic friction, or micro-tribology, has rapidly progressed since the invention of AFM in 1986 [3]. AFM is an experimental method for detecting forces acting between a single microscopic tip apex and an atomically flat surface, and is suitable for observing microscopic friction. As shown in Fig. 1.7a, the atomic force acting between tip and surface is detected as a small deflection of the cantilever, and atomic-scale images of the surface are obtained by scanning the tip. In most experiments of AFM, deflection of the cantilever is measured from the changed reflection angle of a laser beam by the optical detection method. In particular, when the scanning tip contacts with the surface in the repulsive force regime, and the lateral force can be measured, AFM is called Lateral Force Microscopy (LFM) or Frictional Force Microscopy (FFM) [6]. Observed two-dimensional FFM images

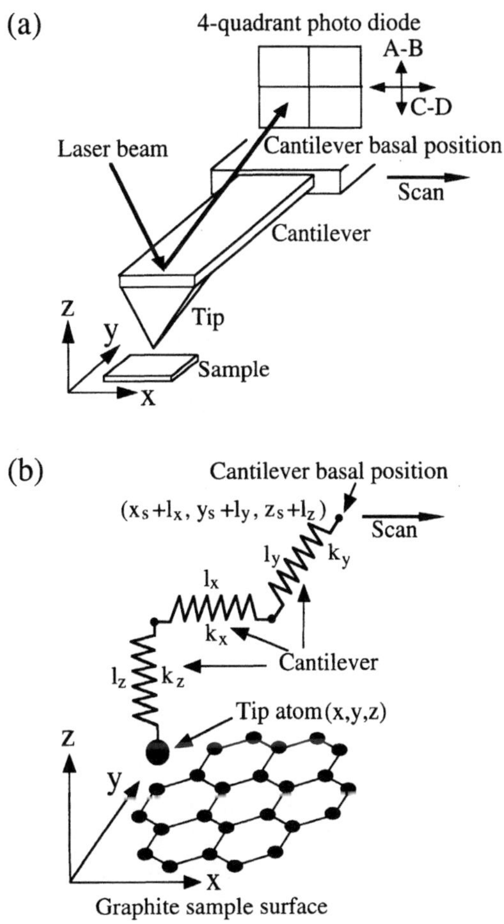

Fig. 1.7. (**a**) Schematic representation of AFM by an optical lever method. (**b**) Model of FFM used in theoretical calculations

yield information about tip atom motion [36–38], which is closely related to the origin of the friction.

Therefore in Sect. 1.3 we address the problem: What does FFM look at? We treat a cleaved graphite surface as a case study. FFM images are interpreted using a theoretical simulation and stable equilibrium condition, and its physical meaning is clarified. Next it is explained how macroscopic conditions such as load, scan direction, anisotropy of the cantilever and surface deformation can influence frictional force images, considering both theoretical and experimental aspects [44–52].

1.3.2 Model and Method of Calculation

In our theoretical simulation, we mimic the tip–surface system of FFM (see Fig. 1.7a) by a single-atom tip connected with a three-dimensional cantilever spring and a monolayer graphite substrate surface as shown in Fig. 1.7b. If the graphite surface is assumed to be rigid, the total potential energy V can be supposed to consist of the harmonic elastic energy of the cantilever V_T, and the microscopic tip–surface interaction V_TS as follows:

$$V\left(x,y,z\right) = V_\mathrm{T}\left(x,y,z\right) + V_\mathrm{TS}\left(x,y,z\right) \ . \tag{1.6}$$

Surface deformation effects can be put into the change in the cantilever spring constant. Then V_T is expressed by

$$V_\mathrm{T}\left(x,y,z\right) = \frac{1}{2}\left[k_x\left(x-x_\mathrm{s}\right)^2 + k_y\left(y-y_\mathrm{s}\right)^2 + k_z\left(z-z_\mathrm{s}\right)^2\right] \ . \tag{1.7}$$

Here (k_x, k_y, k_z) is assumed to be the cantilever spring constant, and (x, y, z) the tip atom position. $(x_\mathrm{s}, y_\mathrm{s}, z_\mathrm{s})$ also denotes the equilibrium tip atom position without the tip–surface interaction. In the simulation we show below, the vertical component k_z is assumed to be 0.25 N/m, which is of the same order as the value realized in most of the experiments. k_x and k_y are treated as parameters. When the natural length of the cantilever is assumed to be (l_x,l_y,l_z), $(x_\mathrm{s}+l_x, y_\mathrm{s}+l_y, z_\mathrm{s}+l_z)$ denotes the basal position of the cantilever. Therefore there is essentially no difference between the cantilever basal position and the tip atom equilibrium position. Hence we shall hereafter call $(x_\mathrm{s}, y_\mathrm{s}, z_\mathrm{s})$ the cantilever basal position.

Next V_TS is expressed as the sum over interatomic potentials:

$$V_\mathrm{TS}\left(x,y,z\right) = \sum_i 4\epsilon\left[\left(\frac{\sigma}{r_i}\right)^{12} - \left(\frac{\sigma}{r_i}\right)^{6}\right] , \tag{1.8}$$

where r_i is the distance between the single atom tip and the ith atom of the graphite surface. Fitting parameters $\epsilon = 0.873 \times 10^{-2}$ eV and $\sigma = 2.49$ Å adopted by Gould et al. [56] from the first principles calculation are used [8].

In the theoretical simulation the tip–surface system is assumed to be kept at absolute zero, and finite temperature effects are neglected. The tip is scanned in the repulsive-force and constant-height mode. The velocity of the cantilever scan ($\simeq$ Å/s) is much smaller than that due to lattice vibration ($\simeq$ 100–1000 m/s). Therefore, when the cantilever comes to a certain scan position, atoms of the tip and substrate surface instantaneously relax within the time scale of the scanning velocity, and the quasi-metastable state on the adiabatic potential energy surface is achieved. Minimization of the total energy V is thus performed for each cantilever basal position (x_s,y_s,z_s) by the molecular relaxation method, which gives the equilibrium position of the tip atom (x, y, z) and the lateral force $F_i(i = x, y)$ acting on the cantilever. The lateral force F_i acting in the $-i$ direction is given by

$$F_i = \left.\frac{\partial V}{\partial x_i}\right|_{(x,y,z)=(x_s,y_s,z_s)} \qquad (i = x, y) \; . \tag{1.9}$$

1.3.3 Mechanism of Atomic-Scale Friction

The Tomlinson mechanism [57] is explained as below for the one-dimensional system for atomic-scale friction. First, with increased load, the corrugation amplitude of the tip–surface interaction V_{TS} increases. For low loads (where $\mathrm{d}^2V_{\mathrm{TS}}/\mathrm{d}x^2 < k$), the total potential energy surface is nearly parabolic and only a single minimum appears, as shown in Fig. 1.8a. However, for high loads (with $\mathrm{d}^2V_{\mathrm{TS}}/\mathrm{d}x^2 \geq k$), the total potential energy surface has several metastable points corresponding to local minima, as shown in Fig. 1.8b. Hence, for the low load (Fig. 1.8a), when the cantilever is scanned, the tip atom shifts continuously (adiabatically), trapped in a minimum of the potential energy surface. However, for the high load (Fig. 1.8b), a tip atom continuously moves for some period of the scan (1)–(3). But it makes a sudden discrete (nonadiabatic) jump from one minimum to another deeper minimum at scan position (3), where the energy barrier between the two minima disappears. Then the elastic energy stored in the cantilever instantaneously dissipates into the tip and surface atoms by the multiple generation of phonons. This energy dissipation originating from localization of the domain of the equilibrium tip atom position is the essence of the Tomlinson friction mechanism. The continuous and discontinuous motion of the tip atom is called stick–slip motion. The process (1)–(4) is caused to repeat by the lattice periodicity L as shown in Fig. 1.8b.

1.3.4 Load Dependence of Frictional Force Images

We find a phenomenon in which the FFM image of a graphite surface changes in a remarkable way with the load for a certain scan direction. First, the characteristics of load dependence are explained through a comparison between

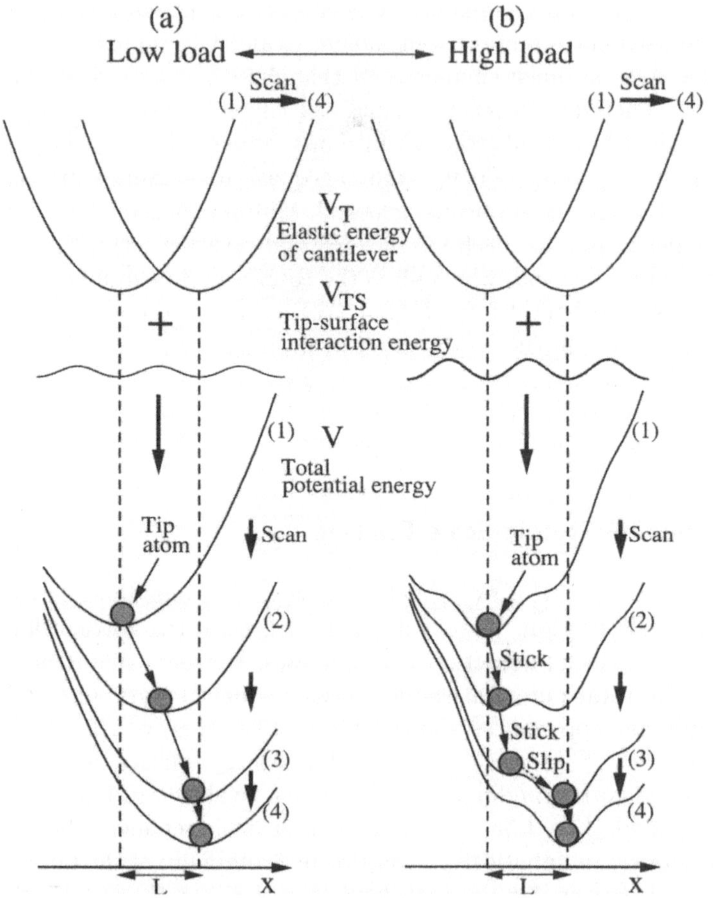

Fig. 1.8a, b. Schematic model of the Tomlinson mechanism. (1)–(4) denote the time evolution of the potential during the cantilever scan along the x direction

theory and experiment. Figures 1.9a–c represent images obtained by the theoretical simulation and experiment. F_x/k_x and F_y/k_y denote, respectively, components parallel and vertical with respect to the scan direction. Here the lateral force F is represented by F/k, the deflection of the cantilever spring. Furthermore, a cantilever is assumed to be an isotropic spring lying in the (x, y) plane, such that $k_x = k_y = 2.5$ N/m. Details of the FFM experiment are described in [36–38]. The average vertical load $\langle F_z \rangle$ is (a) theory 0.675 nN, experiment 44 nN, (b) theory 1.10 nN, experiment 122 nN, and (c) theory 1.40 nN, experiment 327 nN. The theoretical magnitude of $\langle F_z \rangle$ is about two orders smaller than the experimental value. This is due to the fact that a single-atom tip model is adopted in our theoretical simulation.

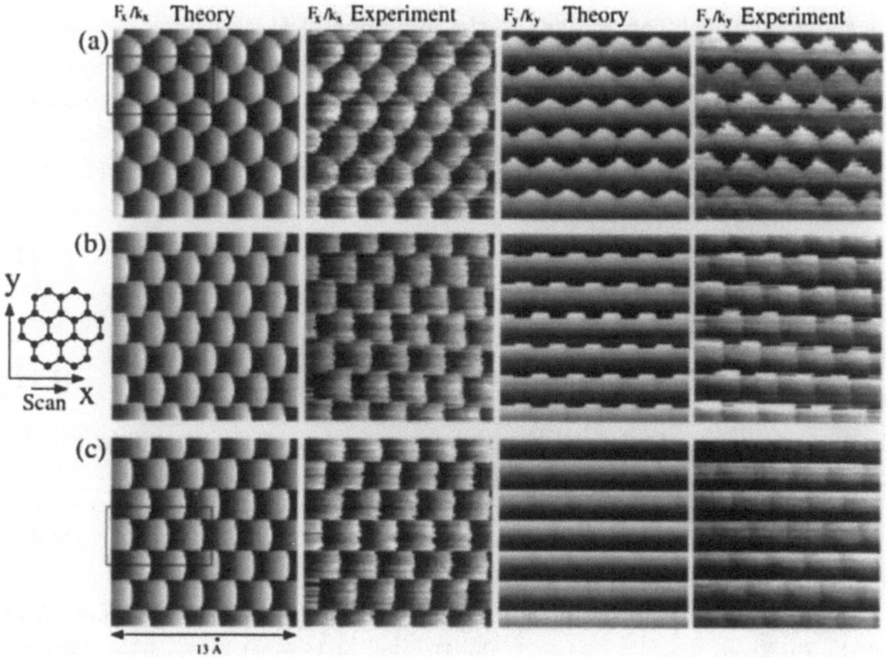

Fig. 1.9. Frictional force images of F_x/k_x and F_y/k_y obtained by theoretical simulation and experiment in the constant-height mode. Average vertical load $\langle F_z \rangle$ is (**a**) theory 0.675 nN, experiment 44 nN, (**b**) theory 1.10 nN, experiment 122 nN, and (**c**) theory 1.40 nN, experiment 327 nN

As shown in Figs. 1.9a–c, the simulated and experimental images (F_x/k_x and F_y/k_y) are in very good agreement with each other for each load (a)–(c). In the F_y/k_y images of Figs. 1.9a–c, we find a characteristic change of boundary between bright and dark areas along the scan direction (x axis). Indeed, a zigzag pattern appears, corresponding to the C-C bond of the graphite lattice, as shown in Fig. 1.9a. However, as the load increases, this zigzag pattern gradually vanishes and becomes a simple uneven pattern as shown in Fig. 1.9b. When the load is further increased, the zigzag pattern completely vanishes and only the straight pattern parallel to the scan direction appears, as shown in Fig. 1.9c. We notice the same load dependence of FFM image patterns in both theory and experiment. We also find an FFM image transition depending on the cantilever stiffness in the constant-height mode, although we do not discuss this problem. Details are described in [44–47].

1.3.5 Method of Analysis of the FFM Image Pattern

We shall give a detailed description of the method of analysis of the FFM image pattern. The concept was presented by Gyalog et al. [40]. First the potential V is converted from a function of the three variables, x, y and z, into a function of two variables, x and y. For that purpose z should be chosen as the point which corresponds to the minimum of V, satisfying

$$F_z = -\frac{\partial V_{\mathrm{TS}}(x,y,z)}{\partial z} = k_z(z - z_{\mathrm{s}}) . \tag{1.10}$$

The solution z of (1.10) can be obtained as $z(x, y; z_{\mathrm{s}})$. V is therefore represented as

$$V(x,y;x_{\mathrm{s}},y_{\mathrm{s}},z_{\mathrm{s}}) = \frac{1}{2}\left[k_x(x-x_{\mathrm{s}})^2 + k_y(y-y_{\mathrm{s}})^2\right] + V'(x,y;z_{\mathrm{s}}) , \tag{1.11}$$

where

$$V'(x,y;z_{\mathrm{s}}) = \frac{1}{2}k_z\left[z(x,y;z_{\mathrm{s}}) - z_{\mathrm{s}}\right]^2 + V_{\mathrm{TS}}[x,y,z(x,y;z_{\mathrm{s}})] . \tag{1.12}$$

In the next step the lateral coordinates x and y of the tip atom can be determined by

$$\nabla_i V = 0 \quad (i = x, y) , \tag{1.13}$$

for each cantilever basal position $(x_{\mathrm{s}}, y_{\mathrm{s}}, z_{\mathrm{s}})$.

If the stable equilibrium condition of the two-dimensional position (x, y) of the tip atom is to be satisfied, the Hessian of V must be positive definite. Hence, the relation

$$|\partial^2 V/\partial x_i \partial x_j| > 0 \;\; (i,j = x,y) \;\; \text{and} \;\; \partial^2 V/\partial x^2 > 0 \tag{1.14}$$

is required in addition to (1.13). Further, the equilibrium condition (1.13) can be rewritten as follows:

$$(x_{\mathrm{s}}, y_{\mathrm{s}}) = \left(x + \frac{1}{k_x}\frac{\partial V'(x,y;z_{\mathrm{s}})}{\partial x}, y + \frac{1}{k_y}\frac{\partial V'(x,y;z_{\mathrm{s}})}{\partial y}\right) . \tag{1.15}$$

(1.15) is the condition on the lateral components of the cantilever basal position $(x_{\mathrm{s}}, y_{\mathrm{s}})$ which can realize the equilibrium tip atom position (x, y).

The quasi-stable equilibrium region of the two-dimensional tip atom position (x, y) is thereby obtained, and mapped onto a domain of the cantilever basal position $(x_{\mathrm{s}}, y_{\mathrm{s}})$. It should be noted that the mapping from (x, y) to $(x_{\mathrm{s}}, y_{\mathrm{s}})$ is monovalent, but the inverse mapping from $(x_{\mathrm{s}}, y_{\mathrm{s}})$ to (x, y) is not necessarily monovalent.

1.3.6 Physical Meaning of Frictional Force Images

Figures 1.10a and b show the stable regions for $\langle F_z \rangle$ = (a) 0.675 nN and (b) 1.40 nN obtained by the method mentioned above. Hatched regions B and C in Figs. 1.10a-1 and b-1 represent regions where tip atoms can exist stably within the (x, y) plane. Other stable regions are given only by boundary curves. These stable regions are hereafter called sticking domains. Solid lines representing honeycomb networks denote C-C bonds in the graphite lattice.

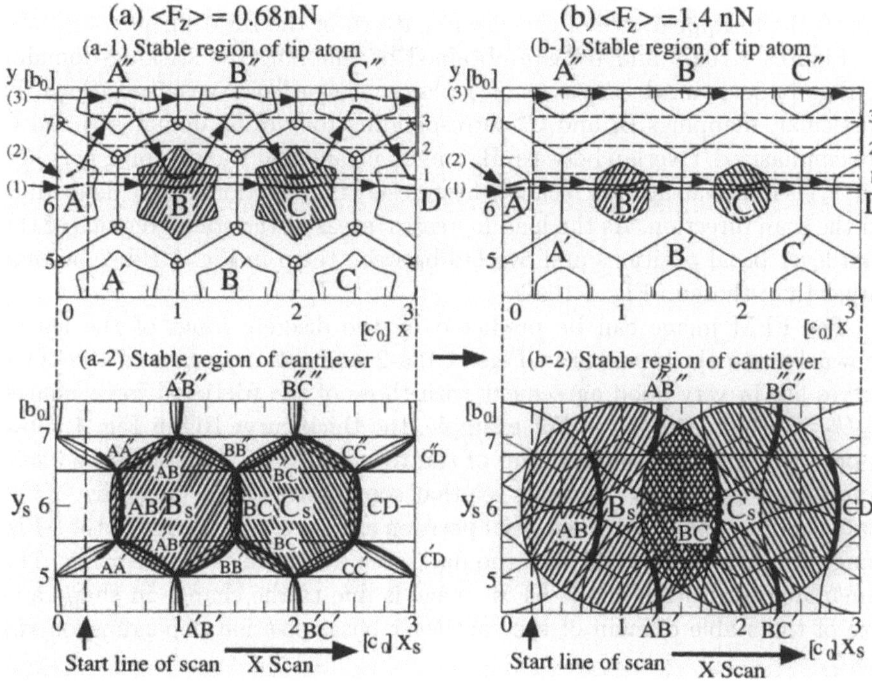

Fig. 1.10. $\langle F_z \rangle$ = (**a**) 0.675 nN and (**b**) 1.40 nN. (a-1) and (b-1) denote sticking domains of the tip atom in the (x, y) plane. (a-2) and (b-2) denote stable regions of the cantilever basal positions in the (x_s, y_s) plane. Regions represented in (**a**) and (**b**) correspond to those within rectangular boxes in Figs. 1.9a and c, respectively. $b_0 = 1.42$ Å and $c_0 = 2.46$ Å

In Fig. 1.10a-1, sticking domains are discontinuously located around hollow sites and on sites of the graphite lattice within the (x, y) plane. However, when the load increases, as shown in Fig. 1.10b-1, only sticking domains around hollow sites appear, and their areas are smaller than those of Fig. 1.10a-1.

Changing the shape of the sticking domains has remarkable effects on tip atom motion. This can be understood by tracing each path of the tip atom motion for a specific scan line. Figures 1.10a-1 and b-1 represent the paths (1), (2) and (3) of the tip atom motion corresponding to the broken lines 1, 2 and 3 of the cantilever scan. Paths (1) and (3) shown in Fig. 1.10a-1 are A $\rightarrow$ B $\rightarrow$ C $\rightarrow \ldots$, and A$''$ $\rightarrow$ B$''$ $\rightarrow \ldots$, which are paths passing the sticking domains around hollow sites located parallel to the x axis scan direction. However, path (2) is A $\rightarrow$ A$''$ $\rightarrow$ B $\rightarrow$ B$''$ $\rightarrow \ldots$, which exhibits a zigzag path intersecting the C-C bonds of the graphite lattice tilted by 30 or 150 degrees from the x axis. This zigzag path produces a zigzag pattern in the frictional force image of F_y/k_y in Fig. 1.9a. However, as the load increases, path (2) in Fig. 1.10b-1 becomes A $\rightarrow$ B $\rightarrow$ C $\rightarrow \ldots$, which is a path passing the sticking domains around hollow sites located parallel to the x axis. This causes the disappearance of the zigzag pattern in the FFM image.

Figures 1.10a-2 and b-2 are obtained by mapping the sticking domains of Fig. 1.10a-1 and b-1 into (x_s, y_s) planes of cantilever basal positions. In particular, mappings B_s and C_s corresponding to sticking domains B and C are emphasized. Overlap between B_s and C_s means that the mapping $(x, y) \mapsto (x_s, y_s)$ is multivalent, which causes hysteresis of the frictional force depending on the scan direction. As the load increases, areas of the stable regions of the cantilever basal positions and overlap between them in Fig. 1.10b-2 become larger than those in Fig. 1.10a-2.

The FFM image can be predicted by the discrete jump of the tip as shown by the thick curves in Figs. 1.10a-2 and b-2. Patterns of the thick curves are in very good agreement with those of the frictional force images F_x/k_x in Figs. 1.9a and c. For example, the thick curve BC in Fig. 1.10b-2 is produced by the discrete jump of the tip atom from the sticking domain B to C in Fig. 1.10b-1. We thus see that some parts of the boundary of the stable region of the cantilever basal position appear as the fringes of the FFM images between the bright and the dark area along the scan direction. The remarkable transition of the FFM image is due to the change in shape and size of the stable domain of the cantilever basal position depending on the load.

Since the slip process of the tip atom motion is an atomic-scale stochastic process, the boundary of the pattern which indicates the position of the slip fluctuates randomly around the thick curves in Figs. 1.10a-2 and b-2 for each x scan. Experimental frictional force images in Fig. 1.9 exhibit such fluctuation features.

1.3.7 Supercell Feature of Images

For FFM images associated with larger loads than the one in Fig. 1.9c, we find a strange supercell feature depending on the start line of the scan. Here we pay attention to the variation of the image pattern in the y direction. As shown in Fig. 1.11a, for the start line AB, the same pattern reappears almost in a cycle of b_0 and $2b_0$. The frame made by the thick straight line indicates the scan region. The b_0 and $2b_0$ periodicity in the y direction corresponds to the graphite atom spacing on the start line AB of the x scan. Thus the frictional force image has a 1×2 supercell symmetry with respect to the unit cell of the graphite lattice. Next, when we choose the start line of the scan A′B′ in Fig. 1.11b, an image of Fig. 1.11b is obtained. The appearance of the long period images is a remarkable feature for the high load case, because the stable region of the cantilever basal position becomes rather large.

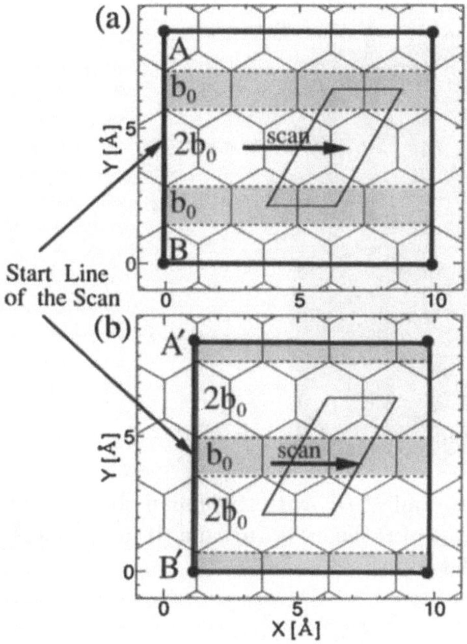

Fig. 1.11. Schematic illustration of two kinds of supercell frictional force image of $F_x(x_s, y_s)$ for $k = 0.25$ N/m. The start line of the scan A′B′ in (**b**) is obtained by shifting AB in (**a**) by 1.23 Å ($= c_0/2$) in the x direction. The masked area has width 1.42 Å ($= b_0$) in the y direction. Rhomboids represented by *solid lines* show 1×2 supercell structures with respect to the unit cell of the graphite lattice

1.3.8 Scan Direction Dependence

For general scan directions, the image can be explained in the same way as for x axis. Frictional force images of Fig. 1.12 are obtained by the y scan of Fig. 1.9 under the condition $\langle F_z \rangle =$ (a) theory 0.7 nN, and (b) experiment 122 nN. It is assumed that $k_x = k_y = 2.5$ N/m. In this case also, theoretical simulation reproduces the experimental result quite well. Although the image patterns of Fig. 1.12 are different from those of Fig. 1.9a, the shape of the boundary of the stable region of the cantilever basal position itself is the same as in Fig. 1.10a-2. Part of the boundary appearing in the observed image pattern changes depending on the scan direction. For this scan direction, no load dependence appears in the image.

Furthermore, Figs. 1.13 are simulated frictional force image patterns obtained by an X scan which deviates by 15° from the x axis in Fig. 1.9. It is assumed that $k_x = k_y = 0.75$ N/m. It is observed that the image pattern deforms as the boundary of the stable region of the cantilever basal position varies.

1.3.9 Anisotropy of the Cantilever

When anisotropy of the cantilever ($k_x \neq k_y$) is considered, the frictional force image pattern changes remarkably. Figures 1.8a and b are obtained for the cantilever with $k_x < k_y$ scanned in the x direction. It is observed that the

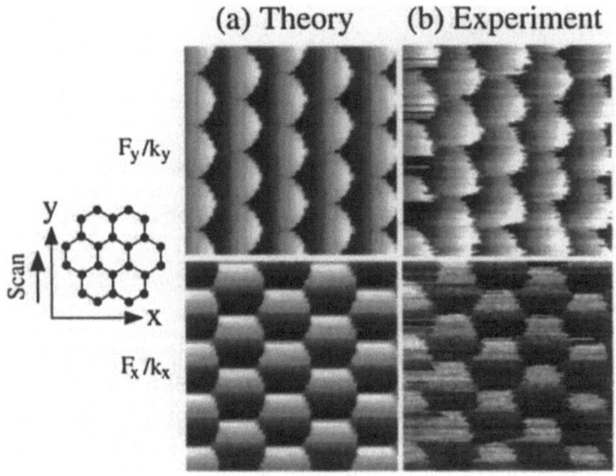

Fig. 1.12. Frictional force images of F_x/k_x and F_y/k_y with the y scan obtained by theoretical simulation and experiment. $\langle F_z \rangle =$ (**a**) theory 0.7 nN and (**b**) experiment 122 nN. The scanning region is 9.9 Å ×9.8 Å

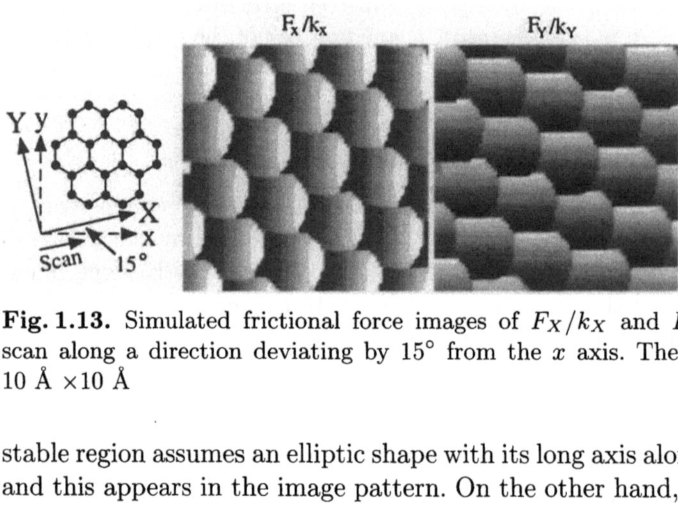

Fig. 1.13. Simulated frictional force images of F_X/k_X and F_Y/k_Y with the X scan along a direction deviating by 15° from the x axis. The scanning region is 10 Å ×10 Å

stable region assumes an elliptic shape with its long axis along the x direction, and this appears in the image pattern. On the other hand, Figs. 1.14c and d are obtained by the y scan of the cantilever with $k_x > k_y$. Contrary to the case for $k_x < k_y$, the stable region assumes an elliptic shape with its long axis along the y direction. The length of the long axis in Figs. 1.14b and d is larger than that in Figs. 1.14a and c, which implies increased cantilever anisotropy. It can be concluded that scan direction and anisotropy have complicated influences on frictional force images obtained in actual experiments.

1.3.10 Surface Deformation

Lastly, we allow the graphite surface to deform, using the surface potential V_S of graphite as follows:

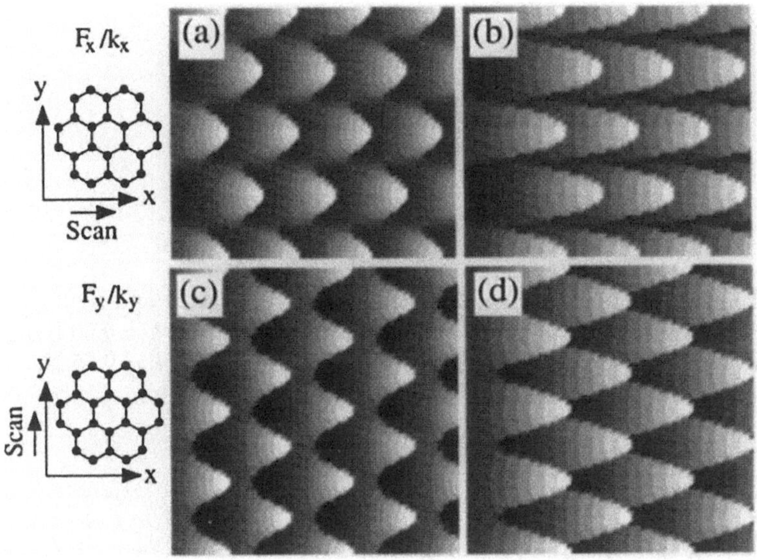

Fig. 1.14. Simulated frictional force images of (**a**), (**b**) F_x/k_x with the x scan and (**c**), (**d**) F_y/k_y with the y scan, obtained using an anisotropic cantilever. Cantilever spring constants are $(k_x, k_y) =$ (**a**) $(0.50, 5.0)$, (**b**) $(0.25, 2.5)$, (**c**) $(10.0, 0.50)$, and (**d**) $(2.5, 0.25)$. The unit is [N/m]

$$V_\mathrm{S} = \frac{1}{2}\sum_{i-j} \mu_\mathrm{r}(r_{ij} - r_0)^2 + \frac{1}{2}\sum_{i-j-k} \mu_\theta r_0^2(\theta_{ijk} - \theta_0)^2 + \frac{1}{2}\sum_{i-(j,k,l)} \mu_\mathrm{p}\left(\delta z_i - \frac{\delta z_j + \delta z_k + \delta z_l}{3}\right)^2 . \quad (1.16)$$

The indices of summation $i-j$, $i-j-k$ and $i-(j,k,l)$ represent nearest-neighbor bonds, bond pairs, and bond triples, respectively. Therefore, the first and second terms correspond to bond-stretching and bond-bending energies, respectively. The third term corresponds to the bending energy of the local planar structure due to the normal displacement of the ith atom from the coplanar position with respect to the three neighboring atoms j, k and l. The parameters of V_S are $r_0 = 1.4210$ Å, $\mu_\mathrm{r} = 41.881$ eV/Å^2, $\theta_0 = 2\pi/3$ rad, $\mu_\theta = 2.9959$ eV/Å^2 and $\mu_\mathrm{p} = 18.225$ eV/Å^2, respectively [44,45]. In this case, the energy $V = V_\mathrm{T} + V_\mathrm{S} + V_\mathrm{TS}$ is minimized.

The lateral force images with the y scan are shown in Figs. 1.15a–c. Figure 1.15b is an image of an elastic surface for $k_x = k_y = k = 0.50$ N/m. Figures 1.15a and c are images of rigid surfaces for $k = 0.50$ N/m and $k = 0.75$ N/m, respectively. As seen in these force images, Fig. 1.15b resembles Fig. 1.15c more closely than Fig. 1.15a. This feature is quantitatively examined in the cross sections of Figs. 1.15a–c at the line indicated by the thick arrow along the y direction. In Fig. 1.16, the lateral force curve with

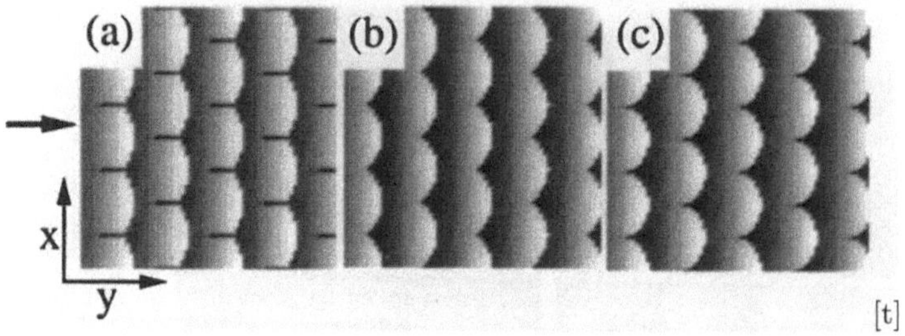

Fig. 1.15. Comparison of lateral force $F_y(x_s, y_s)$ images for elastic and rigid surfaces. The tip is scanned in the y-direction on (**a**) a rigid surface for $k = 0.50$ N/m, (**b**) an elastic surface for $k = 0.50$ N/m, and (**c**) a rigid surface for $k = 0.75$ N/m. The scanning region is 9.9 Å ×9.8 Å

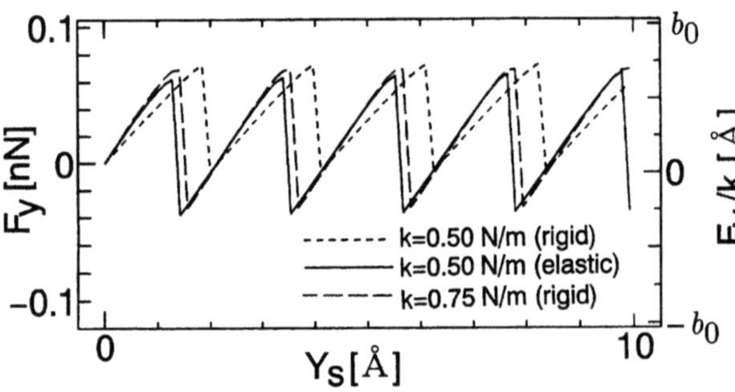

Fig. 1.16. The cross sections of Fig. 1.15 at the line indicated by the *arrow* along the y direction. *Dotted*, *solid* and *broken lines* correspond to (**a**) rigid surface for $k = 0.50$ N/m, (**b**) elastic surface for $k = 0.50$ N/m, and (**c**) rigid surface for $k = 0.75$ N/m, respectively

$k = 0.50$ N/m for an elastic surface shows behavior similar to that with $k = 0.75$ N/m for a rigid surface. Surface deformation thus makes the cantilever effectively stiff. An explanation for this feature can be based on the behavior of the total energy V as follows. The elastic energy of the cantilever V_{T} does not depend on whether the surface is rigid or elastic. On the other hand, spatial variations of the tip–surface interaction potential V_{TS} become reduced when surface deformations are introduced. This is because when the tip atom approaches the potential barrier of V_{TS}, nearby surface atoms will be relaxed to the lower value of V_{TS}. The degree of stick–slip is determined by the ratio of the magnitudes of spatial variations in V_{T} and V_{TS}. Therefore, in the particular case of Fig. 1.15b, the ratio $V_{\mathrm{T}}/V_{\mathrm{TS}}$ for the elastic surface becomes larger than that for the rigid surface, for the same cantilever or the same V_{T}. The enhanced ratio can be realized with a stiffer spring for the rigid surface. It can then be said that the effect of surface elasticity in the x and y directions is included in the cantilever stiffnesses k_x and k_y.

1.4 Dynamic-Mode Atomic Force Microscopy

1.4.1 Importance of the Study of Cantilever Dynamics

In this section we focus on the dynamics of the cantilever motion in dynamic-mode AFM, and reveal various important features relevant to the mechanisms picking up atomistic information from the sample surface via the tip. The reason why knowledge of the macroscopic dynamics of the tip is essential when analysing dynamic-mode AFM images can be explained in the following way. In the case of large amplitude dynamic-mode AFM, the amplitude of cantilever oscillations is of the order of several hundred angstroms, though the tip height above the surface in the contact region is no greater than a few angstroms. Here the contact region means the tip location where the tip–surface interaction becomes significant enough to influence tip dynamics. A model of the interaction force between tip and surface in this regime is shown in Fig. 1.17. It generally includes an attractive region outside and a sharp repulsive part inside. If we mimic the tip and cantilever part as a point mass attached to a harmonic spring, its dynamics can be described by the trajectory in phase space, as seen in Fig. 1.18. In the large area outside the contact region, the trajectory assumes an elliptic form because the motion is the same as that of the harmonic oscillator. But in the contact region, this

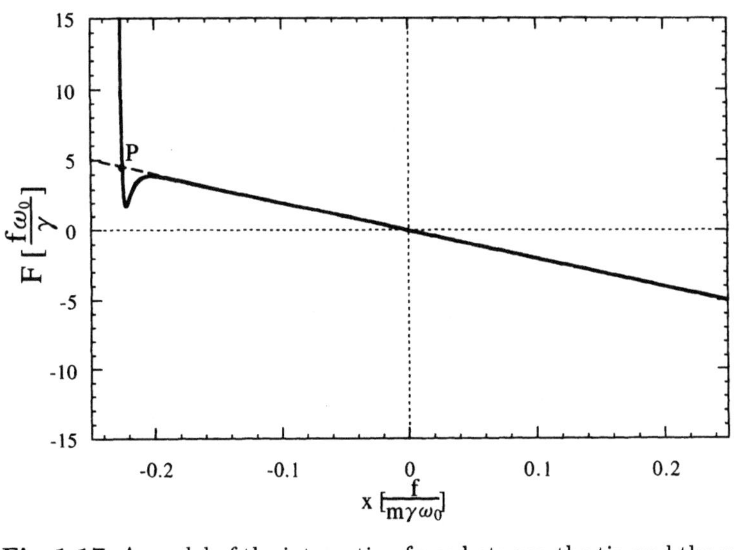

Fig. 1.17. A model of the interaction force between the tip and the surface. Force F and distance x of the mass point, measured from the equilibrium position without the tip–surface interaction, are defined by $F = F(x + d) - kx$ with $F(x + d)$ given by (1.19). The *thin broken line* is the asymptotic line of the force exerted on the mass point from the linear spring. The attractive interaction region is found in front of the crossing point P between the asymptotic and the real force curve where a significant difference appears between the two forces

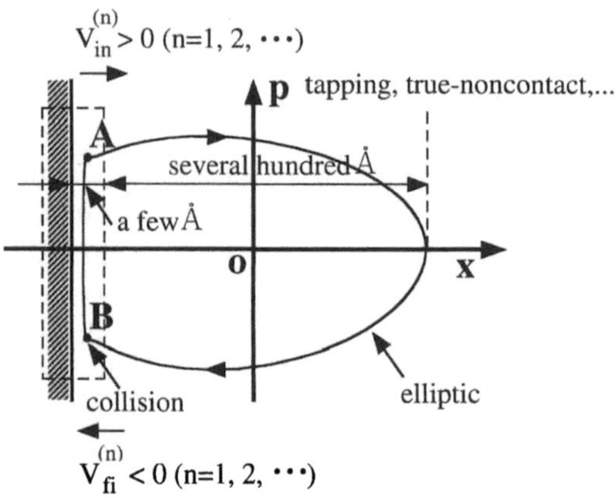

Fig. 1.18. Dynamics of the cantilever described by the trajectory in the phase space. Directions of $v_{\mathrm{fi}}^{(\mathrm{n})} < 0$ and $v_{\mathrm{in}}^{(\mathrm{n})} > 0$ are also shown

elliptic form is truncated by the almost vertical line parallel to the momentum axis. This is due to the sharp repulsive potential of the surface.

The traverse time τ from the lower edge B to the upper edge A of the nearly vertical section of the trajectory is the duration of the tip–surface contact, also envisaged as the tip–surface collision time. The order of this time τ would be estimated as 10–10^2 ns, which is much shorter than the oscillation period, i.e., about 1–10 μs for most cases. However, the tip–surface collision time τ is much larger than the oscillation period of each atom, 10–10^2 fs. Therefore, for the dynamics of individual atoms, the tip motion as a whole in the contact region can be treated as quasi-static. In sharp contrast, the same tip motion in the contact region can be regarded as very rapid, almost instantaneous, compared with the macroscopic oscillation of the cantilever. The important point is that the collision time τ as well as the exact turning position of the tip is determined by the microscopic behavior of the atoms on the surface and the tip in the contact region. Furthermore, such quantities characterize the collision event of tip and surface and thereby govern macroscopic oscillation dynamics of the cantilever.

Thus two issues are involved in the theoretical analysis of dynamic-mode AFM images: the first is to determine the collision time τ, turning point, and dissipation energy in a single collision event from the atomistic dynamics of the contact region. The second issue is to determine the cantilever motion with the scattering condition obtained in this way. From the cantilever motion, the quantity actually observed as the AFM image can be clarified. In the present section we shall focus only on the latter issue and study cantilever dynamics using the simple harmonic spring model.

1.4.2 Model and Method

Choosing the origin of the coordinate $x = 0$ as the equilibrium position of the spring, as shown in Fig. 1.19, the energy of the mass point attached to

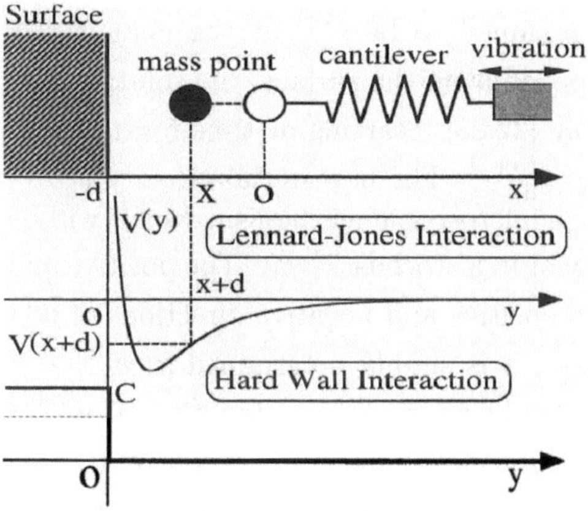

Fig. 1.19. Harmonic oscillator model used in the calculation with Lennard–Jones and hard wall potentials. The distance y of the mass point from the surface is given by $y = x + d$. The definition of x is consistent with Fig. 1.17

the one-dimensional spring and interacting with the surface at the position $x = -d$ is given by

$$E = \frac{p^2}{2m} + \frac{kx^2}{2} + V(x+d) \ , \tag{1.17}$$

where m and k are the effective mass of the cantilever and the cantilever spring constant, and $V(y)$ is the interaction energy between the tip and the surface with separation y where $y = x + d$. The momentum of the tip is $p = m\dot{x}$. If we introduce the external force, $F_{\text{ex}} = f\cos(\omega t - \phi)$, acting on the basal point of the spring, as well as the frictional force $-m\gamma\dot{x}$, the equation of motion is obtained as follows:

$$m\ddot{x} + m\gamma\dot{x} + kx - F(x+d) = f\cos(\omega t - \phi) \ , \tag{1.18}$$

where

$$F(x+d) = -\frac{\mathrm{d}V(x+d)}{\mathrm{d}x} \tag{1.19}$$

is the force due to the tip–surface interaction. The force F in Fig. 1.17 is given by $F = F(x+d) - kx$. In the following sections we investigate the tip motion along the trajectory in the phase space by numerical integration of the equation of the motion (1.18). Two models are assumed for the tip–surface interaction potential $V(x)$. One is the hard wall model,

$$V(x) = \mathrm{C}\Theta(-x) \ , \tag{1.20}$$

where $\Theta(x)$ is the usual step function and C takes a sufficiently large positive value.

The other is the Lennard–Jones potential,

$$V(x) = V_0 \left[\left(\frac{r_0}{x}\right)^{12} - 2\left(\frac{r_0}{x}\right)^{6} \right] . \tag{1.21}$$

In the hard wall case, an analytical expression for the trajectory is easily obtained over the time span in which the tip–surface collision does not take place. If the velocity of the tip is assumed to be $v_{\rm in}^{(n)}$ at the instant $t = t_{\rm in}^{(n)} (n = 1, 2, \dots)$ when the tip is repelled from the surface, the solution $x(t)$ of (1.18) is given in an analytical way [22,23]. Starting at time $t = t_{\rm in}^{(n)}$, the tip returns to the surface at time $t = t_{\rm fi}^{(n+1)}$. The new initial velocity $v_{\rm in}^{(n+1)}$ after the collision is determined by the microscopic mechanism, but it will be given by $v_{\rm fi}^{(n+1)}$ in a deterministic way or in a stochastic way. The positive and negative x directions in Fig. 1.19 are positive and negative directions of $v_{\rm fi}^{(n)}$ or $v_{\rm in}^{(n)}$. For the elastic scattering case, it is simply determined as $v_{\rm in}^{(n+1)} = -v_{\rm fi}^{(n+1)}$. For more general cases it should be given as a function of $v_{\rm fi}^{(n+1)}$ by $v_{\rm in}^{(n+1)} = F(v_{\rm fi}^{(n+1)})$ or stochastically determined by some distribution function depending on $v_{\rm fi}^{(n+1)}$. For the hard wall case the collision time is zero and the new phase of the external force $\phi^{(n+1)}$ of the $(n+1)$th step after scattering therefore changes to

$$\phi^{(n+1)} = \mathrm{mod}\left\{\phi^{(n)} - \omega\left(t_{\rm fi}^{(n+1)} - t_{\rm in}^{(n)}\right), 2\pi\right\} . \tag{1.22}$$

We can thus follow the motion of the tip in the hard wall model by mapping the initial velocity and the phase, $(v_{\rm in}^{(n)}, \phi^{(n)})$ at each successive collision event $n = 1, 2, \dots$.

1.4.3 Features of Cantilever Dynamics in Various Frequency Regimes

The motion of the cantilever we are considering is extremely nonlinear, and various unexpected features appear over its wide frequency range. If the equilibrium position of the tip is rather far from the surface, the tip quite seldom comes into the interaction region at the surface. We might imagine that after a tip–surface collision event, the tip motion would decay into that of the forced harmonic oscillator, the amplitude of which is generally small unless ω is close to the value ω_0. However, this is not necessarily the case, since it may be that the tip collides again and again with surface without disappearing into the small forced harmonic oscillator motion, and sustains the energy of the large amplitude motion with favorable phase-matching to the external force.

Some of the unexpected features of cantilever dynamics can be seen by numerical analysis of (1.18). For example, Fig. 1.20 shows traces of the tip in the phase space for the Lennard–Jones interaction potential. The external force frequency ω is chosen as $0.5\omega_0$, $0.6\omega_0$, and $0.7\omega_0$ for the cases of Figs. 1.20a, b, and c, respectively. For all the cases of Fig. 1.20, the values of γ, d, and V_0 are assumed to be $0.1\omega_0$, $0.175f/m\gamma\omega_0$, and $0.5fr_0$. We choose these parameter values simply because remarkable nonlinear behavior can then be observed.

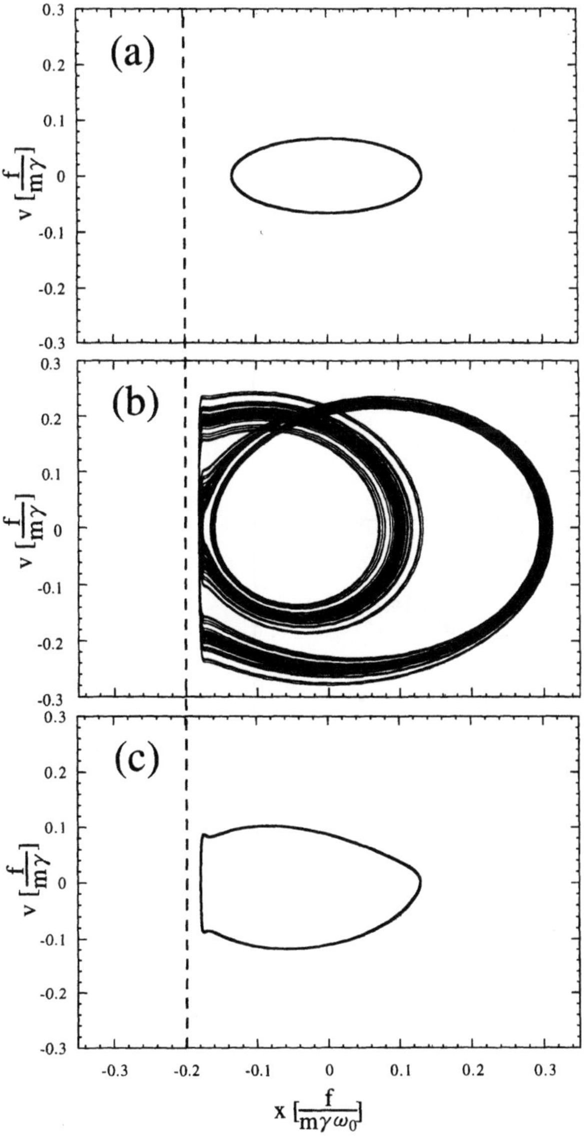

Fig. 1.20. Traces of the tip in phase space for the Lennard–Jones interaction potential when $\omega =$ (**a**) $0.5\omega_0$, (**b**) $0.6\omega_0$ and (**c**) $0.7\omega_0$

Trajectories for the case of $\omega = 0.5\omega_0$ and $0.7\omega_0$ show single period oscillations, but the trajectory for $\omega = 0.6\omega_0$ shows a quasiperiodic oscillation with an almost doubled period. Although the trajectory indicates a double period oscillation, the period of the trajectory ($\simeq 2\pi/\omega$) in Fig. 1.20b is only 67% larger than the cantilever resonance frequency ($\simeq 1.66 \times 2\pi/\omega_0$). These types of multiple period oscillations can be seen below ω_0. Similar behavior also appears for the hard wall interaction case. Fig. 1.20a shows the $(v_{\text{in}}^{(n)}, \phi^{(n)})$ map of the tip for successive collisions ($n = 1, 2, \dots$), when $\omega = 0.7\omega_0$, $\gamma = 0.2\omega_0$, and $d = 0.12 f/m\gamma\omega_0$. The figure indicates that after initial transient steps,

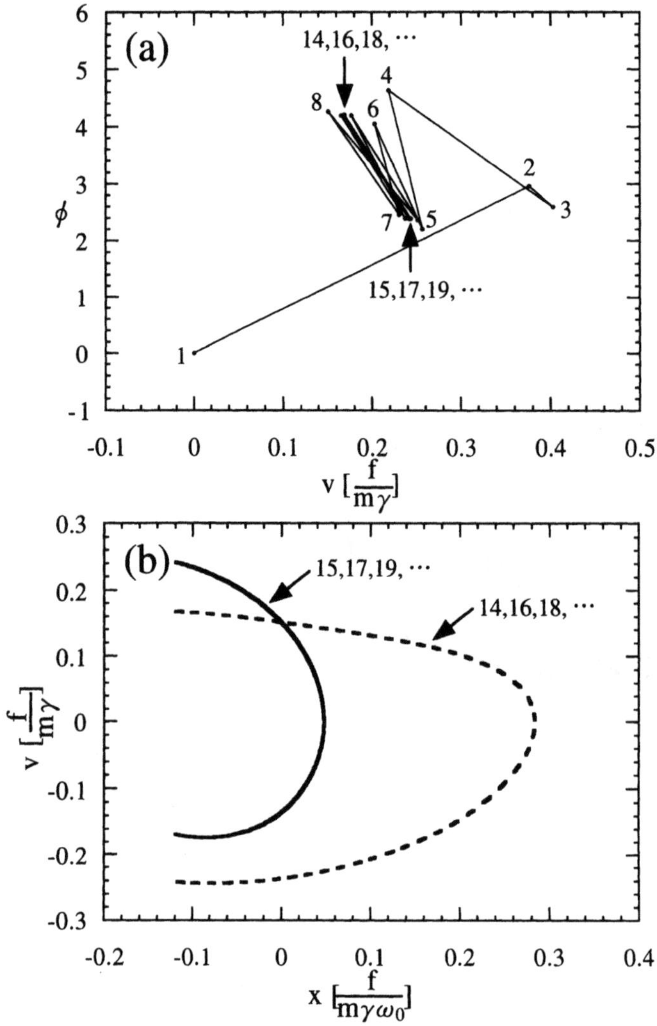

Fig. 1.21. The double period case for the hard wall interaction potential when $\omega = 0.7\omega_0$, $\gamma = 0.2\omega_0$ and $d = 0.12f/m\gamma\omega_0$. (**a**) $(v_{\text{in}}^{(n)}, \phi^{(n)})$ map $(n = 1, 2, \dots)$ and (**b**) traces of the tip in phase space

the tip motion goes into a double period state in which tip–surface collisions invoke the transition between two different kinds of trajectories: for one section of the trajectory the initial outgoing velocity is smaller than the final incoming velocity, and it is reversed for the other kind of trajectory. The bimodal oscillation of the tip in the phase space corresponding to the final state of Fig. 1.21a is shown in Fig. 1.21b.

For the other cases, we found only a simple unimodal oscillation. An example of the unimodal $(v_{\text{in}}^{(n)}, \phi^{(n)})$ map for the successive oscillations starting from the initial step $(v_{\text{in}}^{(1)}, \phi^{(1)}) = (0.0, 0.0)$ is shown in Fig. 1.22a when

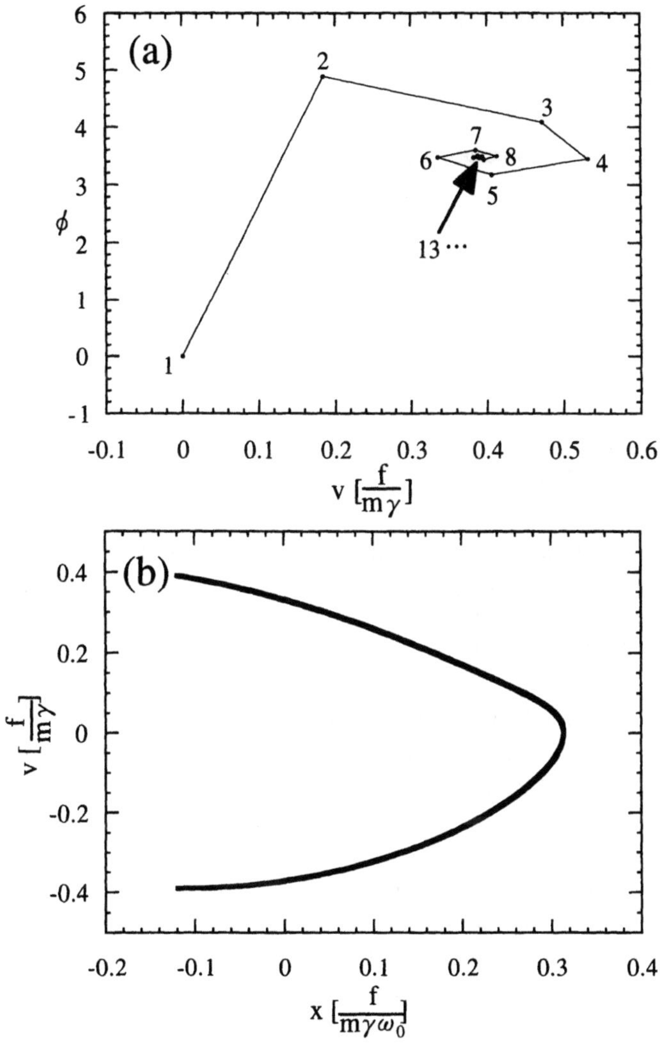

Fig. 1.22. The single period case for the hard wall interaction potential when $\omega = 1.4\omega_0$, $\gamma = 0.2\omega_0$, and $d = 0.12f/m\gamma\omega_0$. (**a**) $(v_{\text{in}}^{(n)}, \phi^{(n)})$ map $(n = 1, 2, \ldots)$ and (**b**) traces of the tip in the phase space

$\omega = 1.4\omega_0$, $\gamma = 0.2\omega_0$, and $d = 0.12f/m\gamma\omega_0$. An example of the unimodal trajectory for the hard wall model in the phase space is also shown in Fig. 1.21b. As indicated in Fig. 1.22a, convergence of the points $(v_{\text{in}}^{(n)}, \phi^{(n)})$ to a fixed point is found to be rapid. In such a case, the tip motion reaches a final stationary state after several transient cycles.

Even if the converged $(v_{\text{in}}^{(n)}, \phi^{(n)})$ point is drastically shifted by some strong perturbation, the $(v_{\text{in}}^{(n)}, \phi^{(n)})$ points resume the same stationary state after passing through several transient steps. The stability of unimodal os-

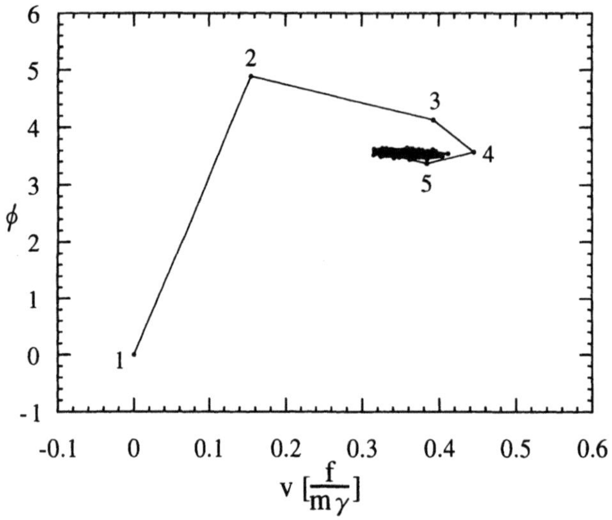

Fig. 1.23. The single period case for the hard wall interaction potential when $\omega = 1.4\omega_0$, $\gamma = 0.2\omega_0$, and $d = 0.12f/m\gamma\omega_0$. The $(v_{\mathrm{in}}^{(n)}, \phi^{(n)})$ map $(n = 1, 2, \ldots)$ for stochastic collision conditions. The repulsion coefficient e is chosen as a uniform random number ranging from 0.8 to 1.0

cillations can also be confirmed by applying stochastic collision conditions. Hence, in the case of Fig. 1.23 for $\omega = 1.4\omega_0$, $\gamma = 0.2\omega_0$, and $d = 0.12f/m\gamma\omega_0$, we generated the repulsion coefficient e by a uniform random number ranging from 0.8 to 1.0 and determined the initial velocity after the $(n+1)$th collision by $v_{\mathrm{in}}^{(n+1)} = -ev_{\mathrm{fi}}^{(n+1)}$. Then, after following a transient path, the $(v_{\mathrm{in}}^{(n)}, \phi^{(n)})$ map shows a scattered distribution in a narrow localized region. This indicates a remarkable stability of the stationary oscillation state. The mode appearing in the higher frequency region is the so-called dynamic touching mode, which will be described in the next section.

1.4.4 Energy Spectra of the Cantilever

The cantilever energy is plotted in Fig. 1.24 as a function of frequency for various tip heights d above the surface. The hard wall model is used, and the friction parameter is set as $\gamma = 0.2\omega_0$. Other parameters are the same as above. If the height d is large enough for the case of Fig. 1.24a, the peak structure approximately exhibits a Lorentz resonance curve typical of the ideal harmonic oscillator. For spectra with the smaller values of d (Fig. 1.24a B–G), the lower slope of the main peak coincides with that for the ideal harmonic oscillator, except for several fine structures in the long tail parts. These fine structures reflect the double, triple and higher multiple period oscillations mentioned in previous sections. The lower slope part deviates abruptly from the normal spectrum above a certain frequency, which depends on the height d. The gradient of the curve in the higher frequency section is considerably

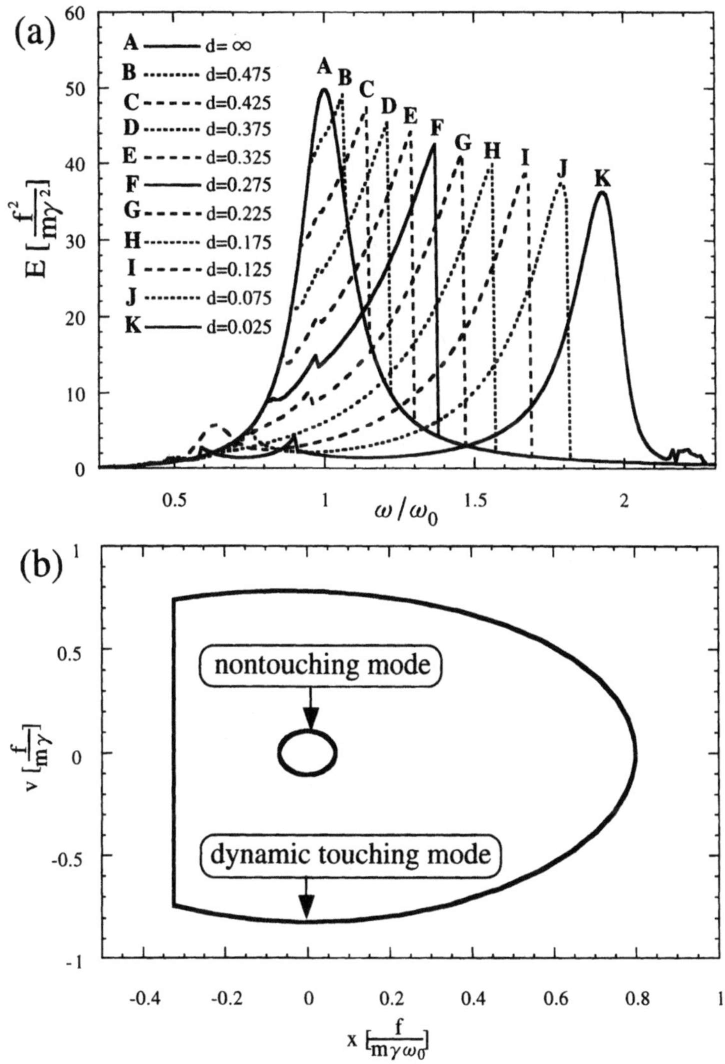

Fig. 1.24. (**a**) Energy spectra of the cantilever as a function of frequency for various tip heights d above the surface, and (**b**) traces of the tip in the phase space for nontouching mode and dynamic touching mode in the case of the hard wall interaction potential

smaller than in the lower frequency section. The upper section corresponds to the touching motion of the tip with the surface. The elliptic trajectory in the phase space shows a truncated and deformed ellipse, as shown in Fig. 1.24b. This is because the amplitude of the tip oscillation exceeds the height. With further increase of the frequency, the spectral curve becomes higher than that of the corresponding normal spectrum, i.e., that for case A ($d = \infty$). However, at a certain frequency the energy abruptly drops to the value of the normal spectrum. Unless the tip equilibrium position is close to the surface, the drop in energy is discontinuous. As a whole the peak is generally shifted to the higher frequency side. This is due to the effect of the strong repulsion from the surface.

In the frequency regime where the spectra take higher values than the normal ones, a 'dynamic bimodal behavior' of the oscillation can be observed. That is to say, the cantilever can assume one of two oscillation modes. One of these is a small-amplitude mode which does not touch the surface and is almost the same as the usual forced harmonic oscillator. The other is called the dynamic touching mode, since the corresponding trajectory shows the deformed elliptic shape truncated by collision with the surface. As shown in Fig. 1.24b, the trajectory of the dynamic touching mode is considerably larger than that of the normal non-touching mode. The dynamic touching mode is a special large-amplitude mode enhanced by favorable phase matching of the tip motion and the external force. The tip gains a large amount of energy owing to acceleration by the force, which compensates for energy dissipation by friction during most of the time period. This is the mode obtained in the previous section by the method of tracing $(v_{\text{in}}^{(n)}, \phi^{(n)})$ for the hard wall model.

Let us now consider what happens if the tip comes close to the surface from a very distant position. The dependence of the cantilever energy on tip height is shown in Fig. 1.25 for several frequencies. If the tip is far from the surface, the tip motion shows the non-touching mode behavior. Thus the energy of the tip is almost constant as a function of tip height. When the tip approaches so closely that it collides with the surface, the oscillation changes to the touching mode and the cantilever energy starts to decrease almost linearly with tip height d. On the other hand, retracting the tip from the

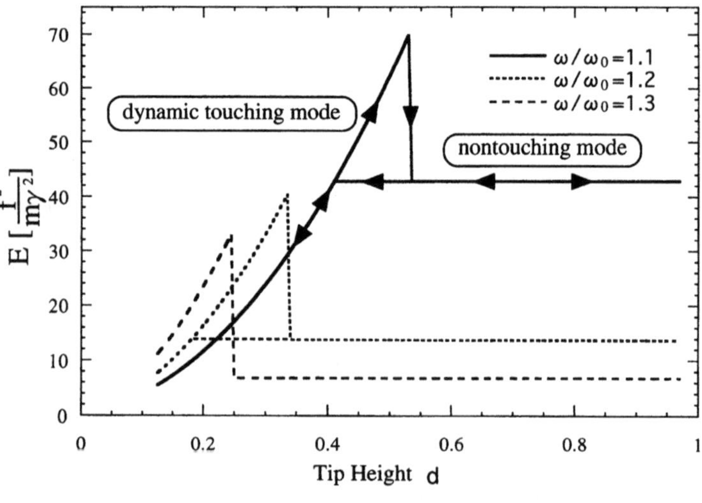

Fig. 1.25. Energy spectra of the cantilever as a function of tip height for various frequencies ω

surface in the touching regime, the oscillation energy increases. It increases beyond the initial value at infinite tip height ($d = \infty$). But when the tip is retracted beyond a certain distance from the surface, the cantilever energy jumps discontinuously from the higher value to the initial value at $d = \infty$. This is nothing other than the transition from the dynamic touching mode to the non-touching mode. Thus the spectrum exhibits hysteresis between the approaching and retracting processes.

1.5 Non-Contact Mode Atomic Force Microscopy

1.5.1 Theory of NC-AFM

Since NC-AFM is a forced vibration system with energy dissipation, a definite understanding of the mechanism of NC-AFM cannot be achieved unless the cantilever dynamics is solved in the form of a true forced vibration with some dissipative mechanism. Therefore, in this section, we present an analytical perturbation theory in order to investigate the detailed relations between the cantilever-forced vibration and the tip–surface interaction potential [24]. As a test case, the theory is applied to a system with a van der Waals tip–surface interaction potential, although it is applicable to any form of potential. We find several important features of the resonance curve. One of these gives a clue to answering the question: What information about the tip–surface potential can be obtained from the frequency shift? The other important feature is the dynamical bistability which exerts a remarkable influence on resonance peak observations in the regime where the turning point of the tip enters the repulsive regime. This feature has already been found in Sect. 1.4 for low Q-values, but we shall discuss finer details, in particular for high Q-value regions. Results from the analytical perturbation method agree fairly well with those obtained by a numerical integration of the equation of motion.

1.5.2 Calculation Method

In the NC-AFM model, a tip with a vertical length h is assumed to be located at the end of the cantilever as shown in Fig. 1.26. If the cantilever basal position is forced to vibrate with amplitude l around $z = u_0$ and driving frequency ω, i.e., $z = u_0 + l\cos(\omega t)$, the equation of motion of the cantilever becomes

$$m\frac{\mathrm{d}^2 z}{\mathrm{d}t^2} + \frac{m\omega_0}{Q}\frac{\mathrm{d}z}{\mathrm{d}t} + k(z - u_0 + h) + \frac{\mathrm{d}V(z)}{\mathrm{d}z} = kl\cos(\omega t) \ , \qquad (1.23)$$

where m is the effective mass of the cantilever and tip, Q is a cantilever Q-value, k is the z-component of the cantilever spring constant, $\omega_0 = \sqrt{k/m}$ is the resonance frequency of the free cantilever, and $V(z)$ represents the tip–surface interaction energy. z indicates the distance between the tip apex and the rigid substrate surface.

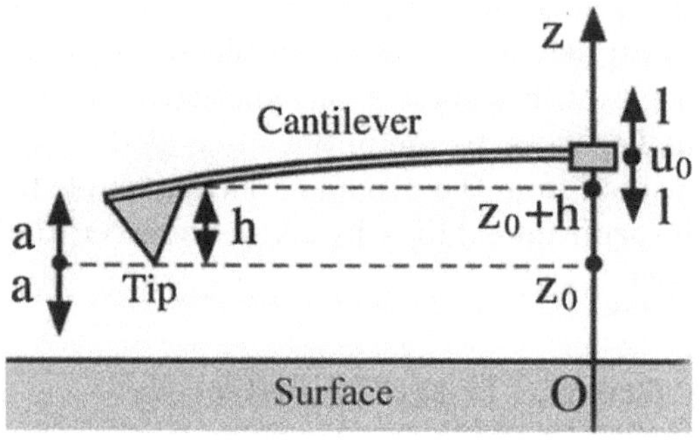

Fig. 1.26. Model of NC-AFM. A tip with height h is forced to vibrate by externally vibrating the cantilever basal position

From the stable equilibrium relation $k(z_0 - u_0 + h) = \mathrm{d}V(z_0)/\mathrm{d}z$, the equilibrium position $z = z_0$ is represented by a function of u_0. Then, on inserting $z = u + z_0(u_0)$ into (1.23), and changing the time variable in (1.23) from t to τ such that $t = \tau/\omega_0$, we obtain

$$\ddot{u} + u = \varepsilon f(u, \dot{u}) + \varepsilon A \cos(\Omega\tau) , \tag{1.24}$$

where

$$f(u, \dot{u}) = -2\mu\dot{u} - \mathrm{C}_0 - \mathrm{C}_1 \frac{\mathrm{d}V[u + z_0(u_0)]}{\mathrm{d}z} , \tag{1.25}$$

$$\mu = \frac{1}{2Q\varepsilon} , \quad \mathrm{C}_0 = \frac{z_0(u_0) - u_0 + h}{\varepsilon} , \quad \mathrm{C}_1 = \frac{1}{\varepsilon k} , \tag{1.26}$$

$$A = \frac{l}{\varepsilon} , \quad \Omega = \frac{\omega}{\omega_0} . \tag{1.27}$$

We shall hereafter use $\dot{u}$ and $\ddot{u}$ to denote $\mathrm{d}u/\mathrm{d}\tau$ and $\mathrm{d}^2u/\mathrm{d}\tau^2$, respectively.

Solution of the differential equation (1.24) for the case: $0 < \varepsilon \ll 1$ and $\Omega = 1 + \mathrm{O}(\varepsilon)$ can be computed using perturbation theory, viz., the Krilov–Bogoliubov–Mitropolsky (KBM) method [58]. If a solution $u(\tau)$ of (1.24) is assumed as $u = a\cos(\Omega\tau - \Phi) + \mathrm{O}(\varepsilon)$ with $\dot{a} = \mathrm{O}(\varepsilon)$ and $\dot{\Phi} = \mathrm{O}(\varepsilon)$, $\dot{a}$ and $\dot{\Phi}$ can be obtained to first order $\mathrm{O}(\varepsilon)$ by solving the following equations:

$$\dot{a} = -\varepsilon\mu a + \frac{\varepsilon}{2} A \sin\Phi \equiv F(a, \Phi) , \tag{1.28}$$

$$\dot{\Phi} = \Omega - 1 + \varepsilon g(a, u_0) + \frac{\varepsilon}{2a} A \cos\Phi \equiv G(a, \Phi) . \tag{1.29}$$

For a certain class of solutions of (1.24), the motion of the cantilever approaches stable simple sinusoidal oscillations, with fixed values of a and Φ. These fixed points can be observed as the critical points of (1.28) and (1.29), which are given by $F(a, \Phi) = G(a, \Phi) = 0$. By deleting Φ at the critical point, we obtain the following frequency–amplitude relation:

$$a = \frac{Ql}{\sqrt{1 + 4Q^2 [\Omega - 1 + \varepsilon g(a, u_0)]^2}} . \tag{1.30}$$

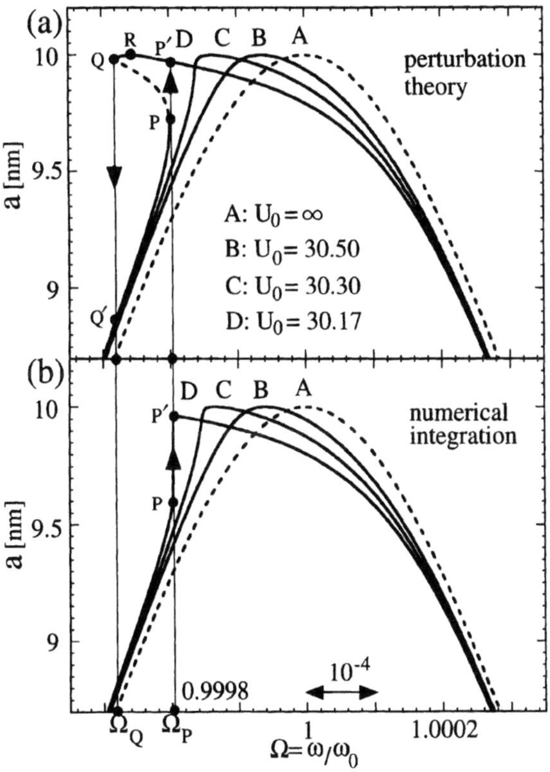

Fig. 1.27. Resonance curves obtained (**a**) by perturbation theory and (**b**) by numerical integration of the equation of motion, for $u_0 =$ (A) ∞, (B) 30.5 nm, (C) 30.3 nm and (D) 30.17 nm. R is a resonance peak of the curve D

This gives the analytical expression for the resonance curve illustrated in Fig. 1.27a. From (1.30), it is easy to see that the maximum value of a is $a_{\mathrm{max}} = Ql$ in the case $\Omega - 1 + \varepsilon g(a, u_0) = 0$. Therefore the frequency shift $\Delta\nu$ is given by

$$\Delta\nu = \frac{\omega_0}{2\pi}\left[\Omega(a_{\mathrm{max}}) - 1\right] = -\varepsilon\nu_0 g(a_{\mathrm{max}}, u_0) \, , \tag{1.31}$$

where $\nu_0 = \omega_0/2\pi$. Thus the characteristic of the resonance curve strongly depends on the Q-value itself, whilst the frequency shift is determined by $a_{\mathrm{max}} = Ql$, the value obtained by multiplying together the two parameters Q and l.

An interesting characteristic of the quantity $g(a, u_0)$ given in (1.29)–(1.31) is expressed by an analytical method as follows:

$$g\,(a, u_0) = -\frac{\mathrm{C}_1}{2\pi a}\int_0^{2\pi} \frac{\mathrm{d}V\left[z_0(u_0) + a\cos\psi\right]}{\mathrm{d}z}\cos\psi \;\mathrm{d}\psi \, , \tag{1.32}$$

as a direct integral of the tip–surface interaction force.

1.5.3 Stability Condition

An equilibrium point is asymptotically stable, when the eigenvalues of the matrix

$$M = \begin{pmatrix} \dfrac{\partial F}{\partial a} & \dfrac{\partial F}{\partial \Phi} \\ \dfrac{\partial G}{\partial a} & \dfrac{\partial G}{\partial \Phi} \end{pmatrix} \tag{1.33}$$

have negative real part. A necessary and sufficient condition for stability is therefore given by

$$\mathrm{tr}\ M = \frac{\partial F}{\partial a} + \frac{\partial G}{\partial \Phi} < 0\,, \tag{1.34}$$

$$\det\ M = \frac{\partial F}{\partial a}\frac{\partial G}{\partial \Phi} - \frac{\partial F}{\partial \Phi}\frac{\partial G}{\partial a} > 0\,, \tag{1.35}$$

where

$$\frac{\partial F}{\partial a} = -\varepsilon\mu\,, \tag{1.36}$$

$$\frac{\partial F}{\partial \Phi} = \frac{\varepsilon}{2} A\cos\Phi\,, \tag{1.37}$$

$$\frac{\partial G}{\partial a} = \varepsilon g(a, u_0) - \frac{\varepsilon}{2a^2} A\cos\Phi\,, \tag{1.38}$$

$$\frac{\partial G}{\partial \Phi} = -\frac{\varepsilon}{2a} A\sin\Phi = -\varepsilon\mu\,. \tag{1.39}$$

The first condition (1.34) is clearly satisfied because $\mathrm{tr}M = -2\varepsilon\mu < 0$. Hence, it is enough to check the second condition (1.35) for stability of the solution of (1.24). The stability condition (1.35) becomes

$$\det M = -\frac{a}{\omega_0}\frac{[\omega/\omega_0 - 1 + \varepsilon g(a, u_0)]}{\mathrm{d}a/\mathrm{d}\omega} > 0\,. \tag{1.40}$$

1.5.4 Application to the Lennard–Jones Potential

As a test calculation, the theory described above will now be applied to a system with a van der Waals tip–surface interaction potential $V(z)$. $V(z)$ is assumed to be an integral of the pairwise Lennard–Jones potential (LJ potential) $v(r)$, with an atom–atom distance r,

$$v(r) = 4\epsilon_{\mathrm{LJ}}\left[\left(\frac{\sigma_{\mathrm{LJ}}}{r}\right)^{12} - \left(\frac{\sigma_{\mathrm{LJ}}}{r}\right)^{6}\right], \tag{1.41}$$

over the region within a spherical tip of radius $R = h/2$ and inside the surface. Then $V(z)$ can be written as

$$V(z) = \int_{\mathrm{T}} \int_{\mathrm{S}} v(r) \rho_{\mathrm{T}} \, \mathrm{d}\boldsymbol{r}_{\mathrm{T}} \, \rho_{\mathrm{S}} \, \mathrm{d}\boldsymbol{r}_{\mathrm{S}}$$
$$= \frac{2}{3}\pi^2 \rho_{\mathrm{T}} \rho_{\mathrm{S}} \epsilon_{\mathrm{LJ}} \sigma^5 \left[R \left\{ \frac{1}{210} \left(\left(\frac{\sigma}{z}\right)^7 + \left(\frac{\sigma}{z+2R}\right)^7 \right) - \left(\frac{\sigma}{z} + \frac{\sigma}{z+2R} \right) \right\} \right.$$
$$\left. + \sigma \left\{ \frac{1}{6} \left(\left(\frac{\sigma}{z+2R}\right)^6 - \left(\frac{\sigma}{z}\right)^6 \right) + \ln\left(\frac{z+2R}{z}\right) \right\} \right] , \tag{1.42}$$

where ρ_{T} and ρ_{S} are number densities of atoms for the spherical tip and sample surface.

Inserting $V(z)$ into (1.32), $g(a, u_0)$ is given by

$$g\left(a, u_0\right) = -\frac{\sqrt{2}}{6} \mathrm{C}_1 \pi^2 \rho_{\mathrm{T}} \rho_{\mathrm{S}} \epsilon_{\mathrm{LJ}} \sigma^4 \left[\frac{R}{\sigma} \left\{ \frac{143}{5 \cdot 2^{11}} \left(\frac{\sigma}{a}\right)^{\frac{3}{2}} \left(\frac{\sigma}{z_0\left(u_0\right) - a} \right)^{\frac{15}{2}} \right. \right.$$
$$+ \frac{67181}{15 \cdot 2^{26}} \left(\frac{\sigma}{a}\right)^9 - \left(\frac{\sigma}{a}\right)^{\frac{3}{2}} \left(\frac{\sigma}{z_0\left(u_0\right) - a} \right)^{\frac{3}{2}} - \left. \frac{1}{8} \left(\frac{\sigma}{a}\right)^3 \right\}$$
$$+ \left\{ \frac{231}{2^9} \left(\frac{\sigma}{a}\right)^{\frac{3}{2}} \left(\frac{\sigma}{z_0\left(u_0\right) - a} \right)^{\frac{13}{2}} - \frac{17493}{2^{22}} \left(\frac{\sigma}{a}\right)^8 \right.$$
$$\left. \left. + 2 \left(\frac{\sigma}{a}\right)^{\frac{3}{2}} \left(\frac{\sigma}{z_0\left(u_0\right) - a} \right)^{\frac{1}{2}} - (3 + 2\sqrt{2}) \left(\frac{\sigma}{a}\right)^2 \right\} \right] , \tag{1.43}$$

where we use a relation, $z_0/a = 1 + (z_0 - a)/a$ and $0 < (z_0 - a)/a \ll 1$.

Parameter values of the potential energy are taken as follows: $\epsilon_{\mathrm{LJ}} = 0.01$ eV, $\sigma_{\mathrm{LJ}} = 0.25$ nm, $R = h/2 = 10$ nm, and $\rho_{\mathrm{T}} = \rho_{\mathrm{S}} = (0.25\ \mathrm{nm})^{-3}$. Those in (1.23) are $Q = 1000$, $k = 50$ N/m, $\omega_0 = 2\pi \times 100$ kHz, and $l = 0.01$ nm. In this case those in (1.24) become $\varepsilon = 0.01$, $\mu = 0.05$, $\mathrm{C}_0 \sim -0.001$ nm, $|\mathrm{C}_1 \mathrm{d}V/\mathrm{d}z| \sim 0.1$–$1.0$, and $A = 1$ nm. Here the large amplitude of the cantilever vibration, of order $Ql = 10$ nm, is considered.

1.5.5 Results

Figure 1.27 shows a comparison between resonance curves for different values of u_0, obtained by (a) perturbation theory given above [see (1.30)] and (b) numerical integration of (1.23). The resonance curve A for $u_0 = \infty$ is shown as a reference. The numerical integration procedure is simply explained. For $t \leq 100Q$ µs, the numerical integration is performed using the Runge–Kutta method with an adapted step size control. For $t \geq 100Q$ µs, when the stationary state is achieved, the time step is fixed as $\Delta t = 0.001$ µs. Then the turning point of the tip can be obtained to high accuracy by a root-finding algorithm such as trisection. Thus the vibration amplitude under the stationary state is given as $a = z_{\max} - z_{\min}$ for each frequency $\Omega = \omega/\omega_0$. The driving frequency ω is increased in steps of $\Delta\omega/\Delta t \simeq 10^{-5}$ Hz.

As the cantilever approaches the surface, the resonance peak shifts towards a lower frequency, i.e., from A to D. This behavior of the resonance curve, in particular, the resonance peak shift in Fig. 1.27b, is extremely well reproduced by perturbation theory, as can be seen from Fig. 1.27a. Resonance curves are represented in (Ω, u_0, a)-space as shown in Figs. 1.28a and b. The projection onto the (Ω, u_0)-plane of the resonance peaks connected by dotted curves in Fig. 1.28 is shown in Fig. 1.29.

In Fig. 1.29 the frequency shift $\Delta\nu$ as a function of cantilever height u_0 is given by the solid curve (perturbation theory) and dots (numerical integration). For the case of the LJ potential in (1.41), and using (1.31) and

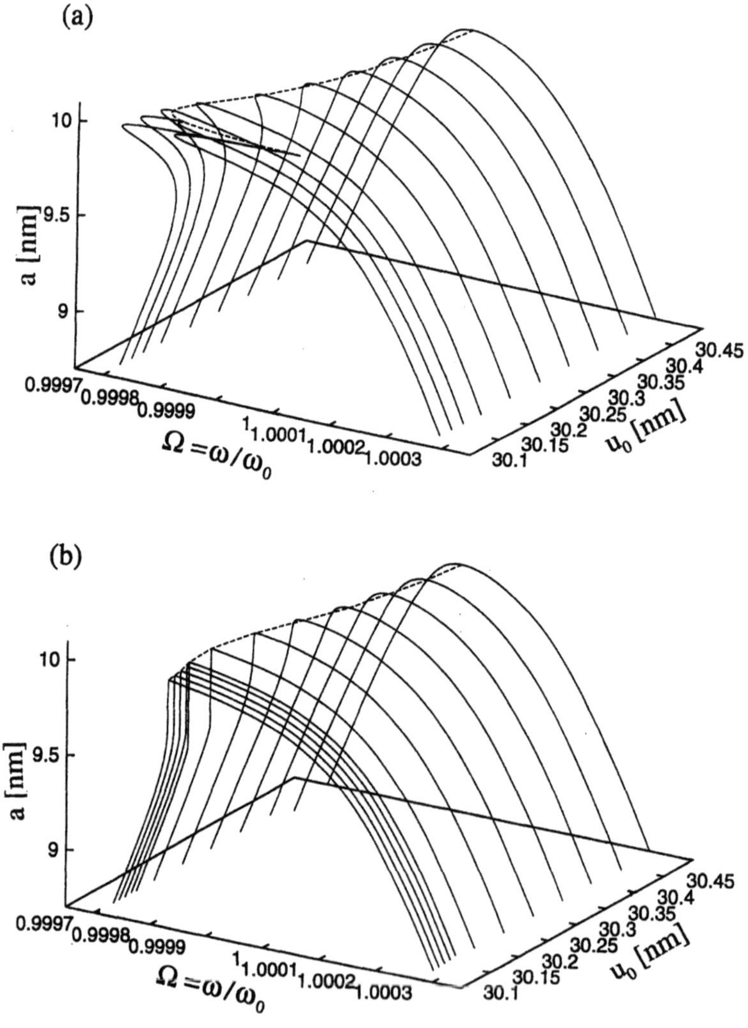

Fig. 1.28. Resonance curves in the (Ω, u_0, a)-space, obtained by (**a**) perturbation theory and (**b**) numerical integration. *Dotted curves* denote the resonance peak

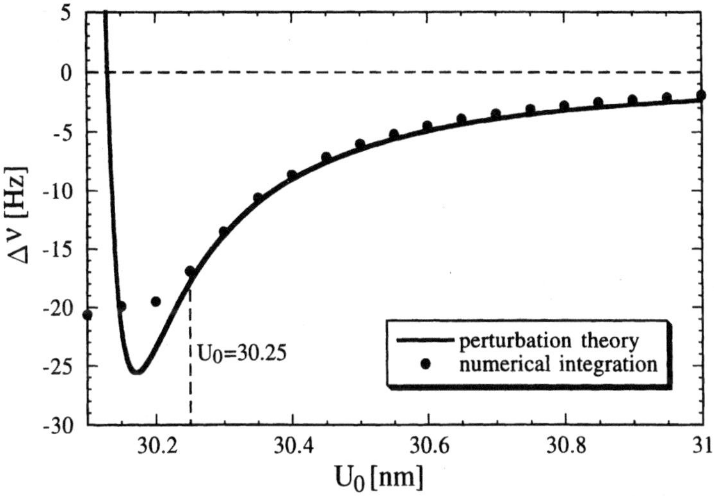

Fig. 1.29. Frequency shift $\Delta\nu$ as a function of cantilever height u_0. The *solid curve* is obtained by perturbation theory and *dots* by numerical integration

(1.32), $\Delta\nu$ can be expressed in a strictly analytical form. The u_0 dependence of $\Delta\nu$ can be roughly estimated from the two dominant terms in $\Delta\nu$ under the condition $R \gg \sigma_{\mathrm{LJ}}$ as follows:

$$\Delta\nu \sim \nu_0 \cdot \frac{\sqrt{2}}{6k}\pi^2 \rho_{\mathrm{T}}\rho_{\mathrm{S}}\epsilon_{\mathrm{LJ}}\sigma_{\mathrm{LJ}}^3 R \left(\frac{\sigma_{\mathrm{LJ}}}{a_{\max}}\right)^{3/2} \times \left[\frac{143}{5\cdot 2^{11}}\left(\frac{\sigma_{\mathrm{LJ}}}{z_0(u_0)-a_{\max}}\right)^{15/2} - \left(\frac{\sigma_{\mathrm{LJ}}}{z_0(u_0)-a_{\max}}\right)^{3/2}\right], \quad (1.44)$$

where $z_0(u_0) \simeq u_0 - 2R$ for $u_0 \gg \sigma_{\mathrm{LJ}}$. The first and second terms in the above equation are derived from the repulsive and attractive terms of the LJ potential, respectively. Therefore, in the region where the attractive interaction is dominant, $\Delta\nu \propto -(u_0 - 2R - a_{\max})^{-3/2}$. When the tip enters the repulsive interaction region, $\Delta\nu$ shows a rather complicated variation based on (1.44). In this way, starting from the van der Waals interaction model, it is demonstrated that $\Delta\nu$ reflects atomic-level information about the tip–surface interaction potential.

As shown in Fig. 1.29, perturbation theory satisfactorily reproduces the results of numerical integration for $u_0 \geq 30.25$. However, for $u_0 < 30.25$, there is a difference between the two. Likewise the shape of the resonance curve D in Fig. 1.27a is rather different from that in Fig. 1.27b in the region $\Omega_{\mathrm{Q}} < \Omega < \Omega_{\mathrm{P}}$. These differences are due to the dynamical bistability, and its mechanism can be well understood within the framework of the present theory. For the resonance curve D in Fig. 1.27a, the amplitude a is a trivalent function of Ω for $\Omega_{\mathrm{Q}} < \Omega < \Omega_{\mathrm{P}}$. Here Ω_{P} and Ω_{Q} are calculated by solving $\mathrm{d}a/\mathrm{d}\Omega = \infty$. Along the dotted curve PQ of Fig. 1.27a, the stability condition

(1.35), det$M > 0$, is not satisfied, and stability cannot be achieved in this region. Thus there exist two different kinds of dynamical stable state in the region $\Omega_Q < \Omega < \Omega_P$. Hence, as Ω is slowly increased, the amplitude a increases until Ω reaches Ω_P. However, if Ω increases further, a is forced to jump from P to P′ in a discontinuous manner. This is the physical origin of the jump from P to P′ in the resonance curve D of Fig. 1.27b. For the case of decreasing Ω, a discrete jump from Q to Q′ is expected.

Consequently, for $u_0 < 30.25$, dynamical bistability produces the 'virtual' resonance peak (P′ in Fig. 1.27b) which is different from the 'real' peak (R in Fig. 1.27a). It is therefore difficult to observe the 'real' resonance peak and frequency shift with good reproducibility in the bistable region.

1.6 Conclusion

In Sect. 1.2, the local density functional method (LDA) for the theoretical calculation of STM images and STS spectra is presented and several examples are discussed. The theoretical images reproduce experimental observations fairly well. It is shown that STM images are determined by electronic states near the Fermi level rather than by the simple atomic structure of the surface. For this reason, a naive interpretation, e.g., assigning bright spots to atoms, often fails. If the top of the tip is formed by a single-atom apex, the effect of the tip is not considered to be so serious as to totally disturb surface images. On the other hand, if the top plane of the apex is formed by more than one atom, abnormal images result. This is confirmed by a simulation in which the tip is tilted. If the tunneling electron stays within a region localised on the atomic scale, electron–electron Coulomb repulsion may cause interesting effects on STM processes, manifested particularly in STS spectra. An example is shown for a hydrogen chemisorbed Si(100) surface [34]. In this context, careful experimental studies and theoretical analysis for weakly adsorbed molecular systems are interesting. Another important aspect of STM is atom manipulation on the surface by the STM tip. Theoretical studies in this direction require a rather different approach from the one discussed here, and this has been treated elsewhere [35].

In Sect. 1.3, we clarify the physical meaning of FFM images and discuss the fundamental feature of atomic-scale friction. First a method for interpreting frictional force images is described, using the stable region of the tip atom and cantilever basal position. Based on this method, the effects of load, scan direction, and cantilever anisotropy on frictional force images are clarified. Furthermore, it is observed that theoretical images reproduce experimental ones quite well, and the effectiveness of the theoretical simulation is shown. If both calculated and experimental results are steadily stored beyond the conventional framework of research (theory and experiment), FFM will become an indispensable tool for understanding the essence of friction in micro-tribology.

In Sect. 1.4, the motion of a realistic cantilever whose tip interacts with the surface by the Lennard–Jones potential or the hard wall potential is analyzed theoretically. If the amplitude of the tip oscillation is much larger than the width of the tip–surface interaction region, various unusual features of tip motion take place depending on the external force frequency. They include pure or quasiperiodic oscillations with a double or multiple period, and cause fine structures in the lower tail of the resonance spectrum. Coexistence of the dynamic touching mode and the non-touching mode is found on the upper frequency side of the normal resonance. If the tip approaches the surface from a very distant region, a discontinuous rise in cantilever oscillation energy should take place, indicating a sudden onset of the dynamic touching mode.

Within the model and parameter range we have investigated, the resonance peak of the spectrum always shifted toward the higher frequency side. This is quite reasonable for the hard wall model or even for the Lennard–Jones potential model assumed only for a single tip apex atom. This is because the range of the attractive tip–surface interaction is rather narrow and the repulsive collision between tip and surface has a dominant effect on cantilever motion. However, if the region of attractive interaction between tip and surface ranges over a much greater distance, we can expect a shift of the resonance peak towards the lower frequency side. Theoretical analysis of the cantilever dynamics for such a situation is performed in Sect. 1.5.

In Sect. 1.5, we present an analytical theory of cantilever forced vibration in NC-AFM. It is demonstrated that the theory is a powerful tool for investigating the frequency shift and other features of the resonance curve including dynamical bistability. Calculated results for the case of a van der Waals interaction agree with those obtained using numerical integration. The frequency shift mechanism is reduced to a simple function $g(a, u_0)$ described by (1.32), which can be obtained from the atomic tip–surface interaction. A transparent and efficient method for simulating NC-AFM images of actual experiments can be obtained through this function. Finally, it is emphasized that our present theory is useful for interpreting a variety of dynamical instabilities which appear in actual experiments and can suggest optimized experimental conditions.

Acknowledgments

The authors would like to thank Prof. K. Kobayashi (Ochanomizu Univ.), Dr. T. Schimizu (Toshiba), Prof. S. Watanabe (Univ. of Tokyo), Dr. T. Uchiyama (JRCAT) and Dr. K. Hirose (NEC) for their collaborative work in Sect. 1.2. The authors would like to acknowledge the cooperation of Prof. K. Kobayashi (Ochanomizu Univ.) and the experimental group of Dr. S. Fujisawa (Mech. Eng. Lab.), Prof. Y. Sugawara (Osaka Univ.) and Prof. S. Morita (Osaka Univ.) for the work presented in Sect. 1.3. The authors also acknowledge the help of Dr. R. Tamura (Univ. of Tokyo), Mr. K. Abe (ISSP, Univ. of Tokyo) and Mr. N. Sato (Univ. of Tokyo) for the study in Sect. 1.4. The series of studies presented here was supported by a Grant-in-Aid from the Ministry of Education, Science, Sports and Culture. Numerical calculations were performed mainly by the Computation Center at the University of Tokyo.

References

1. H.J.Guntherodt and R.Wisendanger (Eds.) Scanning Tunneling Microscopy I, II, III. Springer (1992)
2. M. Tsukada, K. Kobayashi, N. Isshiki and H. Kageshima, Surf. Sci. Reports **13**, 265 (1991).
3. G. Binnig, C. F. Quate and Ch. Gerber, Phys. Rev. Lett. **56**, 930 (1986).
4. T. R. Albrecht, P. Grütter, D. Horne and D. Rugar, J. Appl. Phys. **69**, 668 (1991).
5. Q. Zhong, D. Inniss and V. B. Elings, Surf. Sci. **290**, L688 (1993).
6. C. M. Mate, G. M. McClelland, R. Erlandsson and S. Chiang, Phys. Rev. Lett. **59**, 1942 (1987).
7. J. Bardeen, Phys. Rev. Lett. **15**, 57 (1961).
8. N. Sasaki and M. Tsukada, Phys. Rev. B **52**, 8471 (1995); Jpn. J. Appl. Phys. **34**, 3319 (1995).
9. F. J. Giessibl, Science **267**, 67 (1995).
10. S. Kitamura and M. Iwatsuki, Jpn. J. Appl. Phys. **34**, L145 (1995).
11. R. Erlandsson, L. Olsson and P. Martensson, Phys. Rev. B. **54**, R8309 (1996).
12. N. Nakagiri, M. Suzuki, K. Okiguchi and H. Sugimura, Surf. Sci. **373**, L329 (1997).
13. T. Uchihashi, M. Ohta, Y. Sugawara, Y. Yanase, T. Sigematsu, M. Suzuki and S. Morita, J. Vac. Sci. & Technol. B **15**, 1543 (1997).
14. H. Ueyama, M. Ohta, Y. Sugawara and S. Morita, Jpn. J. Appl. Phys. **34**, L1086 (1995).
15. M. Bammerlin, R. Lüthi, E. Meyer, A. Baratoff, J. Lu, M. Guggisberg, Ch. Gerber, L. Howald and H.-J. Güntherodt, Probe Microscopy **1**, 1 (1997).
16. P. Gleyzes, P. K. Kuo and A. C. Boccara, Appl. Phys. Lett. **58**, 2989 (1991).
17. J. P. Spatz, S. Sheiko, M. Möller, R. G. Winkler, P. Reineker and O. Marti, Nanotechnology **6**, 40 (1995).
18. B. Anczykowski, D. Krüger and H. Fuchs, Phys. Rev. B **53**, 15485 (1996).
19. R. Pérez, M. C. Payne, I. Štich and K. Terakura, Phys. Rev. Lett. **78**, 678 (1997).
20. A. Kühle, A. H. Sørensen and J. Bohr, J. App. Phys.**81**, 6562 (1997).
21. F. J. Giessibl, Phys. Rev. B **56**, 16010 (1997).
22. N. Sasaki, M. Tsukada, R. Tamura, K. Abe and N. Sato, Appl. Phys. A. **66**, S287 (1998).
23. M. Tsukada, N. Sasaki, R. Tamura, N. Sato and K. Abe, Surf. Sci. **401/3**, 355 (1998).
24. N. Sasaki and M. Tsukada, Jpn. J. Appl. Phys. **37**, L533 (1998).
25. J. Tersoff and D. R. Hamann, Phys. Rev. B **41**, 805 (1985).
26. M. Tsukada, K. Kobayashi and N. Isshiki, J. Vac. Sci. & Technol. B **9**, 492 (1991).
27. E. J. van Loenen, J. E. Demuth, R. M. Tromp and R. J. Hamers, Phys. Rev. Lett. **58**, 373 (1987).

28. T. Takahashi, S. Nakatani, N. Okamoto, T. Ichikawa and S. Kikuta, Jpn. J. Appl. Phys. **27**, L753 (1983).
29. M. Katayama, R. S. Williams, M. Kato, E. Nomura and M. Aono. Phys. Rev. Lett. **66**, 2762 (1991).
30. S. Watanabe, M. Aono and M. Tsukada, Phys. Rev. B **44**, (1991) 8330; Appl. Surf. Sci. **60/61**, 437 (1992).
31. T. Yokotsuka, S. Kono, S. Suzuki and T. Sagawa, Surf. Sci. **127**, 35 (1983).
32. S. Watanabe, M. Aono and M. Tsukada, Surf. Sci. **287/288**, 1036 (1993).
33. S. Watanabe, M. Aono and M. Tsukada, Ultramicroscopy **42–44**, 105 (1992).
34. T. Uchiyama and M. Tsukada, Surf. Sci. Lett. **295**, L1037 (1993); J. Vac. Sci. Technol. B **12(3)**, 2205 (1994).
35. K. Hirose and M. Tsukada, Phys. Rev. Lett. **73**, 150 (1994).
36. S. Fujisawa, E. Kishi, Y. Sugawara and S. Morita, Tribo. Lett. **1**, 121 (1995).
37. S. Fujisawa, E. Kishi, Y. Sugawara and S. Morita, Phys. Rev. B **52**, 5302 (1995).
38. S. Morita, S. Fujisawa and Y. Sugawara, Surf. Sci. Rep. **23**, 1 (1996).
39. G. M. McClelland, Adhesion and Friction, M. Grunze and H. J. Kreuzer (Eds.) Springer, Berlin (1990) p.1.
40. T. Gyalog, M. Bammerlin, R. Luthi, E. Meyer and H. Thomas, Europhys. Lett. **31**, 269 (1995).
41. J. Kerssemakers and J. Th. M. De Hosson, Appl. Phys. Lett. **67**, 347 (1995).
42. J. Kerssemakers and J. Th. M. De Hosson, Appl. Phys. Lett. **80**, 623 (1996).
43. J. Kerssemakers and J. Th. M. De Hosson, Appl. Phys. Lett. **82**, 3763 (1997).
44. N. Sasaki, K. Kobayashi and M. Tsukada, Surf. Sci. **357-358**, 92 (1996).
45. N. Sasaki, K. Kobayashi and M. Tsukada, Phys. Rev. B **54**, 2138 (1996).
46. N. Sasaki, K. Kobayashi and M. Tsukada, Jpn. J. Appl. Phys. **35**, 3700 (1996).
47. N. Sasaki, K. Kobayashi and M. Tsukada, NATO-ASI Series, Micro/Nanotribology and its Applications, B. Bhushan (Ed.) Kluwer, Dordrecht 355 (1997).
48. N. Sasaki, M. Tsukada, S. Fujisawa, Y. Sugawara, S. Morita and K. Kobayashi, J. of Vac. Sci. & Technol. B **15** 1479 (1997).
49. N. Sasaki and M. Tsukada, Sci. Rep. RITU, Ser. A **44**, 1 (1997).
50. N. Sasaki, M. Tsukada, S. Fujisawa, Y. Sugawara and S. Morita, Tribol. Lett. **4**, 125 (1998).
51. N. Sasaki, M. Tsukada, S. Fujisawa, Y. Sugawara, S. Morita and K. Kobayashi, Phys. Rev. B **57**, 3785 (1998).
52. N. Sasaki and M. Tsukada, BUTSURI **52**, 704 (1997).
53. H. Hölscher, U. D. Schwarz and R. Wiesendanger, Europhys. Lett. **36**, 19 (1996).
54. H. Hölscher, U. D. Schwarz and R. Wiesendanger, Surf. Sci. **375**, 395 (1997).
55. H. Hölscher, U. D. Schwarz, O. Zwörner and R. Wiesendanger, Phys. Rev. B **57**, 2477 (1998).
56. S. A. C. Gould, K. Burke and P. K. Hansma, Phys. Rev. B **40**, 5363 (1989).
57. G. A. Tomlinson, Philos. Mag. **7**, 905 (1929).
58. N. Bogoliubov and Y. A. Mitropolsky, Asymptotic Methods in the Theory of Nonlinear Oscillations. Gordon and Breach, New York (1961).

2 The Theoretical Basis of Scanning Tunneling Microscopy for Semiconductors – First-Principles Electronic Structure Theory for Semiconductor Surfaces

Takahisa Ohno

Summary. Recent applications of first-principles electronic structure methods to semiconductor surfaces are reviewed. The detailed atomic structures and stabilities of both As-rich and Ga-rich GaAs(001) surfaces are determined from the standpoint of equilibrium energetics and theoretical simulations of scanning tunneling microscopy (STM) images. Microscopic diffusion processes of Si adatoms on a hydrogen- terminated Si(001) surface are studied and the effect of hydrogen termination on Si homoepitaxial growth is discussed. Diffusion constants, that is, preexponential factors and activation energies, of Ga and Al adatoms on an As-rich GaAs(001) surface are evaluated from first-principle calculations of migration potentials. On the basis of these calculated results, stochastic Monte Carlo simulations for a GaAs–AlAs binary system are performed.

2.1 Introduction

The density functional theory proposed by Hohenberg and Kohn [1] provides a powerful tool for predicting properties of condensed matter systems from first principles without using any empirical parameters. The field of first-principles electronic structure theory has been developed since the proposal of the density functional theory and has matured over the last decade. A large variety of theoretical techniques have been invented and successfully applied to many systems including metals, semiconductors, insulators, organic materials, and so on. We are now able to calculate electronic structures of materials and predict their electronic, optical, and magnetic properties with great precision. The static and dynamical structural properties, such as lattice constants, bulk moduli, phonon frequencies, and various reaction processes can also be evaluated.

Designing a material which exhibits a desired property is one of the major goals of materials science. Recently, novel semiconductor structures such as superlattices and quantum wires have been fabricated by using state-of-the-art growth techniques, with the aim of exhibiting novel and exotic properties. It is essential for materials design both to predict properties of the designed material and to find methods for fabricating it. The first-principles electronic structure theory contributes not only to prediction of materials properties, but also to development of methods for fabricating materials by revealing underlying growth mechanisms.

Semiconductor surfaces are of great interest both from a fundamental and an applied point of view. Surfaces are unique systems in that they have many more degrees of structural freedom than bulk crystals, and also in that transport of materials, for example, adsorption and desorption of molecules, frequently occurs on surfaces. Most optoelectronic devices take advantage of the properties of semiconductor surfaces and interfaces. Recent reduction in device size has enhanced the importance of surface and interface properties. In these circumstances, there has been an increasing stimulus to understand the structural and dynamical properties of semiconductor surfaces on a microscopic level. A large variety of surface sensitive experimental techniques such as scanning tunneling microscopy (STM) have been developed. The theory of semiconductor surfaces has also matured concomitantly with experimental developments.

In this paper we will review our recent applications of first- principles electronic structure theory to atomic structures and dynamics of semiconductor surfaces. We first describe the density functional theory in Sect. 2.2. Among several ab initio computational techniques, the pseudopotential method is briefly reviewed. This is widely used for semiconductor materials. Applications to semiconductor surfaces are then discussed in Sects. 2.3 and 2.4. We will describe theoretical investigations of atomic structures and dynamics on semiconductor surfaces. The atomic structures of GaAs(001) surfaces under both As-rich and Ga-rich growth conditions are determined from first principles in Sect. 2.3. The determination of surface structures is a prerequisite to investigations of surface dynamics. Surface diffusion processes of Si adatoms on the hydrogen-terminated Si(001) surface and of Ga adatoms on the As-rich GaAs(001) surface are studied as applications to surface dynamics in Sect. 2.4.

2.2 Computational Methods

Condensed matter systems consist of a number of nuclei and electrons interacting with each other. In order to investigate properties of the system from first principles, we usually assume two major approximations: the adiabatic approximation and the one-electron approximation. The former is an adiabatic separation of the motions of nuclei and electrons. As a result, the problem is reduced to a study of the many-body electronic state, for a given nucleus geometry. The latter is an approximation adopted in solving the many-body Schrödinger equation and the resulting one-electron equation can generally be written as

$$\left\{\frac{\hbar^2}{2m}\nabla^2 + v_{\text{eff}}(\boldsymbol{r})\right\}\psi_i(\boldsymbol{r}) = \varepsilon_i\psi_i(\boldsymbol{r}) \; , \tag{2.1}$$

where ε_i is the eigenvalue of the i-th one-electron state and $\psi_i(\boldsymbol{r})$ is the corresponding wavefunction. $v_{\text{eff}}(\boldsymbol{r})$ is an effective potential derived from the assumed one-electron approximation. Therefore, for a particular material, problems are twofold: how to derive the one-electron potential $v_{\text{eff}}(\boldsymbol{r})$ and how to solve the one-electron equation (2.1).

A local density approximation or its gradient-corrected version to the density functional theory [1,2] is an example of a one-electron approximation which provides an efficient and reliable prescription for practical calculations [3]. It is described briefly in Sect. 2.2.1. Among several ab initio (non-empirical) methods for solving the one-electron equation for condensed matter, mostly crystalline systems, the pseudopotential method is briefly reviewed in Sect. 2.2.2.

2.2.1 Density Functional Theory

According to the density functional theory by Hohenberg and Kohn [1], the non-degenerate ground state of an inhomogeneous electron system is a functional of the electron density $n(\boldsymbol{r})$. For example, the total energy of the ground state E can be given as

$$E = E[n(\boldsymbol{r})] \, . \tag{2.2}$$

A virtue of the density functional theory is that the variational principle is held at the correct ground-state density, namely

$$\frac{\delta E[n(\boldsymbol{r})]}{\delta n(\boldsymbol{r})} = 0 \, , \tag{2.3}$$

under conservation of the total number of electrons

$$\int n(\boldsymbol{r}) \mathrm{d}^3\boldsymbol{r} = N \, . \tag{2.4}$$

The density functional theory is a rigorous theory concerning the ground state of an inhomogeneous electron system but no one knows its correct functional form. However, the functional form is well studied in the case of a homogeneous electon gas system, for which the local density approximation gives the proper description.

Following Kohn and Sham's formulation [2] the total energy can be written formally as a sum of the kinetic energy $T_{\text{s}}[n(\boldsymbol{r})]$ of a *non-interacting* electron system with the same electron density $n(\boldsymbol{r})$, the classical electrostatic energy $E_{\text{HT}}[n(\boldsymbol{r})]$, the exchange-correlation energy $E_{\text{xc}}[n(\boldsymbol{r})]$, which includes all effects of many-body origins, and the interaction energy with the external potential $v_{\text{ext}}(\boldsymbol{r})$:

$$E[n(\boldsymbol{r})] = T_{\text{s}}[n(\boldsymbol{r})] + E_{\text{HT}}[n(\boldsymbol{r})] + E_{\text{xc}}[n(\boldsymbol{r})] + \int v_{\text{ext}}(\boldsymbol{r}) n(\boldsymbol{r}) \mathrm{d}^3\boldsymbol{r} \, , \tag{2.5}$$

$$E_{\mathrm{HT}}[n(\boldsymbol{r})] = \frac{e^2}{2}\int \frac{n(\boldsymbol{r})n(\boldsymbol{r}')}{|\boldsymbol{r}-\boldsymbol{r}'|}\mathrm{d}^3\boldsymbol{r}\mathrm{d}^3\boldsymbol{r}' , \tag{2.6}$$

where the external potential $v_{\mathrm{ext}}(\boldsymbol{r})$ is usually given by nucleus charges.

The variational principle with a local density approximation for the exchange correlation energy

$$E_{\mathrm{xc}}[n(\boldsymbol{r})] = \int \varepsilon_{\mathrm{xc}}[n(\boldsymbol{r})]\, n(\boldsymbol{r})\mathrm{d}^3\boldsymbol{r} , \tag{2.7}$$

where $\varepsilon_{\mathrm{xc}}(n)$ is the exchange-correlation energy density of the homogeneous electron gas of density n, leads to the one-electron equation (2.1) with an effective potential represented as

$$\begin{aligned} v_{\mathrm{eff}}(\boldsymbol{r}) &= v_{\mathrm{ext}}(\boldsymbol{r}) + \frac{\delta E_{\mathrm{HT}}[n(\boldsymbol{r})]}{\delta n(\boldsymbol{r})} + \frac{\delta E_{\mathrm{xc}}[n(\boldsymbol{r})]}{\delta n(\boldsymbol{r})} \\ &= v_{\mathrm{ext}}(\boldsymbol{r}) + e^2\int \frac{n(\boldsymbol{r}')}{|\boldsymbol{r}-\boldsymbol{r}'|}\mathrm{d}^3\boldsymbol{r}' + \left[\frac{\mathrm{d}n\varepsilon_{\mathrm{xc}}(n)}{\mathrm{d}n}\right]_{n=n(\boldsymbol{r})} . \end{aligned} \tag{2.8}$$

Then the electron density is given by summing over occupied states to give the correct number of electrons

$$n(\boldsymbol{r}) = \sum_i^{\mathrm{occ.}} |\psi_i(\boldsymbol{r})|^2 . \tag{2.9}$$

The one-electron equation (2.1) with (2.9) is a non-linear problem with respect to the electron density and may be solved through an iterative procedure (self-consistent field).

2.2.2 Ab Initio Pseudopotentials

Most physical properties of materials depend on the valence electrons to a much greater extent than on the core electrons. The pseudopotential approximation takes advantage of this fact by removing core electrons from the system and including their effects in a pseudopotential which acts on valence electrons. The ab initio pseudopotential is constructed in such a way that its scattering properties for valence electrons are identical to those for valence electrons in the true system including all electrons. Valence wave functions of the pseudopotential have no radial nodes in the core region because of the lack of orthogonality between core wave functions. This means that the pseudopotential is much weaker than the true ionic potential. The weakness of the pseudopotential enables us to expand the wavefunctions using a reasonable number of plane wave basis functions.

The most general form of an ab initio pseudopotential is

$$V_{\mathrm{pseudo}} = \sum_{lm} |lm\rangle\, V_l\, \langle lm| , \tag{2.10}$$

where $|lm\rangle$ are the spherical harmonics and V_l is the pseudopotential for angular momentum l. When the pseudopotential V_{pseudo} acts on the electronic wavefunction, it decomposes the wavefunction into spherical harmonics, each of which is then multiplied by the relevant pseudopotential V_l. A variety of methods for the construction of ab initio pseudopotentials have been proposed [4–7]. These ab initio pseudopotentials are capable of describing the scattering due to the true ion cores in a variety of atomic environments. Ab initio pseudopotential techniques are widely applied to study electronic structures and dynamical properties of a variety of materials, especially semiconductors, as will be described in Sects. 2.3 and 2.4.

2.3 Surface Structures

The surface of a solid no longer has the three-dimensional periodicity of the crystal and surface atoms relax from what would be their ideal bulk positions. Because the bonding in semiconductors is both strong and highly directional, relaxation can occur in such a way that surface atoms move to eliminate their dangling bonds. One prominent example is the (001) surface of silicon, where surface Si atoms relax to form Si–Si dimers. At the surface of a compound semiconductor such as GaAs, there exist dangling bonds of both cation and anion atoms. The charge transfer from cation to anion atoms can induce the surface relaxation of compound semiconductors. We have investigated the atomic structures of GaAs(001) surfaces under both As-rich and Ga-rich growth conditions by using first-principles electronic structure theory.

2.3.1 The GaAs(001) As-Rich Surfaces

TheGaAs(001) surface is very interesting not only in terms of surface physics but also for semiconductor technology of thin-film growth. The GaAs(001) surface manifests a sequence of reconstructions dependent on surface stoichiometry. These range from the As-rich $c(4 \times 4)$ structure to the Ga-rich (4×2) reconstruction. Among them, the As-rich (2×4) surface is of the most technological interest since molecular beam epitaxy (MBE) growth usually begins and ends with this surface under As-rich growth conditions. The atomic structure of the (2×4) surface has been intensively investigated [8–11]. It was reported that this symmetry exists over a range of surface stoichiometry. Electron diffraction measurements actually showed three different (2×4) phases (α, β and γ), although it could not determine the structures definitely since the unit cell was relatively large [9].

The principal model for the (2×4) surface is the vacancy model, which assumes that some As dimers are missing in a (2×4) unit cell. A tight-binding total energy calculation has shown that the (2×4) unit cell with three As dimers and one missing dimer is energetically favorable [8]. This structure obeys the *electron counting model*, which is a guiding principle popularly used to determine low energy surface structures of compound semiconductors. The electron counting model requires a surface structure to be stable when the number of electrons per unit cell will exactly fill all As dangling bonds and empty all Ga dangling bonds. The existence of missing As dimers has been confirmed by recent STM images [10,11].

We have determined the detailed atomic structures and stabilities of As-rich surfaces from the standpoint of equilibrium energetics by using first-principles total-energy calculations [12]. Several reconstruction models for As-rich GaAs(001) surfaces having (2×1), (2×4), $c(2 \times 8)$, $c(4 \times 4)$, and (3×1) symmetries are considered, where As coverage ranges from $\theta = 0.5$, through 2/3, 0.75, 1.0, and finally up to 1.75. The atomic structures of some of the (2×4) surfaces are shown schematically in Fig. 2.1. The structure in Fig. 2.1a contains three As dimers and one dimer vacancy per (2×4) unit cell ($\theta = 0.75$). It is called the β phase. When basic (2×4) units are arranged in anti-phase along the missing-dimer rows, the structure has the $c(2 \times 8)$ symmetry. The structure in Fig. 2.1d is denoted as the $\beta 2$ phase, since there

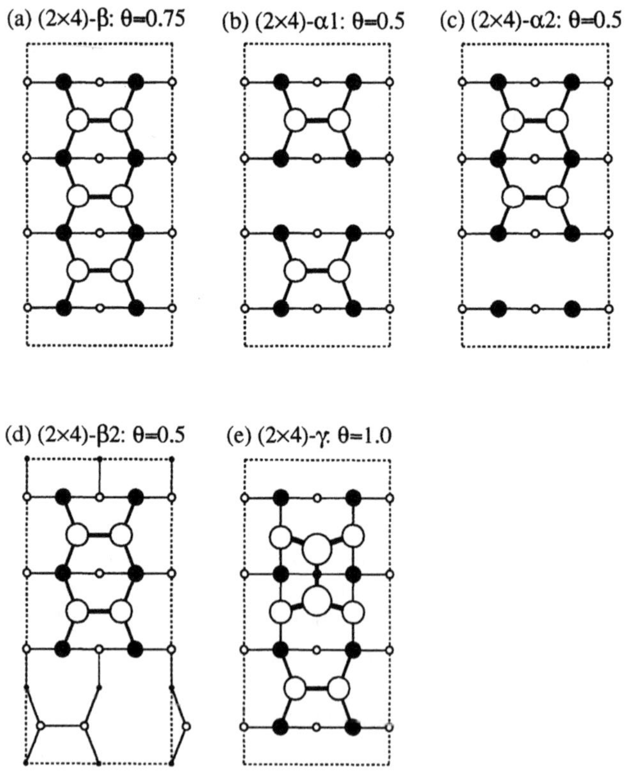

Fig. 2.1a–e. Schematic atomic structures of As-rich GaAs(001)-(2×4) surfaces whose As coverages θ are 0.5, 0.75, and 1.0. *Filled* and *open circles* denote Ga and As atoms, respectively

are actually three As dimers per (2×4) unit cell (that is, two in the top layer and one in the third layer) like in the β phase. The $c(4 \times 4)$ structure consists of three As ad-dimers on top of a complete As monolayer ($\theta = 1.75$). Furthermore, two Ga-rich (4×2) phases are considered: the $(4 \times 2)\beta$ surface consists of three Ga dimers in the top layer and the $(4 \times 2)\beta 2$ surface contains two Ga dimers in the top layer and one in the third layer. The formation energies of the GaAs(001) surfaces are plotted in Fig. 2.2 as a function of the As chemical potential.

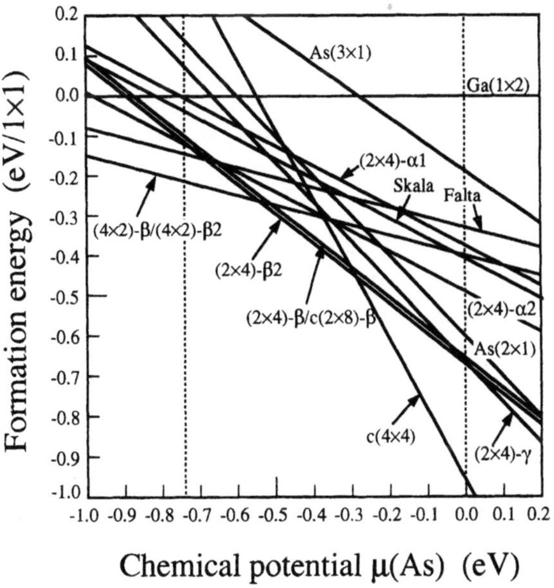

Fig. 2.2. Surface formation energy per (1×1) unit cell for GaAs(001) surfaces. *Vertical dashed lines* indicate the range of the As chemical potential. The *horizontal line* at 0.0 eV is the surface energy of the GaAs(001)-Ga(1×2) surface. The origin of the As chemical potential is taken to be that of bulk As

In this figure the chemical potential is measured relative to that of bulk As, $\mu_{\mathrm{As\,(bulk)}}$. It is noted that the energy difference between the $(2 \times 4)\beta$ and $\beta 2$ surfaces is very small. In fact, the latter is lower in energy by only 0.01 eV per (1×1) unit cell than the former. Both surfaces have three As dimers per (2×4) unit cell and the local bonding environment of each atom in the two surfaces is essentially identical, as shown in Fig. 2.3.

The factor which determines the relative stability of the $(2 \times 4)\beta$ and $\beta 2$ surfaces is of great interest. Since the two surfaces satisfy the electron counting rule, each As dimer-atom is negatively charged and each threefold-coordinated Ga atom is positively charged in the (2×4) unit cell. The electrostatic interactions between these charged atoms favor the $(2 \times 4)\beta 2$ phase [13]. On the other hand, the surface stress induced by the As-dimer formation is more difficult to relax in the $(2 \times 4)\beta 2$ phase because one As-dimer is formed but no As atom is missing in the third layer of the $(2 \times 4)\beta 2$ surface. Thus, the surface stress favors the $(2 \times 4)\beta$ structure. Cancellation of the effects of electrostatic interaction and surface stress leads to the small energy

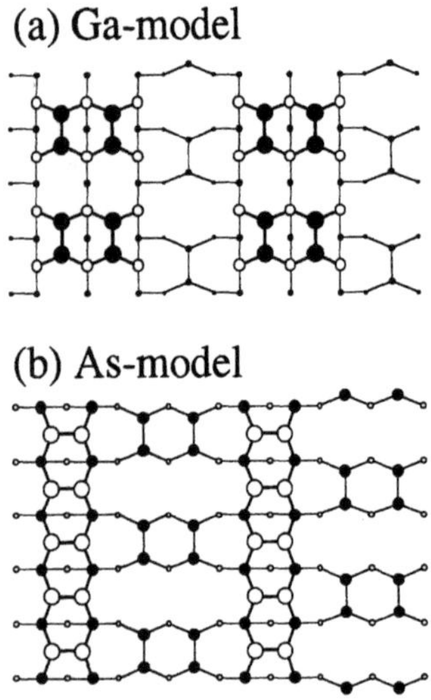

Fig. 2.3. Schematic atomic structures of the Ga-rich GaAs(001)-(4 × 2) surface: (**a**) the Ga model and (**b**) the As model. *Filled* and *open circles* denote Ga and As atoms, respectively

difference between the As-rich $(2 \times 4)\beta$ and $\beta 2$ surfaces. The same holds for the relative stability of the Ga-rich $(4\times 2)\beta$ and $\beta 2$ phases, which have almost the same energy.

Consequently, the $(2\times 4)\beta 2$ surface having two As dimers in the top layer and one in the third layer is found to be the most stable As-rich structure in the range $-0.54 < \mu_{\mathrm{As}} < -0.28$ eV, and may correspond to the well ordered two-As-dimer structure in recent STM images [11]. There is a possibility that the $(2 \times 4)\beta$ surface with three As dimers in the top layer may appear in some MBE growth conditions [9,10], since the energy difference between $(2 \times 4)\beta$ and $\beta 2$ surfaces is very small. The $c(2 \times 8)\beta$ surface has almost the same energy as the $(2 \times 4)\beta$ surface, which indicates that the basic (2×4) units can be arranged both in phase and in anti-phase. When μ_{As} exceeds -0.28 eV, the more As-rich $c(4 \times 4)$ phase becomes stable with respect to the $(2 \times 4)\beta 2$ surface. For $-0.74 < \mu_{\mathrm{As}} < -0.54$ eV the $(2 \times 4)\beta 2$ surface becomes unstable with respect to the more Ga-rich $(4 \times 2)\beta$ or $\beta 2$ surfaces.

2.3.2 GaAs(001) Ga-Rich Surfaces

Little attention has been directed toward the Ga-rich surface reconstructions such as the (4×2) and (4×6) phases. This is mainly due to the difficulty in preparing these Ga-rich surfaces using conventional MBE. The study of structures of Ga-rich surfaces is of increasing importance in order to gain comprehensive understanding of the growth mechanism of the GaAs(001) surface, since the growing front must be terminated alternately with As-rich (2×4) and Ga-rich (4×2) surfaces.

There are two distinctly different models for the Ga-rich (4×2) surface as shown in Fig. 2.3. One is the Ga model [14] consisting of two Ga dimers on the top layer and one Ga dimer at the third layer, which corresponds to the

$(4\times2)\beta2$ surface. The other is the As model [15], which has two As dimers in the top layer and two Ga dimers in the second layer in a unit cell. The STM image of the Ga-rich (4×2) surface shows that the faint features are positioned in between the brighter features along the $[\bar{1}10]$ direction [15,16]. If the faint and brighter features are identified as Ga and As atoms, respectively, the STM image cannot be explained readily by the Ga model since the Ga and As atoms should line up in the Ga model. If STM simply mapped the surface atomic geometry, this would be true. However, STM does not merely map the atomic geometry of the surface but the local density of states near the Fermi level. The correct interpretation of the STM image requires knowledge of the surface local density of states near the Fermi level. In order to resolve the discrepancy in the Ga and As models, we have performed a first-principles total energy calculation [17].

It is found that the As model is energetically very unstable with respect to the Ga model for the allowed range of the As chemical potential. Therefore, the As model can safely be ruled out. The Ga model, that is the Ga-rich $(4 \times 2)\beta2$ surface, is stable in the range $-0.74 < \mu_{\mathrm{As}} < -0.54$ eV. The calculated level charge densities for the Ga model structure are shown in Fig. 2.4. The 91st band is the HOMO (Highest Occupied Molecular Orbital)

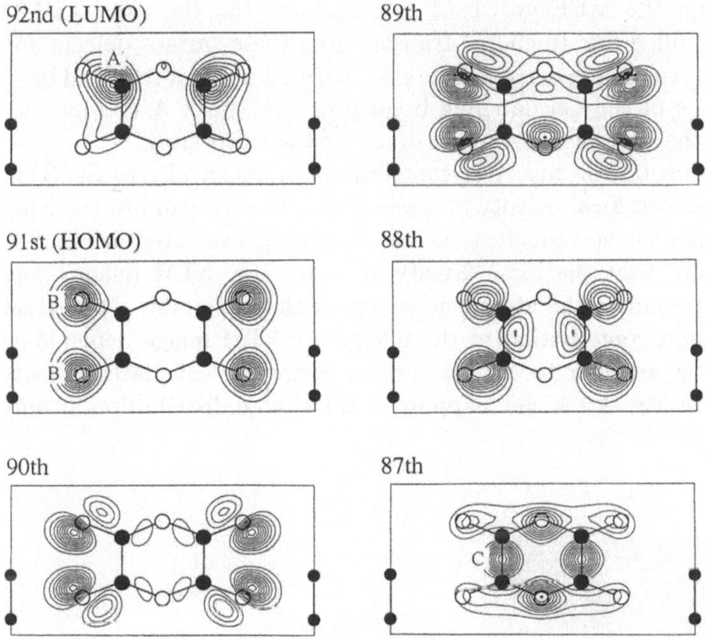

Fig. 2.4. Calculated charge density from the (4×2) Ga model structure at 0.9 Å above the top layer Ga dimer position for the 92nd (LUMO), 91st (HOMO), 90th, 89th, 88th, and 87th bands

band and has a charge localized at the As atoms in the second layer (peak B in the figure). The contribution from the Ga dimer on the top layer becomes noticeable only at the deep 87th band, whose charge distribution peaks at the middle of the Ga dimer (peak C in the figure). As the energy level increases from the 87th band to the 91st band, the contribution from the Ga dimer gradually disappears and the charge derived from the As atoms becomes more dominant. The charge distributions of the 90th, 89th, and 88th bands basically overlap with that of the 91st HOMO band. The 92nd band is the LUMO (Lowest Unoccupied Molecular Orbital) band derived from the Ga dangling bonds and should be basically empty for the ideal semiconducting (4×2) surface.

By comparing the calculated surface energy levels and the filled-state STM imaging condition at about -2 V [15,16], it is determined that all local densities of states between the 87th and 92nd bands contribute to the tunneling current to form the STM image of the (4×2) surface. The 91st HOMO band makes the most significant contribution to the tunneling current because of the lower potential barrier for tunneling, compared with the 90th, 89th, 88th, and 87th bands. Thus, the As atoms in the second layer are imaged as individual brighter protrusions (peak B), whereas the Ga dimer on the top layer is observed as single faint hump (peak C) instead of a pair-like feature.

Since there is a significant presence of surface defects, such as adsorbates and vacancies, on the MBE grown (4×2) surface [16], the 92nd LUMO band is partially filled due to charge transfer from these surface defects and contributes to the tunneling current. The charge distribution of the 92nd band is localized at one of the Ga dangling bond positions (peak A in Fig. 2.4), which is due to the existence of the Ga dimer at the third layer.

Ga dimer atoms on the top layer contribute asymmetrically to the STM image. The integrated local density of states from the 87th band to the 92nd band, proportional to the tunneling current density, is presented in Fig. 2.5. It is clearly shown that the local density of states (the STM image) does not simply correspond to the atomic geometry of the Ga model. The extent of the LUMO band contribution to the filled-state STM image depends on the level of charge transfer to the Ga dimers. Since the calculated density of states shown in Fig. 2.9 is the mapping of the charge distribution at only

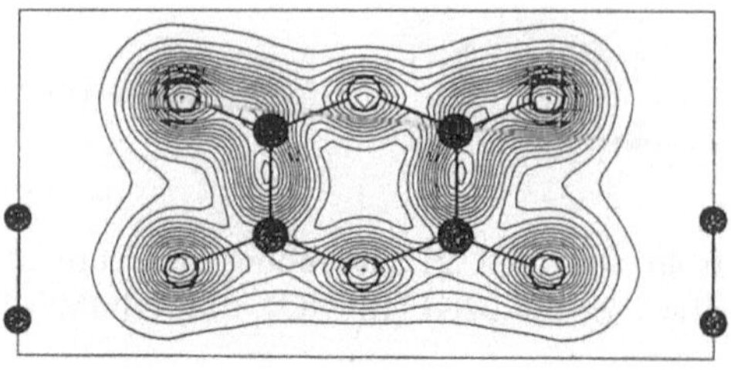

Fig. 2.5. Integrated local density of states from the 87th band to the 92nd band of the (4×2) Ga model

0.9 Å from the surface while the STM image reflects the charge distribution approximately 10 Å away, peaks A and B may shift away from the atomic position to some extent. Acknowledging these effects, the STM image reported [15,16] can be explained well in terms of the (4×2) Ga model.

2.4 Surface Dynamics

The migration process of individual adatoms on a semiconductor surface and their incorporation into growth sites of the surface play important roles in controlling atomic arrangements during epitaxial growth. Understanding these fundamental processes at the atomic level will allow development of methods for manipulating the surface dynamics and offer the exciting possibility of tailoring thin-film growth for specific materials applications. Dynamical properties of surfaces are, however, much more difficult to investigate experimentally, compared with static ones like surface atomic structures. The first-principles electronic structure theory is one of the most powerful methods available to investigate these problems. We will describe our recent studies concerning surface diffusion both of Si adatoms on the hydrogen-terminated Si(001) surface and of Ga adatoms on the As-rich GaAs(001) surface.

2.4.1 Diffusion of Si on H-Terminated Si(001)

Hydrogen termination has attracted considerable interest because of the possibility of changing the growth mode of epitaxial semiconductor films. Adsorption of hydrogen can alter surface properties such as surface energy, adatom diffusion, and nucleation processes, and result in the modification of the morphology of the epitaxial films [18–21]. Recently, the breakdown of Si homoepitaxial growth has been reported in low temperature Si molecular beam epitaxy (MBE) on hydrogenated Si(001) surfaces [18,19]. Three mechanisms have been proposed to explain the disruptive effects of H termination. The first suggests that suppression of surface Si mobility caused by H termination results in defective epitaxial growth [18]. The second claims that the breakdown in epitaxy is correlated with formation of Si dihydrides on the surface [19]. The third mechanism, based on ab initio calculations [22], suggests that surface H atoms hardly segregate from the surface and hinder the incorporation of Si adatoms into growth sites. Experimental results, however, indicate that H atoms do segregate from the Si(001) surface [19]. The role of surface H atoms in Si homoepitaxy has not been satisfactorily understood at the microscopic level. We have theoretically investigated Si adsorption and diffusion on the H-terminated Si(001)-(2×1) surface by using first-principles total-energy calculation techniques [23]. Figure 2.6 shows the structure of the H-terminated Si(001)-(2×1) surface, which corresponds to the H coverage of one monolayer (ML) [24]. Figure 2.7 presents the total

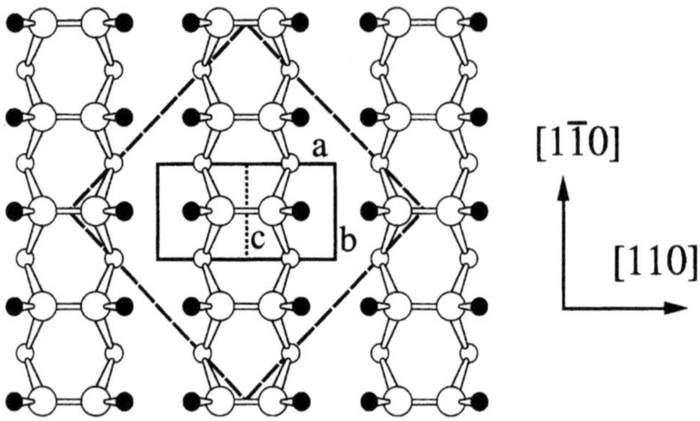

Fig. 2.6. Top view of the H-terminated Si(001)-(2 × 1) surface. *Large* and *small open circles* denote top- and second-layer Si atoms, respectively. *Solid circles* denote the terminating H atoms. The (2 × 1) unit cell is indicated by the *solid line* and the c(4 × 4) supercell by the *dashed line*. *Lines a* and *b* are the boundaries of the (2 × 1) unit cell and *line c* halves the unit cell perpendicular to the Si dimer bond

energy curve as a function of the height of the adsorbing Si atom, when the Si atom is deposited over one Si dimer atom. As the Si atom moves close to the surface, the total energy decreases gradually and then drastically drops by about 1.2 eV. This discontinuous drop in total energy corresponds to the transfer of the H atom from the Si dimer atom to the adsorbing Si atom. The bond between the H atom and the Si dimer atom is broken and a new bond is formed between the H atom and the adsorbing Si atom. The adsorbing Si atom moves further downward and finally forms a bond with the Si dimer atom. In this way, the Si atom adsorbs onto the monohydride Si(001) surface by spontaneously capturing one nearest neighbor H atom, which is consistent with experiments [19]. The geometry of the Si adatom capturing one H atom is hereafter referred to as the H-capture geometry.

Figure 2.8 shows the total energy surface of the Si adatom in the H-capture geometry. The contour map is plotted in the surface region near the *bared* Si dimer atom. The *bared* Si atom means the substrate Si atom from which the terminating H atom is detached by Si adsorption. It is noted that the contour map loses the original (2 × 1) periodicity of the H-terminated Si surface because of H detachment from the surface. There are three almost-degenerate stable sites; the S, M, and E sites. The E site is very close to the dimer center D site.

In these geometries given in Fig. 2.9, the Si adatom binds to one H atom and one Si dimer atom. The Si adatom migrates among these stable sites with an energy barrier of 0.5 eV. The capture of H atoms gives another degree of freedom to Si adsorption on the H-covered Si surface, compared with that on the bare Si surface. Besides the H-capture geometry, there are two other adsorption geometries where the Si adatom captures no and two H atoms, which

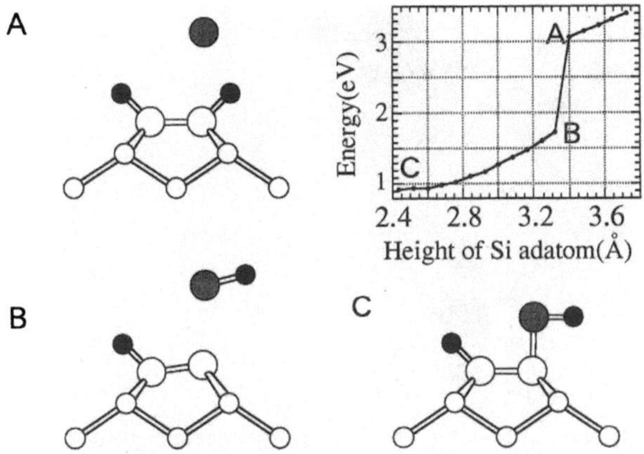

Fig. 2.7. Total energy curve when the Si atom is deposited above one Si dimer atom of the H-terminated Si(001)-(2 × 1) surface, as a function of the height of the deposited Si atom. Three geometries appearing during the Si adsorption are also presented in the *side view*. The *open, solid* and *gray circles* denote substrate Si, H atoms, and deposited Si atoms, respectively

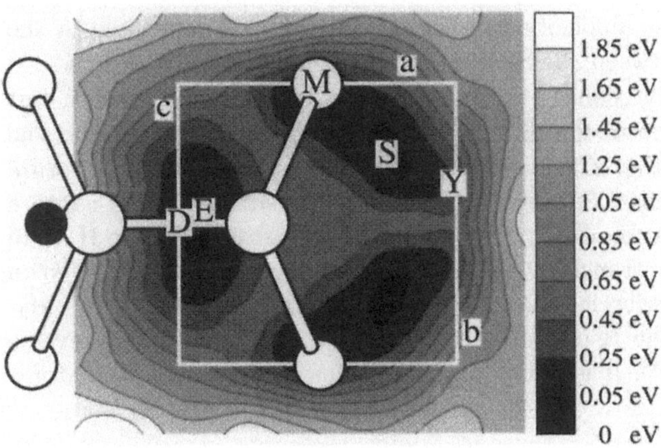

Fig. 2.8. Total energy surface of the Si adatom in the H-capture geometry on the H-terminated Si(001)-(2 × 1) surface. The contour map is plotted near the (1 × 1) surface region containing the Si dimer atom from which the H atom is detached. The (1×1) cell is indicated by the *white line*. *Open* and *solid circles* denote substrate Si and H atoms, respectively. The S, M, and E sites are three almost-degenerate stable adsorption sites. D is the dimer center site. Y is the minimum energy site along the line b. The first contour is plotted at 0.05 eV with respect to the energy of the S site and the contour spacing is 0.2 eV

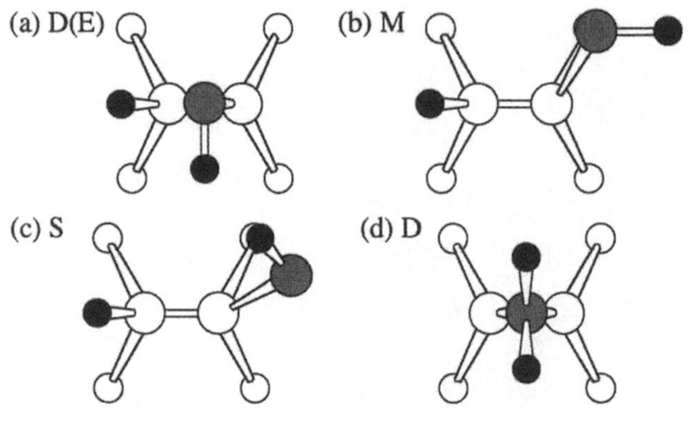

Fig. 2.9. Top views of the stable geometries for the Si adatom on the H-terminated Si(001)-(2 × 1) surface. (**a**), (**b**) and (**c**) are the H-capture geometries at the D (E), M, and S sites, respectively. (**d**) is the H_2-capture geometry at the D site

are denoted by the no- and H_2-capture geometry, respectively. Although the H-capture geometry is the lowest in energy over most of the surface region, the other geometries also play important roles at some positions. The total energy surfaces for these kinds of adsorption geometries are connected by the transfer of H atoms between the Si adatom and the substrate Si atoms.

H mobility not only realizes the most stable adsorption geometry, but assists surface migration of the Si adatom. On the dimer center D site, the H_2-capture geometry (Fig. 2.9d) is more stable by 0.5 eV than the H-capture one (Fig. 2.9a). The initial Si adsorption, however, results in the H-capture geometry instead of the H_2-capture geometry. The migration of one H atom after the adsorption can transform the former geometry to the latter as shown in Fig. 2.10. The barrier for the transformation from the H- to the H_2-capture geometry by the H displacement is 0.2 eV and that for the reverse transformation is 0.7 eV. It is noted that the diffusion barrier of H adatoms

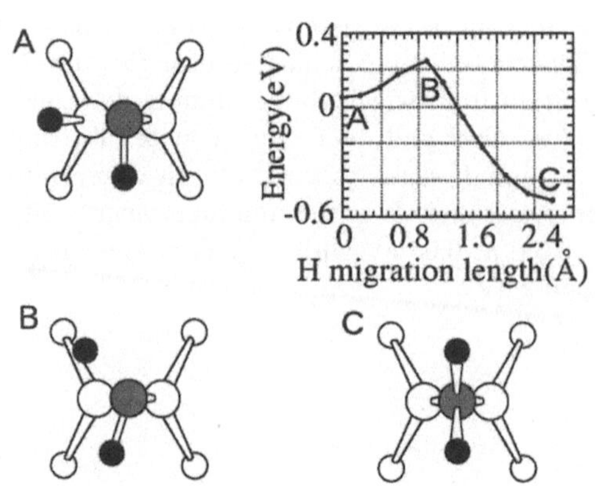

Fig. 2.10. Total energy curve of the H migration during the transformation between H- and H_2-capture geometries at the dimer center D site, as a function of the displacement of the H atom. The geometries appearing during the H migration are also presented in the top view

is measured to be more than 2.0 eV on the Si(001) surface [25]. H mobility is considerably enhanced by Si adsorption on the H-terminated Si(001) surface, since the Si adatom comes so close to the substrate Si atom that one H atom can be transferred between them.

There are two possible migration processes in which the Si adatom in the H_2-capture geometry diffuses from the most stable dimer center D site. One is the process in which the Si adatom migrates from the dimer center while remaining in the H_2-capture geometry. The other is the process assisted by H mobility, in which the Si adatom is transformed to the H-capture geometry by releasing one H atom as shown in Fig. 2.10 and then migrates on the energy surface of the H-capture geometry (Fig. 2.8). The energy barrier for Si diffusion in the H_2-capture geometry is more than 2.0 eV, while the barrier for H release is 0.7 eV and that for Si diffusion in the H-capture geometry is 0.5 eV. Consequently, the Si adatom prefers to migrate from the dimer center in the H-capture geometry by releasing one H atom.

When the Si adatom in the H-capture geometry diffuses into the neighbor (1×1) cell in which the Si dimer atom is H-terminated, the energy abruptly increases as shown in Fig. 2.8. This means that the Si adatom hardly migrates into the next cell as long as it remains in the H-capture geometry. The alternative process of Si migration into the next cell is the following one, assisted by H mobility. When the Si adatom in the H-capture geometry arrives at the boundary a or b line of the (1×1) cell, the transformation to the no-capture geometry occurs by releasing the H atom back to the *bared* Si dimer atom, as shown in Fig. 2.11. Then the Si adatom migrates into the next cell and captures one H atom from the Si dimer atom in the cell to form the H-capture geometry again. The H release has an energy barrier of 1.0 eV (Fig. 2.11) and 1.2 eV at the boundary M and Y sites, respectively. The Si adatom migrates across the boundary c line at the D site, via the H_2-capture geometry by capture and release of one H atom; the Si adatom captures another H from

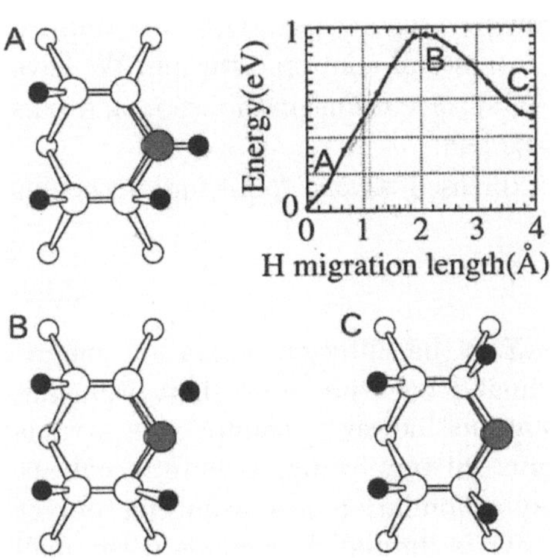

Fig. 2.11. Total energy curve of the H migration during the transformation between the H- and no-capture geometries at the M site, as a function of the displacement of the H atom. The geometries appearing during the H migration are also presented in the top view

the front Si atom of the dimer and then releases one H to the back Si atom. The barrier for this H displacement is 0.7 eV (Fig. 2.10).

Consequently, the Si adatom migrates on the monohydride Si(001) surface by repeated capture and release of H atoms. From the barrier heights of the H movement and the energy difference of 0.5 eV between the H- and H_2-capture stable geometries, the Si diffusion barriers are estimated to be 1.5 eV and 1.7 eV along directions parallel and perpendicular to the dimer rows, respectively. These barriers are very large compared with that on the bare Si(001) surface [26,27]. The Si adatom hardly migrates, apart from the Si dimer atom from which the H atom is detached. The reduction of the surface Si mobility would increase the surface roughness of the epitaxial Si film. Increased surface roughness seems to lead intrinsically to the crystalline to amorphous transition of the Si film on the H- terminated Si(001) surface [18] as well as on the bare surface. As a result, the Si homoepitaxy on the monohydride surface may be disrupted by the reduction of Si migration.

2.4.2 Diffusion of Ga on GaAs(001)

The surface diffusion of Ga adatoms on the As-rich GaAs(001)-(2 $\times$ 4) surface is one of the most important rate limiting processes in MBE growth of GaAs under As-rich conditions, and has been intensively investigated by using the RHEED intensity oscillation technique [28–31]. There is, however, a large discrepancy between reported values of diffusion constants: the activation energy of Ga-adatom diffusion on the As-rich GaAs(001) surface is estimated from 1.3 eV [28] to 4.0 eV [29]. Neither the anisotropy in surface diffusion nor the effects of surface reconstruction on surface diffusion have been clarified. The RHEED intensity oscillations indicate two-dimensional nucleation on terraces of a vicinal GaAs surface and provide macroscopic information about surface diffusion, including the influence of surface defects such as steps and interactions between adatoms themselves. Experimentally it is difficult to obtain information about the pure migration of individual adatoms on the GaAs surface. First-principles theoretical studies are thus required to investigate the microscopic processes of adatom diffusion. We have performed parameter-free calculations of surface diffusion of cation adatoms on the As-rich GaAs(001)-(2 $\times$ 4) surface [32].

The diffusion constant is usually expressed as the following Arrhenius form:

$$D = D_0 \exp(-\Delta E/k_{\mathrm{B}}T) , \tag{2.11}$$

where D_0 is the preexponential and ΔE is the diffusion activation energy. The activation energy ΔE can be evaluated by using static first-principles total energy calculations. These calculations, however, cannot give a value for the preexponential D_0 which contains entropy terms. A more recent approach, first-principles molecular dynamics simulation, is a promising tool for obtaining diffusion constants directly, but is limited to systems with small activation energies at high temperatures. We have calculated the diffusion constant by combining classical transition-state theory with first-principles total energy calculations [33]. The surface diffusion constant D of an adatom is rigorously given by

$$D = \frac{1}{4} \sum_{i,j} |X_i - X_j|^2 n_i \Gamma_{ij} , \tag{2.12}$$

where n_i is the probability for the adatom to be at a stable site X_i and Γ_{ij} is the jump rate from the X_i site to the X_j site. The jump rate can be decomposed into the directional flux Γ_{ij}^0 through a saddle surface separating the X_i and X_j sites and a dynamical correction factor ξ which accounts for immediate return diffusion jumps and correlated multiple jumps. The directional flux Γ_{ij}^0 is expressed by using the transition-state theory [34] as follows:

$$\Gamma_{ij}^0 = \left(\frac{k_\mathrm{B}T}{2\pi\mu} \right)^{1/2} \frac{\int_{S_{ij}} \exp[-E(x)/k_\mathrm{B}T] \mathrm{d}^2 x}{\int_{V_i} \exp[-E(x)/k_\mathrm{B}T] \mathrm{d}^3 x} . \tag{2.13}$$

Here μ is the reduced mass of the diffusing adatom and $E(x)$ is the total energy obtained with the adatom located at x and all other surface atoms relaxed. The three- and two-dimensional integrations are performed, respectively, over the volume V_i centered at the stable X_i site and the saddle surface S_{ij} separating the X_i and X_j sites. In deriving (2.13), it is assumed that vibrational frequencies of surface atoms depend only weakly on the position of the diffusing adatom. We also set the dynamical correction factor ξ equal to unity in the present calculations.

The calculated migration potential of a Ga adatom on the As-rich GaAs (001)-(2 × 4)β surface is presented in Fig. 2.12. A (4 × 4) surface supercell, which contains two (2 × 4)β structures, is used in the calculation as shown

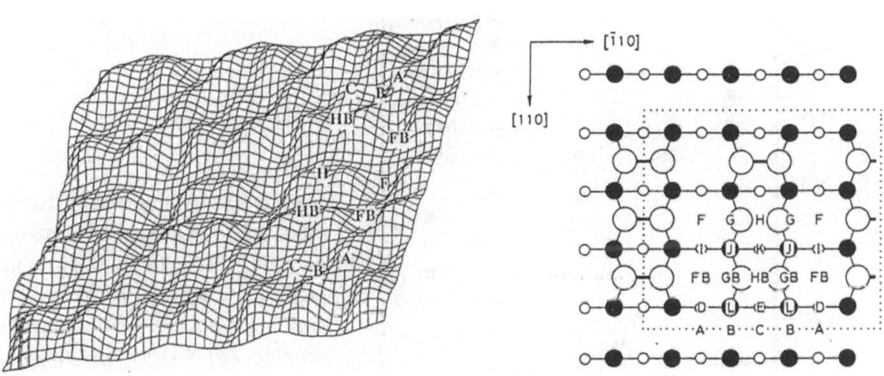

Fig. 2.12. Calculated migration potential of a Ga adatom on the reconstructed GaAs(001)-(2×4)β surface. A (4×4) surface supercell used in the calculation is also shown, in which several adsorption sites are indicated by letters of the alphabet

in this figure, where adsorption sites in the surface unit cell are indicated by letters. The most stable adsorption site of a Ga adatom is the long bridge site (F site) which is located between two As-dimer rows. The other long bridge site in the dimer region (FB site) and the bridge site in the missing dimer region (B site) correspond to local minima on the potential surface, which are 0.43 eV and 0.64 eV higher than the absolute minimum F site. The migration potential surface of an Al adatom has characteristics similar to that of a Ga adatom, while the former has larger undulation than the latter. For calculation of the diffusion constant, we define volumes V_i centered at the local minimum sites X_i and saddle surfaces S_{ij} separating the X_i and X_j sites using a Wigner–Seitz-like construction.

Generally, the diffusion constant will be different for the two principal directions on the surface. These principal directions on the GaAs(001)-(2 ×4) surface are along and perpendicular to the As-missing dimer rows, respectively. The calculated diffusion constants of Ga and Al adatoms for the two principal directions are shown in Fig. 2.13. All diffusion constants follow the Arrhenius behavior. The activation energy and preexponential for diffusion of Ga adatoms are found to be 0.85 eV and 2.1×10^{-2} cm^2/s along the As-missing dimer rows, and 1.05 eV and 2.5×10^{-2} cm^2/s perpendicular to the missing dimer rows. Diffusion of Ga adatoms is anisotropic because of reconstruction of the GaAs(001) surface, and the direction of fast diffusion is parallel to As-missing dimer rows. Al-adatom diffusion exhibits anisotropy similar to that of Ga-adatom diffusion. Activation energies and preexponentials for Al-adatom diffusion are found to be 0.96 eV and 3.3×10^{-2} cm^2/s, and 1.20 eV and 5.4×10^{-2} cm^2/s, respectively, for the directions along and perpendicular to the As-missing dimer rows. The Al adatoms diffuse several times more slowly than the Ga adatoms in the same directions because of

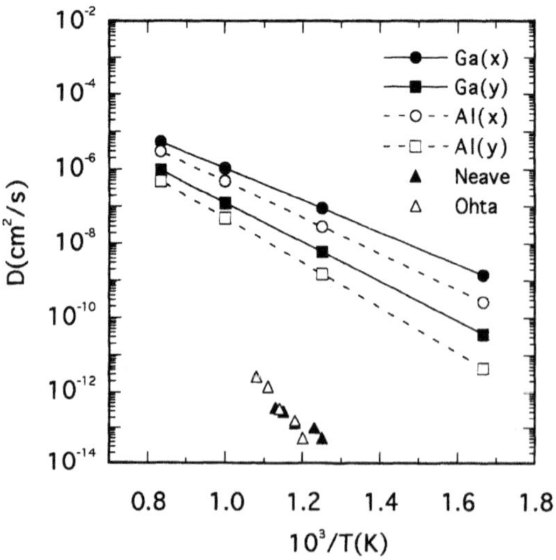

Fig. 2.13. Calculated diffusion constants of Ga and Al adatoms on the GaAs(001)-(2×4) surface, for the two principal directions along (x) and perpendicular (y) to the As-missing dimer rows. Experimental data by Neave et al. [28] and Ohta et al. [31] are also plotted for Ga adatoms

the larger activation energies. Diffusion constants obtained by the RHEED experiments are several orders of magnitude smaller than calculated values. This large discrepancy may arise from macroscopic effects included in the RHEED measurements, such as the influence of surface defects and roughness. It is desirable to determine diffusion constants on the GaAs(001) surface by more microscopic methods that can measure the pure migration of single adatoms, e.g., scanning tunneling microscopy and field ion microscopy.

To investigate the macroscopic dynamical behavior of cation adatoms and the difference between Ga and Al adatoms on the As- stabilized GaAs(001)-(2×4) surface, we performed stochastic Monte Carlo (MC) simulations for a GaAs–AlAs binary system on the basis of calculated results for diffusion constants and migration potentials [35]. In doing so, we took into account the dependence of the calculated migration potential on adatom coverage, which means that the favorable sites for adatoms change from sites on the dimer region to those on the missing dimer region as coverage increases [36].

Figures 2.14a and 2.14b show the ratio of the number n of codeposited Ga and Al adatoms for $Al_{0.5}Ga_{0.5}As$ in the dimer region and missing dimer

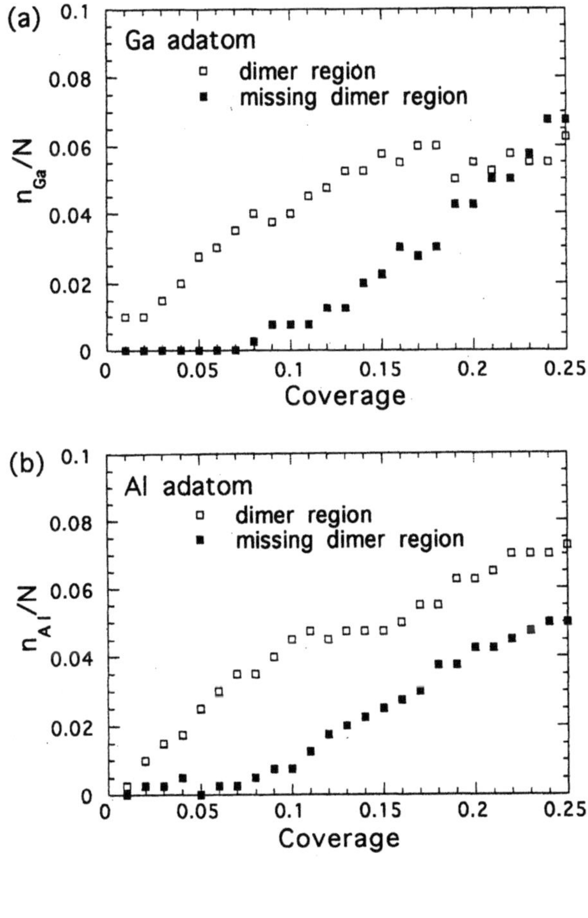

Fig. 2.14. Ratio of the number of (**a**) Ga and (**b**) Al adatoms in the dimer region (*white square*) and missing dimer region (*black square*) to the total number of surface lattice sites as a function of surface coverage. These are obtained by MC simulation at 873 K and growth rate 2 ML/s on the As-stabilized GaAs(001)-(2×4) surface

region to the total number N of surface lattice sites, as a function of adatom coverage θ. It is found from Fig. 2.14 that cation adatoms favorably exist in the dimer region at very low coverages ($\theta < 0.1$) and then the number of cation adatoms in the missing dimer region rises linearly with increased coverage. The number of Ga adatoms in the missing dimer region increases more rapidly than that of Al adatoms. This implies that Ga adatoms migrate faster to the missing dimer region than Al adatoms as coverage increases. This reflects the larger diffusivity of Ga adatoms than Al adatoms on the GaAs(001) surface.

Figures 2.15a and b show the respective atomic arrangements of Ga and Al adatoms at coverage $\theta = 0.10$ and 0.25 on the As-stabilized GaAs(001)-(2×4)

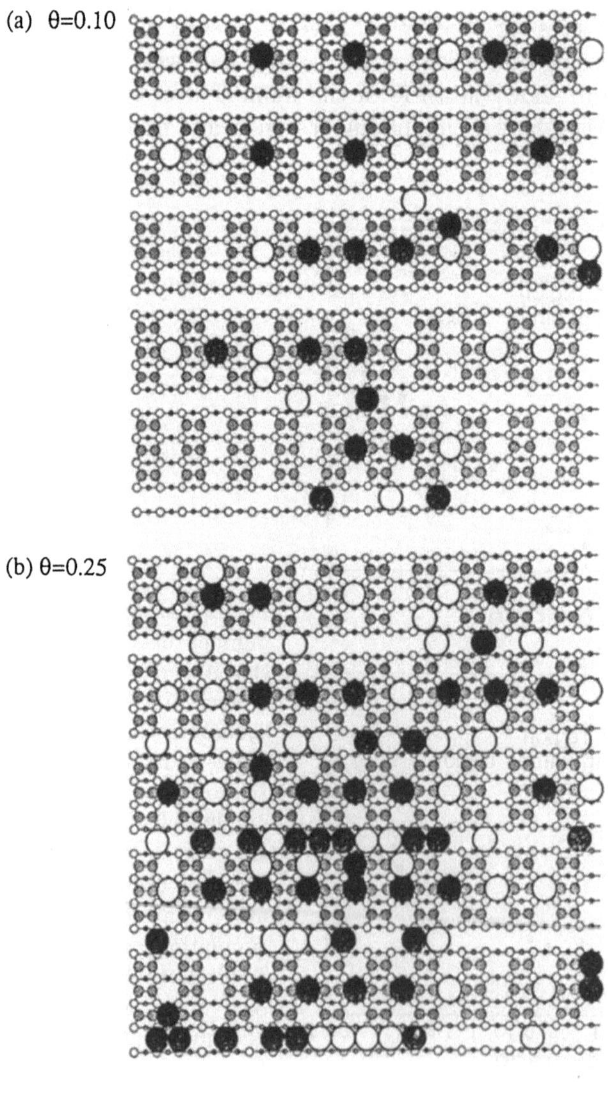

Fig. 2.15. Snapshots of atomic arrangements for $Al_{0.5}Ga_{0.5}As$ relative to adatom coverage (**a**) $\theta = 0.10$ and (**b**) $\theta = 0.25$. *White, black* and *gray circles* denote Ga adatoms, Al adatoms and surface As atoms, respectively

surface. It is clearly seen in Figure 2.15a that randomly impinging cations predominantly occupy lattice sites on As dimers (dimer region) at $\theta = 0.10$. As coverage increases, the favorable lattice sites for adatoms change from sites in the dimer region to those along the missing dimer row as shown in Figure 2.15b. This reflects the coverage dependence of the migration potential.

Acknowledgments

The author would like to give his thanks to J. Nara, T. Sasaki, K. Shiraishi and T. Ito for collaborating on the present work.

References

1. P. Höhenberg and W. Kohn, Phys. Rev. B **136**, 664 (1964).
2. W. Kohn and L.J. Sham, Phys. Rev. B **140**, 1133 (1965).
3. R.O. Jones and O. Gunnarsson, Rev. Mod. Phys. **61**, 689 (1989).
4. G.B. Bachelet, D.R. Hamann, and M. Schluter, Phys. Rev. B **26**, 4199 (1982).
5. L. Kleinman and D.M. Bylander, Phys. Rev. Lett. **48**, 1425 (1982).
6. N. Troullier and J.L. Martins, Phys. Rev. B **43**, 1993 (1991).
7. D. Vanderbilt, Phys. Rev. B **41**, 7892 (1990).
8. D.J. Chadi, J. Vac. Sci. Technol. **A5**, 834 (1987).
9. H.H. Farrell and C.J. Palmstrom, J. Vac. Sci. Technol. **B8**, 903 (1990).
10. M.D. Pashley, K.W. Haberern, W. Friday, J.M. Woodall, and P.D. Kirchner, Phys. Rev. Lett. **60**, 2176 (1988).
11. T. Hashizume, Q.K. Xue, J. Zhou, A. Ichimiya, and T. Sakurai, Phys. Rev. Lett. **73**, 2208 (1994).
12. T. Ohno, Phys. Rev. Lett. **70**, 631 (1993).
13. J.E. Northrup and S. Froyen, Phys. Rev. **B50**, 2015 (1994).
14. D.K. Biegelsen, R.D. Bringans, J.E. Northrup, and L.E. Swartz, Phys. Rev. B **41**, 5701 (1990).
15. S.L. Skala, J.S. Hubacek, J.R. Tucker, J.W. Lyding, S.T. Chou, and K.Y. Cheng, Phys. Rev. B **48**, 9138 (1993).
16. Q. Xue, T. Hashizume, J.M. Zhou, T. Sakata, T. Ohno, and T. Sakurai, Phys. Rev. Lett. **74**, 3177 (1995).
17. T. Ohno, Surf. Sci. **357–358**, 265 (1996).
18. D.P. Adams, S.M. Yalisove, and D.J. Eaglesham, Appl. Phys. Lett. **63**, 3571 (1993).
19. M. Copel and R.M. Tromp, Phys. Rev. Lett. **72**, 1236 (1994).
20. A. Sakai and T. Tatsumi, Appl. phys. Lett. **64**, 52 (1994).
21. J.E. Vasek, Z. Zhang, C.T. Salling, and M.G. Lagally, Phys. Rev. B **51**, 17207 (1995).
22. T. Ogitsu, T. Miyazaki, M. Fujita, and M. Okazaki, Phys. Rev. Lett. **75**, 4226 (1995).
23. J. Nara, T. Sasaki, and T. Ohno, Phys. Rev. Lett. **79**, 4421 (1997).
24. J.J. Boland, Phys. Rev. Lett. **65**, 3325 (1990).

25. K. Sinniah, M.G. Sherman, L.B. Lewis, W.H. Weinberg, J.T. Yates, Jr., and K.C. Janda, Phys. Rev. Lett. **62**, 567 (1989).
26. G. Brocks, P.J. Kelly, and R. Car, Phys. Rev. Lett. **66**, 1729 (1991).
27. Y.-W. Mo, J. Kleiner, M.B. Webb, and M.G. Lagally, Phys. Rev. Lett. **66**, 1998 (1991).
28. J. H. Neave, P. J. Dobson, B. A. Joyce, and J. Zhang, Appl. Phys. Lett. **47**, 100 (1985).
29. J. M. Van Hove and P. I. Cohen, J. Crystal Growth **81**, 13 (1987).
30. T. Nishinaga, T. Shitara, K. Mochizuki, and K. I. Cho, J. Crystal Growth **99**, 482 (1990).
31. K. Ohta, T. Kojima, and T. Nakagawa, J. Crystal Growth **95**, 71 (1989).
32. T. Ohno, K. Shiraishi, T. Ito, Mater. Res. Soc. Symp. Proc. **326**, 27 (1994).
33. P.E. Blochl, C.G. Van de Walle, and S.T. Pantelides, Phys. Rev. Lett. **64**, 1401 (1990).
34. G. Vineyard, J. Phys. Chem. Solids **3**, 121 (1957).
35. T. Ito, K. Shiraishi, and T. Ohno, Appl. Surf. Sci. **82/83**, 208 (1994).
36. K. Shiraishi, T. Ito, and T. Ohno, Solid State Electron **37**, 601 (1994).

3 Atomic Structure of 6H-SiC (0001) and (000$\bar{1}$)

L. Li and T. Sakurai

Summary. In this paper we present a critical review of the current status of surface studies on 6H-SiC (0001) and (000$\bar{1}$). The article is divided into three major sections. The first consists of a summary of surface preparation procedures and the experimental methods used for the work covered by this article. This is followed by a discussion of the progress which has been made in the determination of the atomic structure and composition of the numerous surface phases of clean 6H crystals. Finally we describe work on surface phonons of 6H-SiC (0001) and adsorption of Ga on the SiC surface as related to developments in the use of 6H-SiC as substrate for growth of III-nitrides.

3.1 Introduction

Recent progress in experimental investigation of silicon carbide (SiC) surfaces has been stimulated by the increasing interests in applications of SiC in microelectronic devices involving high-power, high-temperature, and high-frequency operation [1]. The potential advantages of SiC in device applications stem not only from its favorable thermal and electronic properties, but also from its excellent structural properties. These make it the most promising substrate material for the growth of group III nitrides, most notably GaN [2–4].

Of over 170 polytypes of SiC [5], i.e., crystal structures differing only in the stacking sequence of the closed-packed layers, the cubic 3C and hexagonal 6H polytypes have been the most intensely studied. As shown in Fig. 3.1, the stacking sequences for 3C- and 6H-SiC are given by ABCAB...and ABCA′C′B′AB..., respectively. Here A, B and C indicate the three possible lateral positions of the Si-C bilayers, while the primed and unprimed letters indicate a change in the bond orientation. For example, the A and A′ bilayers have equivalent spatial positions, but their bonds are rotated 180 degrees with respect to each other. The dimensions of the 3C unit cell in the ⟨111⟩ direction and the 6H unit cell in the ⟨0001⟩ direction are 7.55 Å and 15.18 Å, respectively, i.e., the distance between the Si-C bilayer is 2.52 Å. The nearest-neighbor distance on the hexagonal {111} and {0001} surface is 3.08 Å. Experimentally, the 3C-SiC {111} and 6H-SiC {0001} surfaces exhibit many similarities, as expected from the closely related stacking sequences. The 3C-SiC {111} and 6H-SiC {0001} surfaces are polar. The ideal (111) and (0001) surfaces are terminated by silicon atoms while the $(\bar{1}\bar{1}\bar{1})$ and $(000\bar{1})$ surfaces are terminated by carbon atoms.

The consequence of the polar character of these surfaces is that a single complete pseudomorphic adlayer cannot exist on either surface, since it would possess three dangling bonds per atom. Stoichiometry of the surface can be varied by preparation methods, producing a variety of surface structures depending on the exact termination. The driving force for forming these different reconstructions is to diminish the number of dangling bonds associated with surface atoms. This is clearly demonstrated by the (3×3) reconstruction found on both (0001) and $(000\bar{1})$ surfaces in which the number of dangling bonds is reduced from twenty seven to one per unit cell [6,7]. Another feature of bonding in SiC is that it is a mixture of ionic and covalent bonding. The partially ionic character results from the difference in electronegativity between silicon and carbon [8]. The valence charge density of SiC has been calculated, and the charge transfer is estimated to be from Si to C [9,10]. This should lead to an easily discernible effect on the STM experiment. For example, the bonding in GaAs is also polar. The STM images of

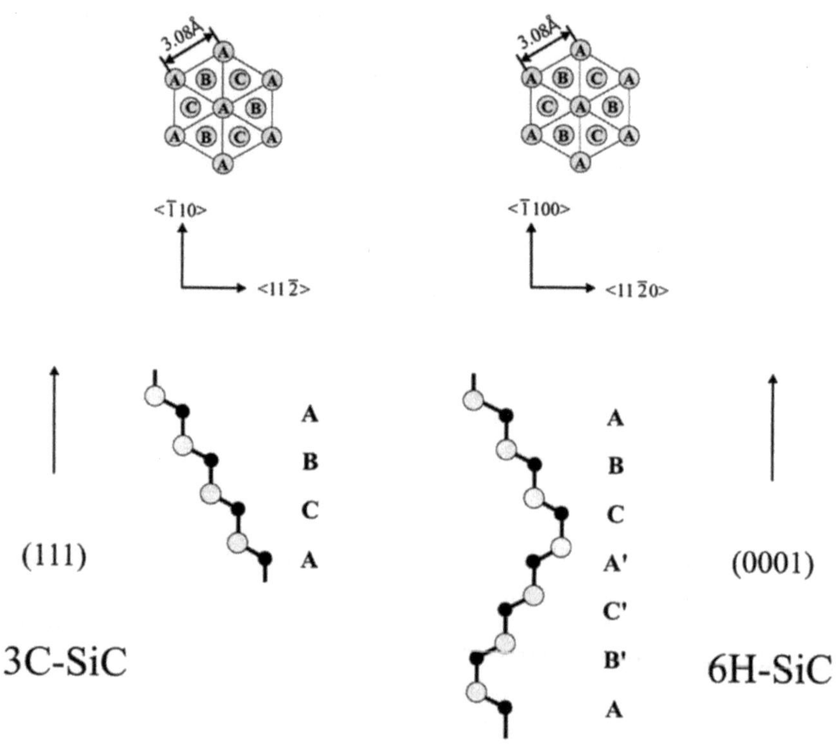

Fig. 3.1. *Top* and *side* views of ball-and-stick models for 3C- and 6H-SiC

the GaAs(110) surface exhibit contrast reversal between the filled and empty state STM images [11]. This has been explained by the fact that the surface structure of GaAs(110) results in filled states being mainly localized on As atoms, while empty states are concentrated around the Ga. This causes As atoms to appear bright in the filled state image, with Ga atoms appearing dark. On the other hand, in the empty state image, it is the Ga atoms that appear bright, while the As atoms appear dark. The detailed calculations cited above show that the SiC bond is in fact more polar than GaAs. As expected, contrast reversal is seen on a number of SiC reconstructions. This is dramatically illustrated on the $(\sqrt{3} \times \sqrt{3})$ phase of the (0001) surface [12,13]. Here a complete contrast reversal is observed between the filled and empty state images. In the filled state image a bright honeycomb structure is seen. On the other hand, in the empty state image the dark spots in the center of the honeycomb network become bright, while the honeycomb structure itself becomes dark. The resulting bright spots in the empty state image thus form centered hexagons.

The extant literature of the surface structure and electronic properties of 6H-SiC surfaces is very extensive. It includes Low Energy Electron Diffraction (LEED) [6,14,17,19], X-ray Photoelectron Spectroscopy (XPS) [20–22], Scanning Tunneling Microscopy (STM) [6,7,12,13,18,24–27], Electron Energy Loss Spectroscopy (EELS) [9,15,17,23], as well as detailed band structure measurements using photoemission on a number of different surfaces of different SiC polytypes [28–30]. There are also several theoretical calculations for proposed models of the surface structure [7,31–34]. In this article we will focus on the STM and High Resolution Electron Energy Loss Spectroscopy (HREELS) studies of the 6H-family of surfaces, in particular (0001) and (000$\bar{1}$). We begin with a description of the surface preparation procedure. This is followed by a discussion of the structures of the various phases observed on these surfaces. Next, the results of our HREELS study of the surface phonons of 6H-(0001) are presented. Finally, we conclude by enumerating gallium adsorption experiments on 6H-(0001) and (000$\bar{1}$) surfaces.

3.2 Surface Preparation

The preparation of high quality clean surfaces of SiC in ultrahigh vacuum (UHV) presents some unique challenges when compared with elemental semiconductors such as silicon. In the latter case, clean well-ordered surfaces can be achieved by simply repeatedly flashing the sample to high temperatures. A second method that has been successful for elemental semiconductors is cycles of sputtering and annealing. However, application of these relatively simple methods to SiC provides much less satisfactory results. Von Bommel et al. [14] showed that atomically clean ordered surfaces of 6H-SiC could be obtained by annealing in UHV, but problems with surface precipitation of graphite were encountered. This was attributed to sublimation of silicon that produces excess carbon on the surface. Upon cooling, this surface carbon rearranges to form graphite, which is the lowest energy phase. Another distinct disadvantage is that with this preparation method, the surface phase transformations can proceed only in the direction of successively Si-poorer compositions. Ion bombardment suffers from the problems encountered when SiC surfaces are prepared by annealing alone. Since the bombarded surfaces are badly damaged, high-temperature annealing is required in order to restore surface order.

A new approach is needed in order to avoid graphitization of the surface and to provide precise control of surface stoichiometry. Thus the required method must remove surface oxide, while at the same time either preventing the depletion of silicon or providing a constant source of silicon to replace that which is lost by evaporation when the sample is heated to high temperature ($> 900°C$). In order to achieve this, Kaplan and Parrill [35] proposed removing the surface oxide by in situ chemical reactions. They reported removal of all the oxide from both 6H and 3C samples after several minutes of exposures at 970°C to a gallium flux in the range of 0.1 to 1.0 equivalent monolayer s^{-1}. The Ga atoms impinging on the surface react to form GaO, which is volatile at this temperature. The reaction product contains only gallium and oxygen, thus avoiding silicon depletion at the surface. A similar approach using a Si beam has been the most successful of all surface preparation methods tried to date. The removal of SiO_2 by Si beam etching has been applied to the cleaning of both Si and SiC surfaces [12,15,36]. However, a fairly tight temperature window exists in which this process is effective. For temperatures below 850°C, a layer of polycrystalline silicon grows because the reaction rate of the impinging silicon atoms with the SiO_2 layer is too slow. At temperatures between 900°C and 1000°C, the arriving Si atoms convert surface oxide to SiO which is volatile in this temperature range, thus removing the oxide without otherwise disturbing the surface. Contaminant carbon is similarly converted to SiC, thereby becoming part of the crystal structure. At this point the stoichiometry of the surface can be manipulated. The surface silicon/carbon ratio can be changed in one of three ways: (1) by adding Si, (2) by desorbing silicon by annealing, and (3) by adding carbon [12,15,21–23]. In this way the entire range of surface phases can be obtained and studied.

3.3 Surface Structure of 6H-SiC (0001) and (000$\bar{1}$)

In the following, we will review the structures of the surface phases of 6H-SiC(0001) and 6H-SiC(000$\bar{1}$). In spite of the polar nature of these two surfaces, i.e., (0001) is silicon terminated, while (000$\bar{1}$) is carbon terminated, a similar family of phases is formed on each. Work on the (0001) surface will be discussed first, and this will be followed by a description of the structures of the (000$\bar{1}$) surface.

3.3.1 The (0001) Surface

The existence of four distinct (0001) surface phases is now well established. In order of increasing Si content, the observed phases are $(\sqrt{3}\times\sqrt{3})$, $(2\sqrt{3}\times 6\sqrt{3})$, (3×3), and (7×7). In the case of the (7×7) and (3×3) phases, widely accepted structural models have been proposed. However, the details of the $(2\sqrt{3}\times 6\sqrt{3})$, $(\sqrt{3}\times\sqrt{3})$ phases are not yet completely understood, and are still the subject of some debates in the literature. Although a more complete understanding of the (3×3), and (7×7) surfaces exists, in the discussion following we begin with the Si-deficit $(\sqrt{3}\times\sqrt{3})$ surface. The reason for this is that under optimal cleaning conditions, i.e., those which result in the best surface ordering, $(\sqrt{3}\times\sqrt{3})$ is typically the first surface phase obtained. All other Si-rich phases can then be produced by changing the silicon to carbon ratio of the surface through deposition or sublimation of silicon.

$(\sqrt{3}\times\sqrt{3})$ Phase. The $(\sqrt{3}\times\sqrt{3})$ phase is the most commonly observed and thus the most widely studied surface structure of SiC. For example, Kaplan et al. [15] reported the $(\sqrt{3}\times\sqrt{3})$ structure in their pioneering LEED work. However, no specific structural model was proposed. More recently, Owman et al. [18] proposed a structural model based on their STM experiments on samples prepared by annealing at 950°C in UHV. In this model, 1/3 ML of adatoms resides on top of a bulk-like Si-C bilayer as shown in Fig. 3.2. The adatoms may occupy either the four-fold coordinated sites above second layer atoms (T_4) or else set in the threefold coordinated hollow sites above the fourth layer atoms (H_3). In either case the adatoms bond to three surface

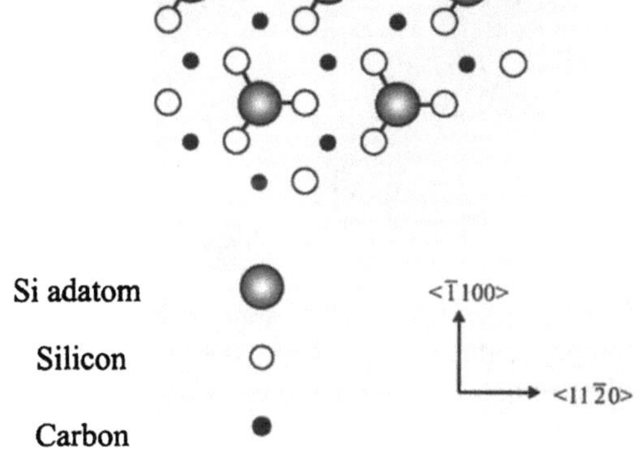

Fig. 3.2. Ball-and-stick adatom model of the $(\sqrt{3}\times\sqrt{3})$ reconstruction of Si-terminated 6H-SiC(0001) surface (from [18])

atoms, thus eliminating their dangling bonds. This configuration results in one dangling bond per unit cell. Assuming that the adatom model is correct, there are two issues which need to be addressed in order to obtain a complete understanding of the structure of this phase. The first is whether the adatoms occupy the T_4 or H_3 sites. Secondly, the chemical identity of the adatoms, i.e., Si or C, must be determined. STM observations show only a single domain, i.e., adatoms occupying exclusively T_4 or H_3, but not a mixture of both. Total energy calculations have shown that the T_4 site is more favorable than the H_3 site [33,34]. Unfortunately, the second issue, namely the chemical identity of the adatoms, cannot be resolved using STM.

A further consequence of the adatom model is that the surface must be metallic. The total energy calculations for this model indicate that there are occupied states at the Fermi level [33,34]. However, recent photoemission studies by Mårtensson et al. [28–30] indicate that the $(\sqrt{3} \times \sqrt{3})$ surface has semiconducting properties, i.e., no occupied states at the Fermi level. The authors of the study also found a surface state located at 1.3 eV below the Fermi level. In order to explain its origin, they proposed a different structural model that consists of Si adatoms on top of a modified carbon layer. This surface state then arises from the bonds associated with the Si adatoms, but the detailed structure was not given.

Further STM studies were conducted by Li et al. [12,13,27] that led to a different model. The sample preparation method was different from that of Owman et al. [18]. The sample was annealed in UHV in a flux of Si evaporated from a silicon slab approximately five centimeters from the SiC surface. They observed a marked bias dependence in the images of the $(\sqrt{3} \times \sqrt{3})$ structure, indicating that the surface is semi-conducting. An atom resolved image of the

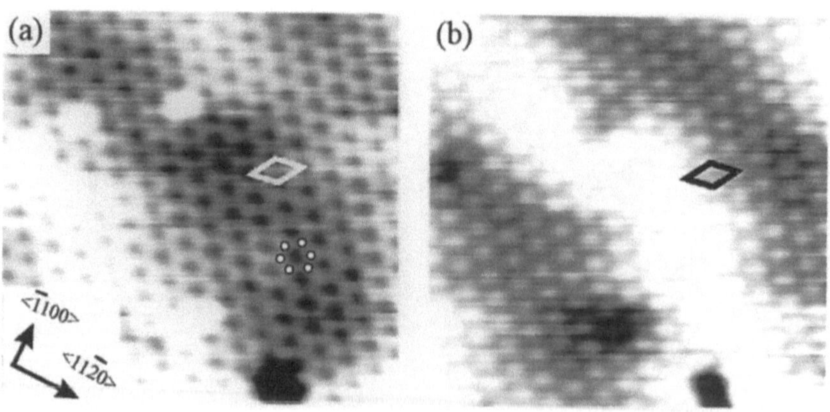

Fig. 3.3. Simultaneously recorded STM images taken with sample bias of (**a**) −1.2 V (filled state) and (**b**) +1.2 V (empty state) in the 6H-SiC(0001) $(\sqrt{3} \times \sqrt{3})$ structure. Image size = 70 Å ×70 Å. The unit cell is indicated by *white* and *black lines*, respectively (from [12])

($\sqrt{3}\times\sqrt{3}$) structure of the surface is shown in Fig. 3.3. The images were taken simultaneously at a sample bias of -1.2 V on the forward scan and $+1.2$ V on the back scan. By switching the bias voltage this way, we were sampling the filled states in image (a) and the empty states in (b) over the same area on the surface. It is evident that contrast in images (a) and (b) undergoes a reversal. The bright spots forming hexagons in image (b) become dark in (a) and the dark regions surrounding the bright spots in (b) become bright, forming honeycombs in (a). The measured unit cell spacing is 5.4 ± 0.1 Å, in good agreement with the expected value of 5.3 Å for the ($\sqrt{3} \times \sqrt{3}$) reconstruction. The measured corrugation is 0.2 ± 0.1 Å.

This STM results shown in Fig. 3.3 is apparently inconsistent with the Si adatom model since a metallic surface should not show bias dependence. Based on these STM results, a variant of the surface vacancy model for this ($\sqrt{3} \times \sqrt{3}$) was proposed. One such model is reproduced in Fig. 3.4.

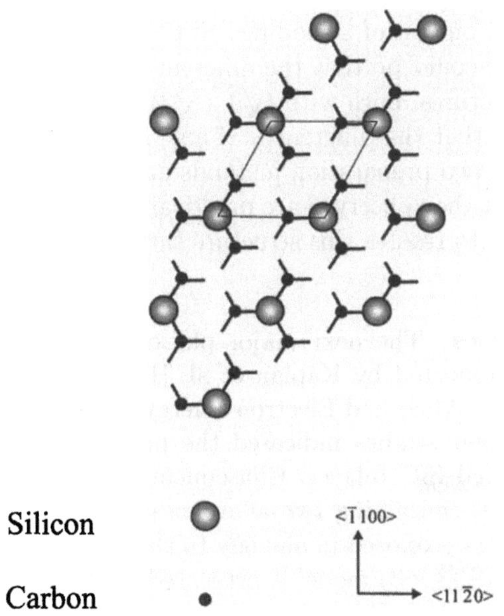

Fig. 3.4. Ball-and-stick vacancy model of ($\sqrt{3} \times \sqrt{3}$) reconstruction of Si-terminated 6H-SiC(0001) surface. The unit cell is indicated by the *black line*

The vacancies are created in the outer Si-C bilayer because it is missing two thirds of its Si atoms. The remaining Si atoms produce the ($\sqrt{3}\times\sqrt{3}$) periodicity. The missing Si atoms cause the surface to relax, with the remaining Si atoms moving inwards, closer to the surface C atoms. The bonding configuration of the Si changes from sp^3 to primarily sp^2. Furthermore, due to the ionic character of the Si-C bonds, charge transfer is expected from Si to C. Intuitively, this leads to the observed contrast in the STM images as follows. The electron density in the remaining p-type orbital of the sp^2 hybridized silicon atoms is withdrawn into the bonds now formed between the three unsaturated bonds of the C atoms, which in turn are formed upon removal of the silicon atoms. There are six unsaturated C bonds per unit cell, leading to two C-C bond complexes. Since there is one silicon atom and two C-C bond complexes per unit cell, assuming complete charge transfer from the Si to the C, this results in an additional one-half of an electron per C-C bond complex. Since the p-orbital of the remaining silicon at the surface is depleted of charge, the sites of the silicon atoms appear dark in the filled state image. At the same time, since the C-C bond complex contains a significant amount of charge, these sites appear bright in the filled state image. Nonetheless, calculations are required in order to understand fully the details of the electronic structure of this model so that a better comparison can be made with the STM data.

It is clear that the structure of the $(\sqrt{3} \times \sqrt{3})$ phase remains controversial. On one hand STM data of Owman et al. [18] is more consistent with the simple adatom model with Si on top of an intact Si-C bilayer. On the other hand, photoemission and the STM studies performed by Li et al. [12,13,28–30] support a more complex model with Si on top of a modified Si-C bilayer. One possible cause for this disagreement could be that the different preparation methods create distinct surface structures, both with $(\sqrt{3} \times \sqrt{3})$ periodicity. Recently, Starke et al. [37] reported that the differences, if any, between the two structures obtained using these two preparation methods are small, because the LEED spot intensities that they observed are practically the same. More investigation will be necessary to resolve this structure satisfactorily.

(3×3) and $(2\sqrt{3} \times 6\sqrt{3})$ Phases. The next major phase of the 6H-(0001) surface is the (3×3), first reported by Kaplan et al. [15] based on LEED, Auger Electron Spectroscopy (AES) and Electron Energy Loss Spectroscopy (EELS) measurements. Their studies indicated the presence of a silicon adlayer on top of the terminal SiC bilayer. Consequently, a Dimer Adatom Stacking fault (DAS) model containing two adatoms per unit cell with different stacking orientation was proposed in analogy to the (7×7) reconstruction of Si(111) [11]. The DAS (3×3) model is presented in Fig. 3.5. There are two basic features in this model. First, there are two adatoms in the unit cell. Second, the unit cell is divided into two parts by a stacking fault. Studies of this surface phase have recently been conducted by several different groups using a variety of techniques [6,7,12,13,21,23]. These studies have led to some general agreement with regard to the structure of this phase. In the following we will discuss STM results and summarize the current understanding of the structure of the (3×3) phase. The images shown in Fig. 3.6(a) and (b) are empty-state and filled-state images taken simultaneously on the same area of the (3×3) surface. Regular arrays of bright spots, one per unit cell, form the (3×3) periodicity in both images. However, there are some spots that are less bright, giving the appearance of recessed 'atoms'

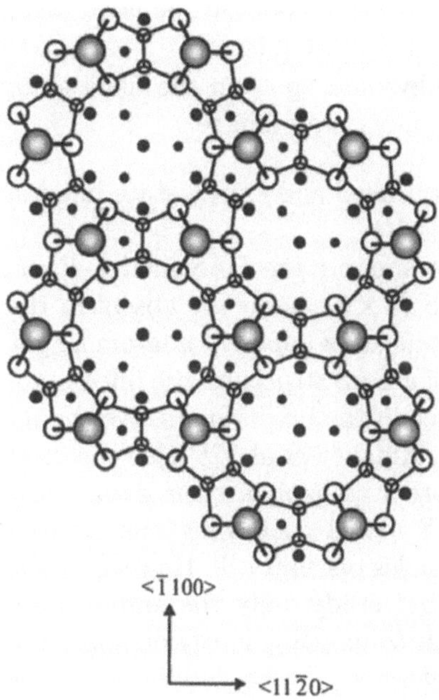

Fig. 3.5. Ball-and-stick DAS model of the Si-rich (3 × 3) structure of the Si-terminated 6H-SiC(0001) surface (from [15])

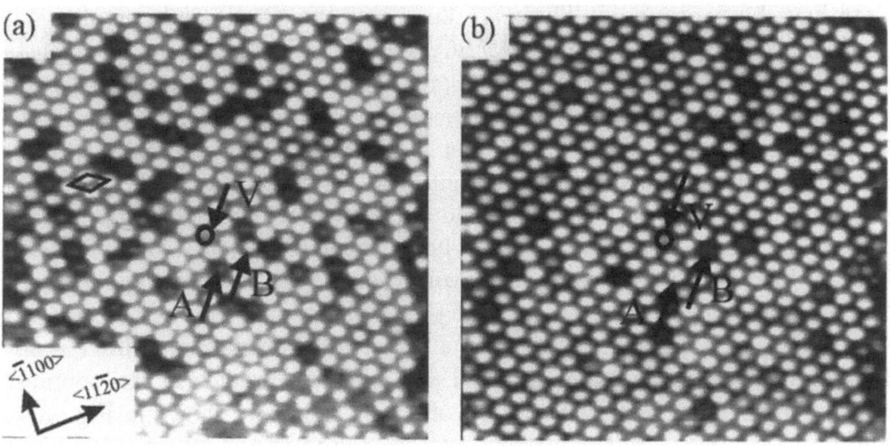

Fig. 3.6. Simultaneously recorded STM images taken with sample bias of (**a**) +2.5 V (empty state) and (**b**) −2.5 V (filled state) the 6H-SiC(0001) 3 × 3 structure. Image size = 200 Å ×200 Å. The unit cell is indicated by the *black line.* Three types of defects A, B and V are indicated by *arrows*

or 'clusters' while others are dark, resembling vacancies. While the majority of the bright spots are common to both images, hardly any of the less bright or dark spots occupy the same sites in the two images. We can identify the following sites that change polarity on the (3 × 3) surface:

- those marked (A), which are occupied by bright spots in the filled state image, and become dark spots in the empty state image;
- those marked (B), which are occupied by dark spots in the filled state image, and become bright spots in the empty state image.

If a dark spot occupies the same site in both filled and empty state images, then we consider it to be a vacancy (marked V).

It is clear that the STM data does not support the DAS model. First, the very prominent corner holes of the DAS (3×3) model are absent in the images. Second, contrary to the filled-state images of apparent adatom height difference on faulted and unfaulted halves of a DAS structure, the filled-state image of Fig. 3.6b shows no height difference in the bright spots. We should point out that in a recent STM experiment, Kulakov et al. [25] also observed only a single protrusion per unit cell. In order to explain their data, these researchers proposed a variant of the DAS model containing one adatom cluster on a silicon adlayer with two corner holes per unit cell. However, their model also includes a stacking fault. In that study, only the empty state image was presented, thus it was not possible to unambiguously demonstrate the presence or absence of the stacking fault.

Based on the STM data, the DAS model and its variants can with confidence be ruled out. Two other models that are consistent with the extant data have been proposed. These are known as the simple and complex silicon or carbon tetramer models [6,7,26]. The simple tetramer model consists of either a silicon or a carbon tetramer adsorbed in a 3×3 pattern on top of a Si-C bilayer. The difference between the simple and complex tetramer model is that in the complex tetramer model there is an additional layer of silicon psuedomorphically adsorbed onto the top Si-C bilayer (see Fig. 3.7). Thus the remaining task becomes one of distinguishing between the simple and complex tetramer models. Recent holographic LEED data combined with density functional theory strongly favors the complex tetramer model [6,7]. The adatom cluster saturates nine of the 27 dangling bonds of the adlayer. The remaining dangling hybrids form a threefold coordination within the adlayer. Only one single dangling bond per unit cell remains at the topmost atom. This model gives good account of the STM images shown in Fig. 3.6a and b. The majority of the bright spots are due to the single dangling bond left per unit cell. The spots A could arise from a carbon atom substituted for silicon in the adatom position. The spots B could arise from substitution of the silicon trimer by a carbon trimer.

The next phase is $(2\sqrt{3} \times 6\sqrt{3})$. This is a transition phase between the (3×3) and $(\sqrt{3} \times \sqrt{3})$ [23]. It is produced either by desorbing silicon from the (3×3) phase or adding silicon to the $(\sqrt{3} \times \sqrt{3})$ phase. Two images of this phase are shown in Fig. 3.8. In the large-scale image, two phases can be seen oriented 30 degrees from one another. The (3×3) phase appears on the island in the lower left-hand corner, while the predominant $(2\sqrt{3} \times 6\sqrt{3})$ phase is present on the rest of the surface. A high-resolution image of the $(2\sqrt{3} \times 6\sqrt{3})$

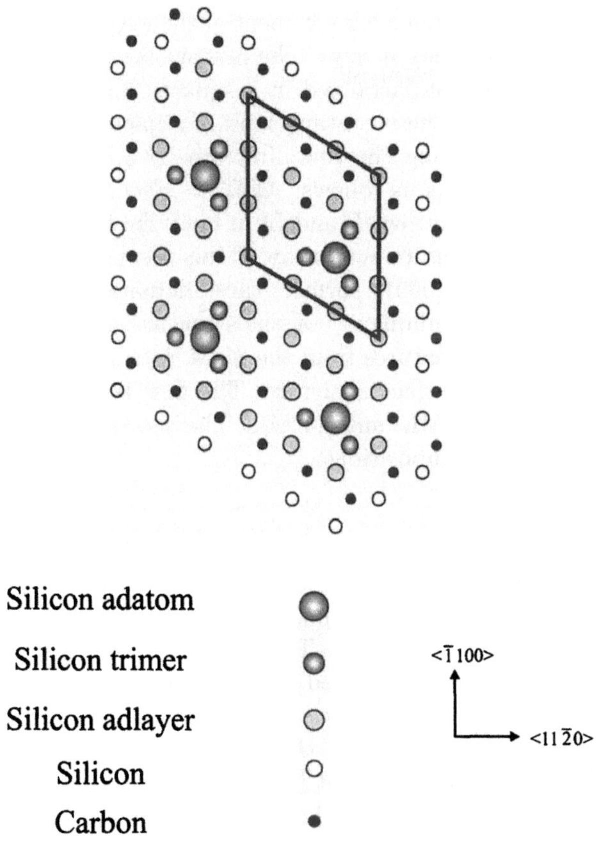

Fig. 3.7. The tetramer model proposed for the Si-rich (3 × 3) structure of the Si-terminated 6H-SiC(0001) surface (from [7]). The unit cell is indicated by a *black line*

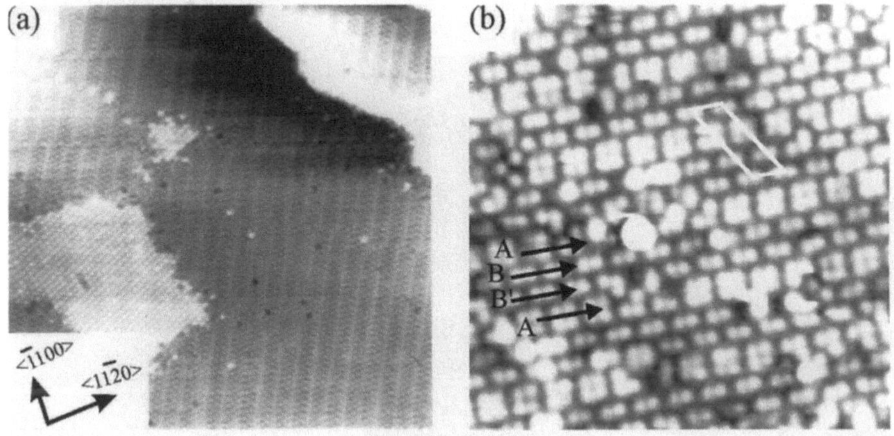

Fig. 3.8. STM images of the $(2\sqrt{3} \times 6\sqrt{3})$ reconstruction of the Si-terminated 6H-SiC(0001) surface. (**a**) Filled state (−2.0 V) and (**b**) empty state (+2.0 V). Image sizes are 650 Å ×650 Å and 120 Å ×120 Å, respectively

structures is shown in Fig. 3.8b. The atomic arrangement of the $(2\sqrt{3}\times6\sqrt{3})$ structure can be seen to form three types of rows. The first we label as type A, containing four adjacent bright spots arranged in a square. The second and third we label as type B and B′. These contain pairs of adjacent bright spots. However, B′ is phase-shifted along the row direction by 5.3 Å with respect to type B. The row sequence is as follows: ABB′A. Also note that row A appears higher in the image than rows B and B′ in both the filled and empty state images. Due to the degenerate symmetry of this reconstruction from six-fold symmetry of the 6H-SiC(0001) surface, three domains can be seen on large terraces. Because of the complexity of this structure, no model has yet been proposed. However, two features from the STM data should be included in any future model of the surface structure. The first is that the silicon coverage should be between 1 ML and 1.4 ML. The second is that rows A and B should have different compositions.

(7 × 7) Phase. The most silicon-rich phase on the 6H-SiC(0001) surface occurred after deposition of more than two monolayers of silicon on the $(\sqrt{3}\times\sqrt{3})$ phase [12,27]. A filled state STM image is shown in Fig. 3.9. The well-known Si(111)-(7 × 7) is evident [11]. The most interesting feature of this surface is the large triangular shaped steps, in contrast to the relatively straight steps on the $(\sqrt{3}\times\sqrt{3})$ surface. The origin of this interesting step structure is believed to be strain in the silicon film caused by the 20% lattice mismatch between Si(111) and SiC(0001). Multiple steps are visible in the image, with each single atom step being approximately 3.1 Å high.

Fig. 3.9. Filled-state (−2.0 V) STM image of the (7 × 7) reconstruction formed on the 6H-SiC(0001) surface. Image size = 360 Å ×360 Å

3.3.2 The (000$\bar{1}$) Surface

The C-terminated (000$\bar{1}$) surface also displays a rich variety of structures. The phase most often reported in previous LEED studies is $(1 \times 1)/(\sqrt{3} \times \sqrt{3})$ [15,35], but (2×2) and (3×3) have also been reported [10]. In order of decreasing Si content, the phases observed by Li et al. [8,38] in their STM studies are (7×7), $(3 \times 3)/(2 \times 2)$, $(2\sqrt{3} \times 2\sqrt{3})$. It should be noted that the $(2\sqrt{3} \times 2\sqrt{3})$ phase has not been observed in previous LEED studies. After initial cleaning by silicon beam etching, the $(2\sqrt{3} \times 2\sqrt{3})$ is the first surface phase obtained. Upon increasing the amount of silicon, the surface rearranges to $(3 \times 3)/(2 \times 2)$. As in the case of the (0001) surface, when the Si coverage is above 2 ML, a (7×7) structure is produced.

$(2\sqrt{3} \times 2\sqrt{3})$ Phase. The $(2\sqrt{3} \times 2\sqrt{3})$ phase is observed on the surface immediately after cleaning by silicon beam etching. The simultaneously obtained filled and empty state images are shown in Fig. 3.10. Images show a marked contrast reversal between filled and empty state images. In the filled state image two types of rows are observed. The first consists of adjacent bright spots (labeled A in the image), while the second has alternating bright and dark spots (labeled B in the image). These two rows occur alternately, with B-type rows being phase shifted with respect to one another. Thus the rows appear in the following sequence: ABAB′, where the prime indicates the phase shift. This creates a centered hexagonal pattern, with the center of the hexagon being dark. The side of the hexagon is 5.3 Å. In the empty state image, all bright spots become dark, while dark spots become

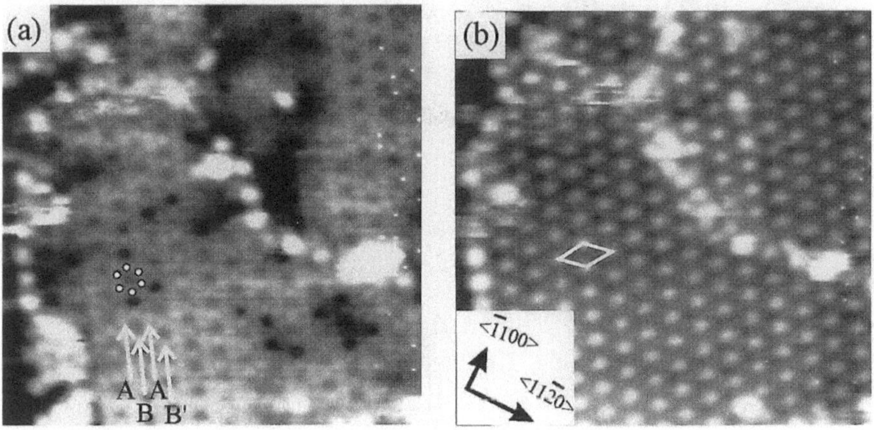

Fig. 3.10. Simultaneously recorded STM images taken with sample bias of (**a**) −1.6 V (filled state) and (**b**) +1.6 V (empty state) the 6H-SiC(000$\bar{1}$) $(2\sqrt{3} \times 2\sqrt{3})$ structure. Image size = 150 Å ×150 Å. The unit cell is indicated by *white lines* (from [12]). *Open white circles* mark the centered hexagonal structure

bright. That is to say, there is complete contrast reversal. This results in row A becoming completely dark while row B still consists of alternating dark and bright spots, but is shifted by 5.3 Å along the row direction. Bright spots in the empty state image also create a centered hexagonal structure. However, the side of the hexagon is doubled to 10.6 Å, giving rise to a $(2\sqrt{3} \times 2\sqrt{3})$ pattern. An alloy model was proposed to explain the features seen in the STM image [12]. This model consists of C adatoms bonded to a terminal C-Si bilayer. The C adatoms are arranged in a $(\sqrt{3} \times \sqrt{3})$ pattern with the adatoms setting at the four-fold coordinated sites (T_4). This is shown in Fig. 3.11. One third of the C adatoms are replaced with Si, resulting in a $(2\sqrt{3} \times 2\sqrt{3})$ structure. It was assumed that the source of carbon was from bulk segregation [20], while silicon arises from silicon beam used to clean the surface. The contrast reversal observed at the corners of the unit cell was simply explained by charge transfer from substitutional silicon adatoms to underlying carbons. This could possibly explain why a $(2\sqrt{3} \times 2\sqrt{3})$ pattern is not observed by LEED. The LEED is sensitive to the geometry, but cannot differentiate between chemical species. Thus to LEED, this alternate arrangement of silicon and carbon atoms at T_4 sites will appear to be $(\sqrt{3} \times \sqrt{3})$.

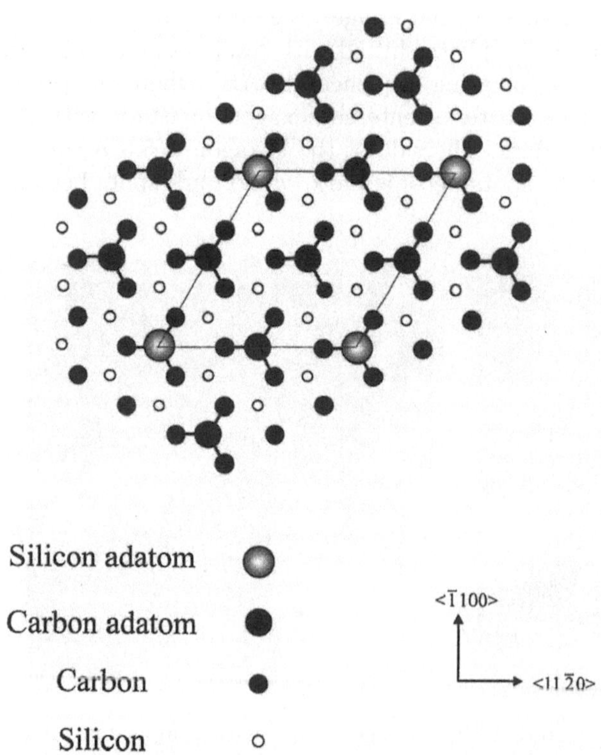

Fig. 3.11. Ball-and-stick alloy model of 6H-SiC(000$\bar{1}$) $(2\sqrt{3} \times 2\sqrt{3})$. The unit cell is indicated by a *black line* (from [12])

However, with STM, which is more sensitive to variations in local electronic structure, it is possible to distinguish between different kinds of species and thus show that the surface periodicity is in fact $(2\sqrt{3} \times 2\sqrt{3})$. One shortcoming of this model is that it fails to explain the contrast reversal seen in row A. Those rows consist of carbon adatoms bonded to the carbon terminal layer of the surface. In a heuristic interpretation this results in a non-polar bond.

In order to overcome this deficiency of the model, a vacancy model has been proposed to explain the STM result [38]. This vacancy model is shown in Fig. 3.12. The vacancies are formed by removal of 1/6 ML of carbon atoms from the terminal carbon layer. These are arranged in a $(2\sqrt{3} \times 2\sqrt{3})$ pattern. This creates two types of sites on the surface. The first consists of a centered hexagon in which the center is a vacancy. Within the vacancy there is a dangling bond complex consisting of three silicon-dangling bonds. The second is also a centered hexagon, but the center is simply another carbon atom with one dangling bond. In addition, the carbon atoms forming the ring of the hexagon can be in two different chemical environments. Four of the carbons in any given ring are associated with threefold coordinated second layer silicon atoms. The remaining two are bonded to four-fold coordinated silicon atoms. Thus those carbons bonded to the threefold coordinated silicon atoms have additional charge density available for tunneling.

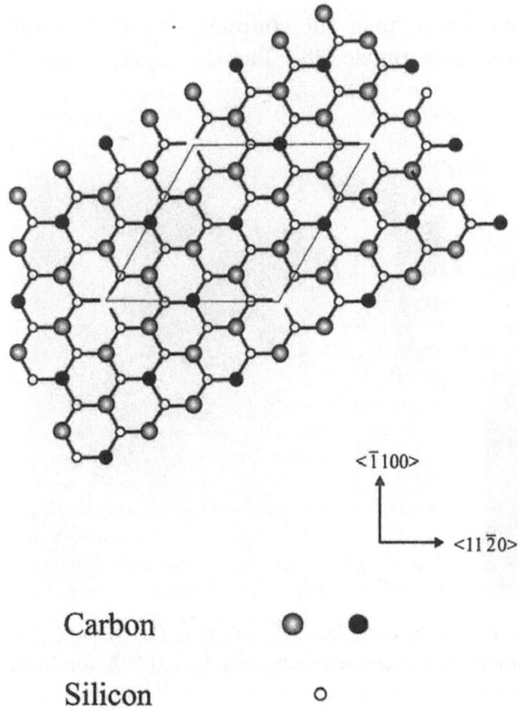

Fig. 3.12. Ball-and-stick vacancy model of the 6H-SiC(000$\bar{1}$) $(2\sqrt{3} \times 2\sqrt{3})$. The unit cell is indicated by a *black line*

The features seen in the STM images can be explained by this model in the following way. Dark spots in the filled state image (Fig. 3.10a) are due to these carbon vacancies. As noted above, since the carbon atoms withdraw charge from the silicon atoms, this leaves the silicon dangling bonds relatively charge deficient. Thus there is less electron density available for tunneling. Therefore these hexagonal complexes appear dark in the filled state image and bright in the empty state image. This results in a contrast reversal of rows B and B′. All bright spots in the filled state image correspond to centered hexagons of carbon with carbon atoms at the center. So for row A, since the carbon atoms have dangling bonds which are relatively charge-rich due to charge transfer from the silicon dangling bond complexes, they should appear bright in the filled state image (Fig. 3.10a). This model agrees well with STM data and the $(2\sqrt{3}\times 2\sqrt{3})$ LEED pattern should be expected. However, in all previous LEED studies of 6H-SiC(000$\bar{1}$) surfaces, no $(2\sqrt{3}\times 2\sqrt{3})$ pattern has been reported. In order to firmly establish this new structure, further investigation is necessary.

$(2\times 2)/(3\times 3)$ and (7×7) Phases. When more than one monolayer of Si is deposited on the $(2\sqrt{3}\times 2\sqrt{3})$ surface, a disordered (2×2) structure develops first, followed by the (3×3). They are shown in Figs. 3.13a and b, respectively.

The (3×3) phase is basically the same as the (3×3) phase on the silicon terminated surface. Nonetheless, long-range ordering is not as good as in the Si-terminated surface. We therefore expect the complex tetramer model shown in Fig. 3.7 to apply to this surface phase [38]. The (2×2) structure is

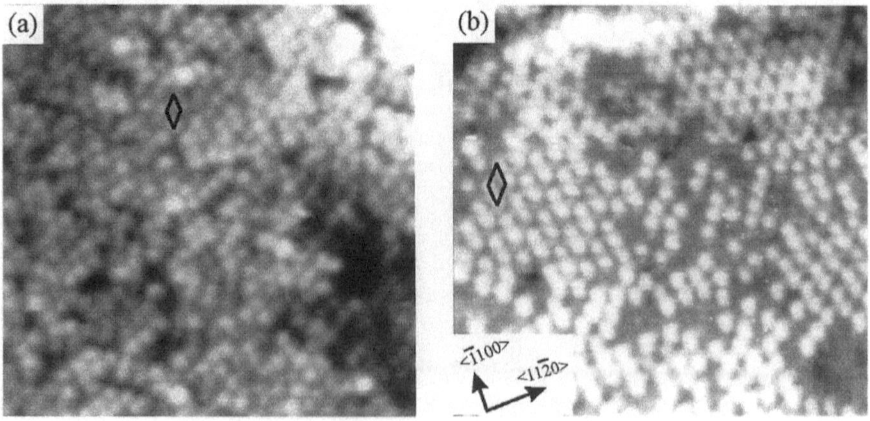

Fig. 3.13. Filled-state STM images of the Si-rich (**a**) (2×2) and (**b**) (3×3) reconstructions of the 6H-SiC(000$\bar{1}$) surface. Image sizes are 160 Å ×160 Å for both images (from [12])

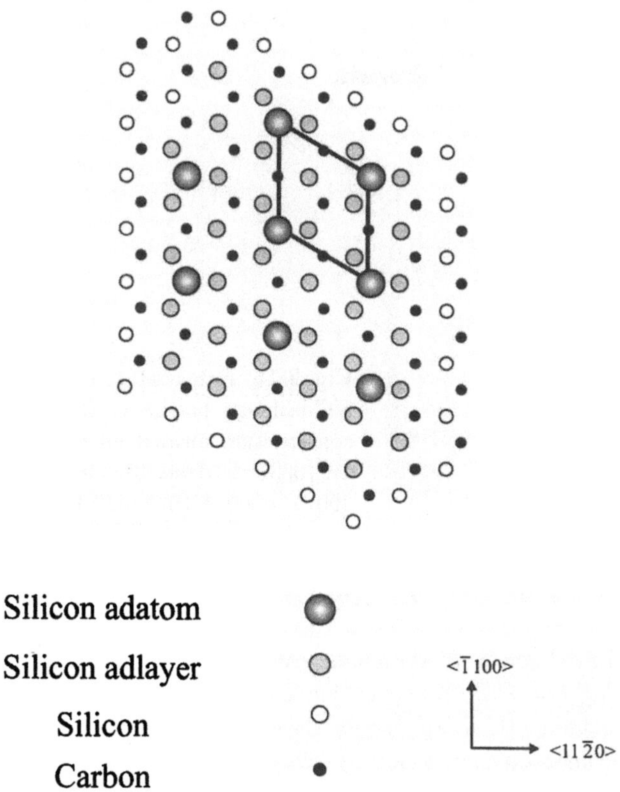

Fig. 3.14. Adatom model of the (2×2) 6H-SiC(000$\bar{1}$) surface. The unit cell is outlined by the *black line*

even less well ordered than the (3×3) surface, with a relatively high density of defects appearing as dark spots in the STM image. There is one protrusion per unit cell and no bias dependence is observed. A possible model is shown in Fig. 3.14 [38]. It consists of simple adatoms arranged in a 2×2 pattern on top of a layer of silicon pseudomorphically grown on the top Si-C bilayer. This gives a surface silicon coverage of 1.25 ML. The model accounts both for the silicon coverage and for the lack of bias dependence in the images. This phase has been not extensively investigated. Only one previous LEED study has reported the existence of a (2×2) structure [14]. More work is therefore needed to confirm this structure and provide further evidence to establish the proposed model.

When silicon coverage is above 2 ML, the (7×7) reconstruction is formed. An STM image of the (7×7) structure is shown in Fig. 3.15. Large-scale images of the surface indicate that the domain size has substantially increased. However, to accommodate the lattice mismatch with the underlying SiC surface, additional steps are formed (as indicated in the figure). Because the initial density of steps on this surface is lower, only one additional step formed by the lattice mismatch is seen in the image.

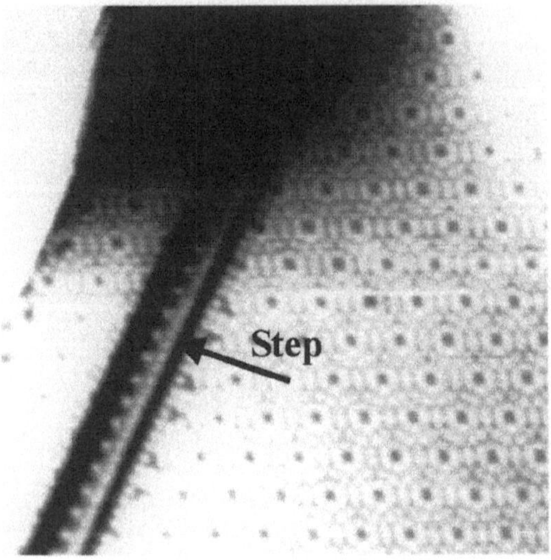

Fig. 3.15. Empty-state (+2.0 V) STM image of the (7 × 7) reconstruction formed on the 6H-SiC(000$\bar{1}$) surface. Image size = 300 Å ×300 Å (from [12])

3.4 Surface Phonons of 6H-SiC (0001)

Fuchs–Kleiwer phonons have been studied on a variety of polar semiconductor surfaces, such as GaAs, InP and GaP. However, because of the relatively deep sampling depth of the electrons, the frequency of these modes is rather insensitive to the microscopic near-surface structure [39]. That is to say, the FK frequency is approximately the same, irrespective of surface orientation. Surface phonons of 6H-SiC(0001) have been studied by several groups [13,19,23]. In those studies, with the exception of that done by Li et al., surfaces carried a significant amount of surface contamination. Li et al. [13] studied two different surface phases of the 6H-(0001) surface using a system that combined STM and High Resolution Electron Energy Loss Spectroscopy (HREELS). In that experiment, using the STM to confirm the surface structure, clean ($\sqrt{3} \times \sqrt{3}$) and (3 × 3) surfaces were prepared. The reader is referred to previous sections of this monograph for an in-depth discussion of the properties and structure of the various phases of 6H-SiC(0001). One important point that should be noted here is that the different surface phases have very different Si content. On the ($\sqrt{3} \times \sqrt{3}$) surface, the Si/C ratio is approximately 0.33 while the (3 × 3) surface consists almost entirely of Si. In addition, surface corrugation is very different. The corrugation of the ($\sqrt{3} \times \sqrt{3}$) surface is approximately 0.2 Å, and that of the (3 × 3) is 2.5 Å. Thus, if the FK mode is significantly affected by the detailed surface structure, these two surfaces should be sufficiently different to perturb the frequency of the mode. The ($\sqrt{3} \times \sqrt{3}$) surface exhibited a high electron reflectivity leading to a relatively high HREELS signal (200 000 counts/s in the elastic peak); while the reflectivity of the (3 × 3) surface appeared to be significantly lower, since it was only possible to obtain an elastic peak signal of about 100 000 counts/s.

The HREELS spectra taken from the two surfaces are shown in Fig. 3.16. The main FK phonon line and two replicas are clearly visible. The position of the phonon on the $(\sqrt{3} \times \sqrt{3})$ surface is 110 meV. On the (3×3) surface it is 115 meV. These values agree with previous studies that were performed on

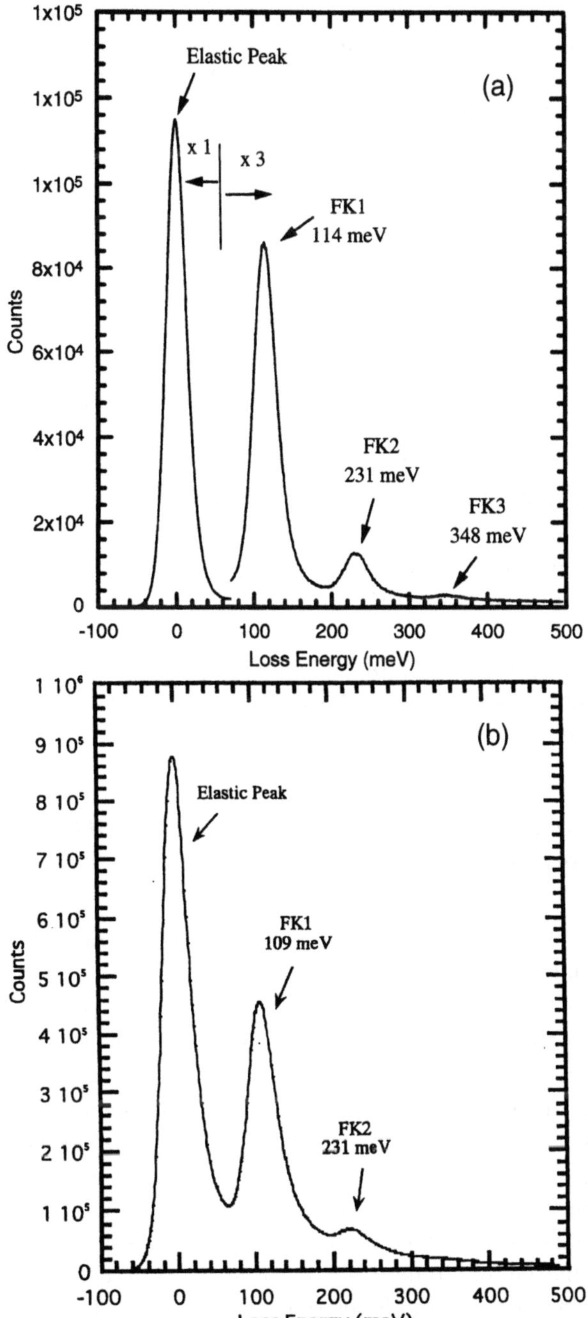

Fig. 3.16. HREELS spectra of the clean 6H-SiC(0001) $(\sqrt{3} \times \sqrt{3})$ and (3×3) surfaces. $E_p = 7$ eV (from [13])

surfaces with a significant amount of contamination [19,23]. The spectrum of the (3 × 3) surface after exposure to approximately 7 L of O_2 was also measured. It suffices here to note that exposure to this amount of O_2 results in a coverage of approximately 30 percent. The position of the FK1 peak is 112 meV. Thus adsorption of < 1 ML of O_2 did not result in any significant change in FK frequency. In addition it was not possible to see any features due to adsorbed O_2 because of the very strong FK phonon peaks. Given the resolution of the data on these surfaces, the shift in the FK1 peak was considered to lie within the limits of experimental error.

In order to gain some further insight into scattering processes, a simple estimate can be made for the effective probing depth of the electrons. This is defined by the electron momentum parallel to the surface [39,40]:

$$|Q_{\text{parallel}}|^{-1} \sim \left[k_0 \frac{h\omega_{\text{so}}}{2E_0} \sin\theta \right]^{-1} , \tag{3.1}$$

where k_0 is the wave vector of the incoming electron with energy E_0. The scattering angle is θ and ω_{so} is the frequency of the surface optical phonon. For SiC, the probing depth $|Q_{\text{parallel}}|^{-1}$ is estimated to be approximately 180 Å. The SiC bilayer height is about 2.5 Å. Therefore, 180 Å amounts to approximately 72 bilayers. Thus it is clear that, as expected, the electron probing depth is much deeper than the first few bilayers which are affected by the details of surface morphology. Finally, because of the relatively low carrier mobility (compared with InAs or InP), no coupled phonon/plasmon mode is expected.

In summary, all studies of FK phonons of SiC surfaces have yielded approximately the same frequency, indicating that FK phonons are insensitive to near-surface structure and stoichiometry.

3.5 Effect of Surface Polarity for Gallium Adsorption onto 6H-SiC Surfaces

One very important problem that needs to be overcome in the growth of group III nitrides is the lack of a lattice-matched substrate. Because 6H-SiC has a relatively small mismatch in both lattice parameter and thermal expansion coefficient, it should be easier to produce single crystal epitaxial layers on this substrate. Nonetheless, when used as a substrate material for GaN growth, 6H-SiC displays strong polarity dependence [41]. On the silicon-terminated (0001) surface, two-dimensional GaN growth results in a smooth film. In contrast, three-dimensional growth occurs on the $(000\bar{1})$ carbon terminated surface, producing hexagonal pyramids of GaN on the SiC surface. It has been suggested that this marked difference in film morphology arises from differences in the lattice mismatch due to interfacial charge [42]. Furthermore, GaN growth studies have shown that pre-deposition of gallium on the SiC surface prior to growth leads to marked improvement in the quality of the GaN epitaxial layer [43].

A study of gallium adsorption was performed on both the Si-terminated 6H-SiC(0001) ($\sqrt{3} \times \sqrt{3}$) and C-terminated 6H-SiC(000$\bar{1}$) ($2\sqrt{3} \times 2\sqrt{3}$) surfaces by Li et al. [44]. The structure of the Ga-terminated 6H-SiC surface showed strong polarity dependence. On the Si-terminated ($\sqrt{3} \times \sqrt{3}$) surface, the Ga atoms arranged themselves into parallel rows in three different domains oriented at 120 degrees with respect to each other at 1 ML coverage; while on the C-terminated ($2\sqrt{3} \times 2\sqrt{3}$) surface, sets of two concentric rings formed an overall ($4\sqrt{3} \times 4\sqrt{3}$) reconstruction at 1 ML coverage.

Results of Ga adsorption on the C-terminated ($2\sqrt{3} \times 2\sqrt{3}$) surface will be described first. At one monolayer coverage, the overlayer consists of sets of two concentric interlocking rings of Ga atoms, as shown in Fig. 3.17. The outer rings consist of 12 distinct spots in the STM image. Thus the original threefold symmetry of the substrate is maintained. In addition, there is an inner ring of six atoms, with a vacancy in the center. Overall, the Ga atoms form a ($4\sqrt{3} \times 4\sqrt{3}$) structure. Some striking features are observed in the image. First, there is a contrast difference between inner and outer rings, with atoms in the inner ring appearing lower or darker than those of the outer ring. The exact amount of the contrast depends on sample bias. In addition, higher contrast is observed in the empty-state image. Second, the atoms of the inner ring appear alternately darker and brighter relative to one another. This is seen in both empty and filled state images, but is more pronounced in the empty state image. The observed contrast could be explained by the structural model outlined in the following paragraph (see Fig. 3.18). To construct this model, one starts first with the unreconstructed 6H-SiC(000$\bar{1}$) surface.

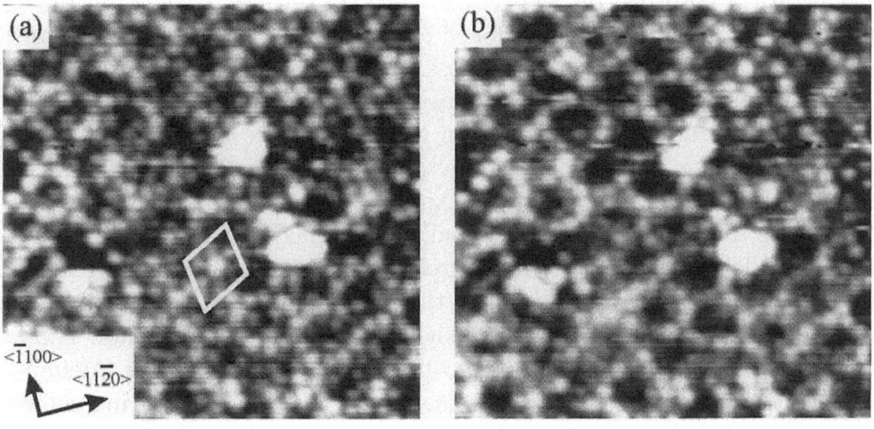

Fig. 3.17. Simultaneously recorded STM images taken with sample bias of (**a**) −1.6 V (filled state) and (**b**) +1.6 V (empty state) of 1 ML of Ga adsorbed on the ($2\sqrt{3} \times 2\sqrt{3}$) C-terminated 6H-SiC(000$\bar{1}$) surface. Image size = 165 Å ×165 Å. The unit cell is indicated by the *white line* (from [44])

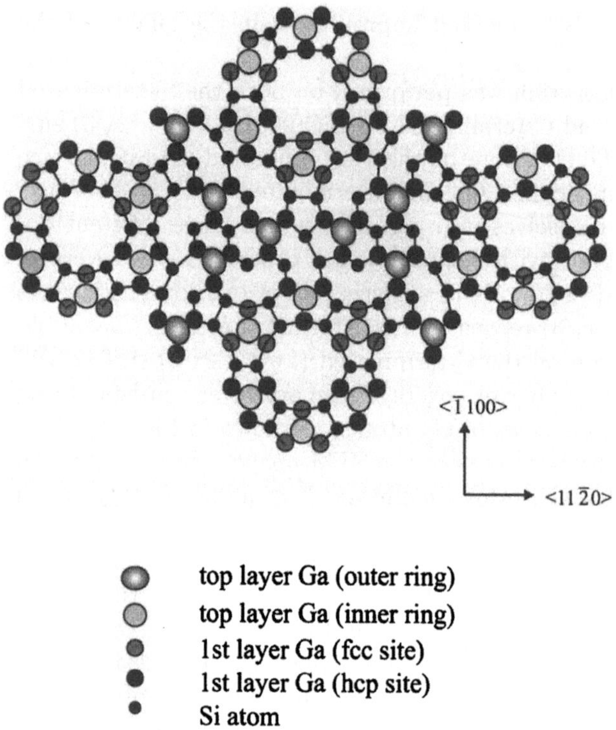

Fig. 3.18. Ball and stick model for the $(4\sqrt{3}\times 4\sqrt{3})$ structure of one monolayer of Ga adsorbed on the $(2\sqrt{3}\times 2\sqrt{3})$ surface (from [44])

The silicon atoms in the top SiC bilayer are symbolized by the smallest black circles in the figure. The first step is to replace the top bilayer C atoms with Ga atoms. This is indicated by medium sized black and dark gray circles. The substitution of Ga for the C atoms in the first layer is analogous to the behavior of Ga adsorbed on Si(111) and Ge(111) surfaces [45,46]. On those surfaces, the Ga atoms substitute for Si (Ge) respectively, forming a domain wall structure. The driving force for those structures is relaxation of strain induced by the large lattice mismatch between Ga and Si (Ge). It is reasonable to assume that this would occur, given the large lattice mismatch between SiC and Ga. There are two distinct sites for adsorbed gallium: hcp and fcc. The type of site is determined by stacking sequence. Fcc and hcp sites are analogous to the 'unfaulted' and 'faulted' sites formed on the Si(111)-(7×7) surface [11]. The medium sized black circles in the figure show 'faulted' sites, while medium sized gray circles show the 'unfaulted' sites. The large gray and shaded gray circles in the figure show the 2nd layer Ga atoms. These atoms form the 'ring' structure that is observed in the STM image. As indicated in the caption, shaded gray circles show atoms in the outer ring, while large gray circles show those of the inner ring. Note that all Ga atoms in the outer ring lie on 'unfaulted' sites, while those of the inner ring alternate between 'unfaulted' and 'faulted' sites. There are two additional similarities with the Si(111)-(7×7) structure. First, there are corner holes in which the silicon atom in the top bilayer is missing. Second, the silicon atoms in the top bilayer are dimerized.

In a manner analogous to that of the Si(111)-(7 $\times$ 7) surface, the contrast observed in the STM image for Ga/SiC-6H(0001) can be explained by differing electronic structure at the various bonding sites. On the Si(111)-(7 $\times$ 7) surface, due to the differing local electronic structure, corner adatoms appear slightly higher than central adatoms in the filled state image [11]. In the Ga/SiC case, atoms of the inner and outer rings are analogous to the corner and center adatoms, respectively, and it is thus reasonable to believe that one could observe a contrast difference between the two rings. Furthermore, three of the Ga atoms that form the inner ring sit in faulted sites while the other three sit in unfaulted sites. Thus the electronic structure of these two types of Ga atoms should be different, plausibly explaining the fact that three of the atoms of the inner ring appear relatively bright, while the other three look relatively dark. The geometric height of all of the atoms in the second layer is the same in this model. However, the apparent height difference seen in the images is due to the different local electronic structure of 'faulted' and 'unfaulted' sites.

Turning now to the Si-terminated ($\sqrt{3} \times \sqrt{3}$) surface, one monolayer coverage of Ga results in an ordered structure. The adsorbed gallium forms a row structure consisting of domains oriented 120 degrees with respect to each other. The image shown in Fig. 3.19 illustrates the three possible domains of the row structure. The direction of the rows is oriented 30 degrees away from the $\langle\bar{1}100\rangle$ direction of the clean surface. Some row meandering is also visible, but the change in direction of the rows is always 120 degrees. The spacing between rows was measured to be 9.6 Å. In addition to the primary order, i.e., formation of rows on the surface, some fine structures can be seen

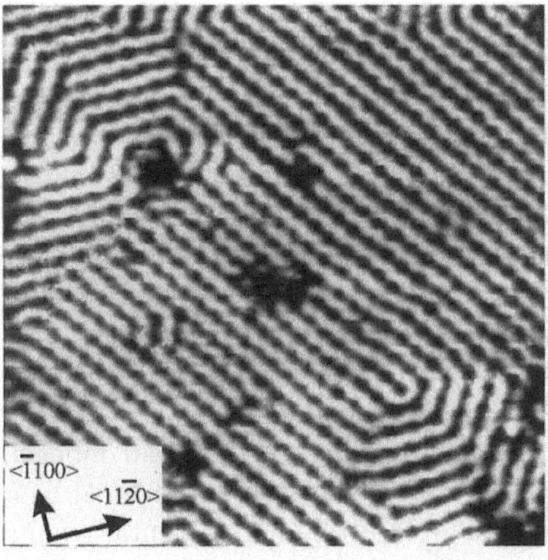

Fig. 3.19. Empty state image of the Ga row structure obtained on the ($\sqrt{3} \times \sqrt{3}$) Si-terminated 6H SiC(0001) surface. Sample bias = +1.0 V. Image size = 260 Å $\times$260 Å (from [44])

in the image. The rows are not perfectly straight, but rather zigzag slightly in a regular manner, so that the average direction is still a straight line. In addition, the position of the row does not depend on the polarity of the bias voltage. This indicates that the positions of the maxima seen in STM images correspond to the physical positions of the Ga atoms. However, since the internal structure of the rows was not resolved, a detailed structural model for this surface was not proposed.

In previous MBE studies of GaN on 6H-SiC, it was found that termination of the surface with gallium significantly improves the quality of the GaN layer [42]. On both the (0001) and $(000\bar{1})$ surfaces, the STM revealed that adsorption of elemental gallium results in termination of the surface by gallium. On the C-terminated surface, the gallium displaces surface carbon atoms, forming a direct Ga-Si bond. Growth of GaN on this surface should be quite efficacious, since simple models of the growth indicate that a gallium-terminated surface is invariably obtained. A consequence is that film and substrate surfaces will have opposite polarity, i.e., a (0001) SiC surface yields a $(000\bar{1})$ GaN surface. In the case of the silicon-terminated surface, the row structure formed by adsorbed gallium should allow enhanced diffusion along rows, a property that should also be favorable to growth.

3.6 Conclusions

Several methods have been employed to obtain clean, well-ordered SiC surfaces. Among these, in situ silicon beam etching has been the most successful. Several reconstructions have been observed in STM and LEED experiments on samples prepared using this method. On the (0001) surface, in order of increasing Si content, the phases observed by STM are $(\sqrt{3}\times\sqrt{3})$, $(2\sqrt{3}\times6\sqrt{3})$, (3×3), and (7×7). However, the $(2\sqrt{3}\times6\sqrt{3})$ and (7×7) phases have not been seen using LEED. In the case of the (7×7) and (3×3) phases, widely accepted structural models have been proposed. However, the details of the $(2\sqrt{3}\times6\sqrt{3})$, $(\sqrt{3}\times\sqrt{3})$ phases are not yet completely understood, and are still the subject of some debate in the literature. Currently, the two possible models are the vacancy model and the adatom model for the $(\sqrt{3}\times\sqrt{3})$. Further study is required with a variety of techniques. Studies of the FK phonons of SiC surfaces have been performed and have yielded approximately the same frequency for the (3×3) and $(\sqrt{3}\times\sqrt{3})$ structures, indicating that FK phonons are insensitive to near-surface structure and stoichiometry. Further experiments using vibrational spectroscopy of adsorbed hydrogen should be very illuminating, providing insight into the chemical identity of surface species. Moreover, additional theoretical calculations from the new models, to determine energetics and band structure of the surface, are critical to a detailed and definitive understanding of the interesting experimental results presented here.

On the (000$\bar{1}$) surface, in order of decreasing Si content, the phases observed are (7×7), $(3 \times 3)/(2 \times 2)$ and $(2\sqrt{3} \times 2\sqrt{3})$. Both an alloy and a vacancy model have been proposed for the $(2\sqrt{3}\times 2\sqrt{3})$ reconstruction. Upon increasing the amount of silicon, the surface rearranges to $(3 \times 3)/(2 \times 2)$. The structure of the (000$\bar{1}$)-(3×3) is essentially the same as that of the (0001)-(3×3) surface. On the other hand, the (2×2) is not well understood. Nonetheless, an adatom model has proposed which explains the features observed in STM images. Finally, when the Si coverage is above 2 ML, the well known (7×7) structure is produced. While the (3×3) and (7×7) are well characterized, our knowledge of the structure of (2×2) and $(2\sqrt{3}\times 2\sqrt{3})$ is not yet complete. Further studies, both experimental and theoretical, are necessary in order to provide a full understanding of the electronic and structural properties of these surfaces.

The structure of the Ga terminated 6H-SiC surface shows strong polarity dependence. On the Si-terminated $(\sqrt{3} \times \sqrt{3})$ surface, Ga atoms arrange themselves in parallel rows in three different domains oriented at 120 degrees with respect to each other at 1 ML coverage. On the C-terminated $(2\sqrt{3} \times 2\sqrt{3})$ surface, sets of two concentric rings are formed, producing an overall $(4\sqrt{3} \times 4\sqrt{3})$ reconstruction at 1 ML coverage. A structural model for the $(4\sqrt{3}\times 4\sqrt{3})$ is proposed. The key ingredient of the model is the substitution of top layer carbon atoms by Ga. If the interface is maintained during N adsorption, i.e., the Ga and N atoms do not undergo an exchange reaction, this substitution will result in a polarity reversal between the SiC substrate and the GaN film.

References

1. MRS Bulletin, Vol. **22**, (1997).
2. H. Morkoc, S.N. Mohammed, Science **267**, 51 (1995).
3. S. Nakamura, M. Senoh, S. Nagahana, N. Iwasa, T. Yamada, T. Matsushita, H. Kiyoku and Y. Sugimoto, Jpn. J. Appl. Phys. **35**, L74 (1996).
4. T.W. Weeks, Jr., M.D. Bremser, K.S. Ailey, E. Carlson, W.G. Perry and R.F. Davis, Appl. Phys. Lett. **67**, 401 (1995).
5. G. Pensl and R. Helbig, Festkörperprobleme/Adv. Solid State Phys. **30**, 133 (1990).
6. K. Reuter, J. Bernhardt, H. Wedler, J. Schardt, U. Starke and K. Heinz, Phys. Rev. Lett. **79**, 4818 (1997).
7. U. Starke, J. Schardt, J. Bernhardt, M. Franke, K. Reuter, H. Wedler, K. Heinz, J. Fürthmuller, P. Käckell and F. Bechstedt, Phys. Rev. Lett. **80**, 758 (1998).
8. J.C. Phillips, Rev. Mod. Phys. **42**, 317 (1970).
9. A. Garccia and M.L. Cohen, Phys. Rev. B **47**, 4215 (1993), 4221 (1993).
10. M. Sabisch, P. Krüger and J. Pollmann, Phys. Rev. B **51**, 13367 (1995).
11. Scanning Tunneling Microscopy, J.A. Stroscio and W.J. Kaiser (Eds), Academic Press (1993).
12. L. Li, Y. Hasegawa, I.S.T. Tsong and T. Sakurai, J. de Physique IV **6**, 167 (1996).

13. L. Li, C. Tindall, O. Takaoka, Y. Hasegawa and T. Sakurai, Surf. Sci. **385**, 60 (1997).
14. A.J. van Bommel, J.E. Crombeen and A. van Tooren, Surf. Sci. **48**, 463 (1975).
15. R. Kaplan, Surf. Sci. **215**, 111 (1989).
16. S. Nakanishi, H. Tokutaka, K. Nishimori, S. Kishida and N. Ishihara, Appl. Surf. Sci. **41/42**, 44 (1989).
17. U. Starke, Ch. Bram, P.-R. Steiner, W. Hartner, L. Hammer, K. Heinz and K. Möller, Appl. Surf. Sci. **89**, 175 (1995).
18. F. Owman and P. Mårtensson, Surf. Sci. **330**, L639 (1995).
19. J. Schardt, Ch. Bram, S. Müller, U. Starke, K. Heinz and K. Möller, Surf. Sci. **337**, 232 (1995).
20. L. Muehlhoff, W.J. Choyke, M.J. Bozack, J.T. Yates Jr. J. Appl. Phys. **60**, 2842 (1986).
21. T.M. Parrill and V.M. Bermudez, Solid State Commun. **63**, 231 (1987).
22. L.I. Johansson, F. Owman and P.Mårtensson, Phys. Rev. B **53**, 13793 (1996).
23. H. Nienhaus, T.U. Kampen and W.Mönch, Surf. Sci. **324**, L328 (1994).
24. I.S.T. Tsong, J. Amer. Ceram. Soc. **76**, 269 (1993).
25. M.A. Kulakov, G. Henn and B. Bullemer, Surf. Sci. **346**, 49 (1996).
26. L. Li and I.S.T. Tsong, Surf. Sci. **351**, 141 (1996).
27. L. Li, Y. Hasegawa, T. Sakurai and I.S.T. Tsong, J. Appl. Phys. **80**, 2524 (1996).
28. L.I. Johansson, F. Owman and P. Mårtensson, Phys. Rev. B **53**, 13803 (1996).
29. L.I. Johansson, F. Owman and P. Mårtensson, Surf. Sci. **360**, L478 (1996)
30. L.I. Johansson, F. Owman and P. Mårtensson, Surf. Sci. **360**, L483 (1996).
31. P. Badziag, Surf. Sci. **236**, 48 (1990).
32. P. Badziag, Surf. Sci. **337**, 48 (1995).
33. J.E. Northrup and J. Neugebauer, Phys. Rev. B **52**, R17001 (1995).
34. M. Sabisch, P. Krüger and J. Pollmann, Phys. Rev. B **55**, 10561 (1997).
35. R. Kaplan and T.M. Parrill, Surf. Sci. **165**, L45 (1986).
36. M. Tabe, Jpn. J. Appl. Phys. **21**, 534 (1982).
37. U. Starke, M. Franke, J. Bernhardt, J. Schardt, K. Reuter and K. Heinz, Proceedings of the ICSCIII-N97 (in press).
38. L. Li, C. Tindall and T. Sakurai (to be published).
39. L.H. Dubois and G.P. Schwartz, Phys. Rev. B **26**, 794 (1982).
40. D.L. Mills, Prog. Surf. Sci. **8**, 143 (1977).
41. T. Sasaki and T. Matsuoka, J. Appl. Phys. **64**, 4531 (1988).
42. S.Y. Ren and J.D. Dow, Appl. Phys. Lett. **69**, 251 (1996).
43. S. Strite, M.E. Lin, and H. Morkoc, Thin Solid Films **231**, 197 (1993).
44. L. Li, C. Tindall, Y. Hasegawa and T. Sakurai, Appl. Phys. Lett. **71**, 2776 (1997).
45. D.M. Chen, J.A. Golovchenko, P. Bedrossian and K. Mortensen, Phys. Rev. Lett. **61**, 2687 (1988).
46. E. Artacho, P. Molin-Matá, M. Böringer, J. Zegenhagen, G.E. Franklin and J.R. Patel, Phys. Rev. B **51**, 9952 (1995).

4 Application of Atom Manipulation for Fabricating Nanoscale and Atomic-Scale Structures on Si Surfaces

T. Hashizume, S. Heike, T. Hitosugi and K. Kitazawa

Summary. Scanning Tunneling Microscopy/Spectroscopy (STM/STS) and atom manipulation have been used to fabricate and characterize nanoscale and atomic-scale structures on Si(111)–7×7 and hydrogen-terminated Si(100)–2×1–H surfaces. We first fabricate 0.2–0.6 nm deep and 10 nm wide trench lines by applying a high voltage and/or a large tunneling current between the STM tip and the Si(111)–7×7 surface. When the trench lines form a closed figure such as a circle or a square, the apparent height of the surface area inside the closed trench becomes 0.1–0.3 nm lower than outside. This phenomenon is explained by a naturally formed Schottky barrier between metallic surface states and substrate, and by electronic conduction through the surface states. Atom manipulation is then applied on a hydrogen-terminated Si(100)–2×1–H surface to extract hydrogen atoms one by one and to fabricate atomic-scale dangling bond (DB) structures. Electronic structures of DB wires are studied using STS. The wires composed of both unpaired DBs and paired DBs show a finite density of states at the Fermi level and do not show semiconductive band gaps. The results agree well with first-principles theoretical calculations. The hydrogen-terminated Si(100)–2×1–H surface and its interaction with thermally deposited Ga atoms and C_{60} molecules is investigated. They migrate on the hydrogen-terminated area and preferentially adsorb on DBs. Several methods for manipulating hydrogen atoms (detaching, attaching and moving) are also tested. Atomic-scale Ga wires on the Si surface are fabricated for the first time by thermally depositing Ga atoms on the DB wires.

4.1 Introduction

Three decades ago, Feynman gave a lecture entitled 'There's Plenty of Room at the Bottom' and discussed switching devices on atomic and molecular scales [1]. The idea became more realistic when Scanning Tunneling Microscopy and Spectroscopy (STM/STS) [2] and its use for atom manipulation [3–5] became available. One of the important applications of atom manipulation is to realize nanoscale electronic devices which overcome the limitations of present semiconductor technology [6]. So far, such surface modification techniques have mainly been used to fabricate nanoscale structures on the surface. One exception is to change surface properties, reported by

Eigler et al. [7], in which a circular 'quantum corral' was fabricated by manipulating Fe atoms on the Cu(111) surface, and standing waves of electrons confined within the surface states were formed. Although they demonstrated a property change of the surface by atom manipulation, it may not be easy to apply to electronic devices because of the metal substrate they used.

While the Si(111)–7×7 surface [8] shows metallic surface electronic states, the hydrogen-terminated Si(100)–2×1–H surface exhibits a semiconductive band gap and is one of the promising substrates for atomic-scale device fabrication. The surface structure has been systematically studied by STM and the 2×1–H monohydride surface structure is now well understood [9,10]. Hydrogen atoms on the surface can be desorbed using the STM tunneling current and nanoscale DB patterning has been demonstrated [11,12]. An atomically flat surface is routinely obtained. The bulk conductivity can be made small enough for characterizing physical properties of the atomic structures (conductivity, for instance) either by using low dopant-level samples or by reducing the temperature of the sample. Surface states that result from the DBs of Si dimers are passivated by hydrogen adsorption. The surface is less reactive and can be kept clean (even for a couple of weeks) in an Ultra High Vacuum (UHV). Reactivity to some of the metal atoms is low enough for stable adsorption, and surface mobility of these metal atoms is expected to be high [13]. Adsorption of Ga on the clean Si(100)–2×1 surface was studied by STM and formation of Ga dimer rows was reported on the surface after annealing it at 750 K [14,15].

In the present review, we report on fabrication and characterization of nanoscale and atomic-scale structures on Si(111)–7×7 and hydrogen terminated Si(100)–2×1–H surfaces using STM/STS and atom manipulation. Surface properties of the Si(111)–7×7 surface are drastically modified when part of the surface is surrounded by an insulating trench pattern, which is formed by a large tunneling current between the tip and the sample at a relatively high bias voltage. When the trench forms a completely closed loop, the surface inside the loop is imaged lower than the outside by 0.1–0.3 nm, and further STM manipulation is impossible or difficult inside the trench. However, when the loop is not complete, properties of the inside region are the same as those of regular surfaces; apparent height difference is not observed and trench patterns can be fabricated inside the trench. Atomic-scale DB structures are fabricated on the hydrogen-terminated Si(100)–2×1 surface by extracting hydrogen atoms with impinging electrons and/or electric field. A row of fabricated DBs forms a one-dimensional atomic structure that we call a DB wire. Although this DB wire is chemically too reactive to be used in practical atomic-scale devices, we expect it to be an instructive example of an atomic-scale structure on an insulating or semiconducting surface. We analyze our results based on recent first-principles calculations [13,16]. Placing DB wires on the semiconductive Si surface results in the formation of electronic states across the Fermi energy located in the energy gap region of the substrate. The DB wires play a further important role in fabricating more promising atomic wires. Thermally deposited Ga atoms preferentially adsorb on the DBs and DB wires on the surface and this selectivity is used to demonstrate a novel method of fabricating atomic-scale Ga wires on the Si(100) surface by adsorbing Ga atoms on to the DB wire.

4.2 Experimental Aspects

All experiments were carried out in UHV-STM systems. Rectangular Si(111) and Si(100) samples of 2×15 mm^2 were cut from As-doped n-type 5 and 7 to 13 mΩ·cm silicon wafers, respectively, and were set on sample holders made of tantalum. After outgassing by heating at 650°C for 12 h, the samples were annealed at 1250°C for 10 s to remove the oxide layer and were slowly cooled down to obtain Si(111)–7×7 and Si(100)–2×1 surfaces, which were checked by STM observation. To prepare hydrogen-terminated Si(100)–2×1–H surface, the clean surface was exposed to an atomic hydrogen flux of typically 3×10^{-2} ML/s for 5 min (1 ML here is defined as the number of Si atoms on the bulk-terminated ideal Si(100) surface: 6.78×10^{14} atoms/cm^2), while the sample was kept at approximately 390°C [11]. The atomic hydrogen was formed by decomposing hydrogen molecules with a heated tungsten filament. Sample temperatures for annealing were measured with an optical pyrometer and those for hydrogen termination were estimated by extrapolating the measured temperature–annealing-power curve using the Stefan–Boltzmann radiation law.

These sample preparations were performed in the sample preparation chamber with pressure in the low 10^{-10} torr range. The samples were then transferred to the STM chamber. The chamber pressure was maintained below 7×10^{-11} torr during STM measurements. Scanning tips were prepared by electrochemical etching of tungsten wires and were heat-treated by electron bombardment in the preparation chamber for the Si(111)–7×7 experiment. For fabricating DBs on the Si(100)–2×1–H surface, a field ion microscope (FIM) was used to clean and shape the apex of the tip [17,18]. We also used sharp needle structures on the sample surface. These were made by applying ramped voltages (typically of 10.0 V) while keeping the tunneling current constant, in order to counterimage the scanning tip with atomic resolution (Needle Formation and Tip Imaging, or NFTI method) [19].

The STM images were taken at a sample bias voltage V_s ranging from -3.0 to $+3.0$ V and a constant tunneling current I_t ranging from 10 to 1000 pA. The trenches on the Si(111) 7×7 surface were formed by moving the tip on the surface at a constant tunneling current I_t ranging from 0.5 to 300 nA and a tip bias voltage V_t ranging from $+3.0$ to $+5.0$ V. Individual hydrogen atoms on the Si(100)–2×1–H surface were manipulated by applying a pulsed voltage to the sample (typically, a sample bias voltage $V_s = +2.9$ V, a tunneling current $I_t = 400$ pA, and a pulse duration ranging from 100 to

300 ms were used). The DB wires were fabricated by repeatedly addressing the tip to desired positions and applying the pulse voltages. DB wires are also fabricated by desorbing the hydrogen atoms using the tunneling current from the STM tip [11]. A Ga dispenser made of a tungsten filament and a small amount of Ga metal was placed 2 cm away from the sample and heated to 600–650°C in order to deposit Ga on the sample. All experiments were carried out at room temperature.

4.3 Property Changes in the Si(111)–7×7 Surface

Figures 4.1a, b and c show typical gray-scale filled-state STM images of the Si(111)–7×7 surface after fabricating (a) line-, (b) circle- and (c) spiral-shaped trench patterns. The apparent depth of the trench measured from the surface level at the outside of the trench ranges between 0.2–0.4 nm and the width of the trench ranges over 8–15 nm. The atomic structure of the trench

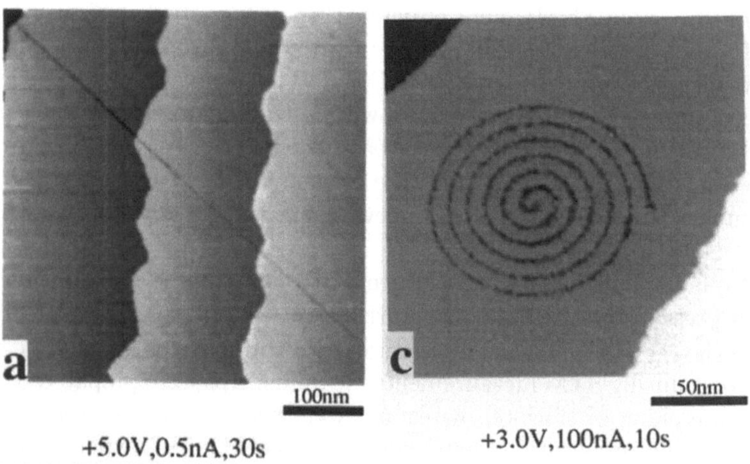

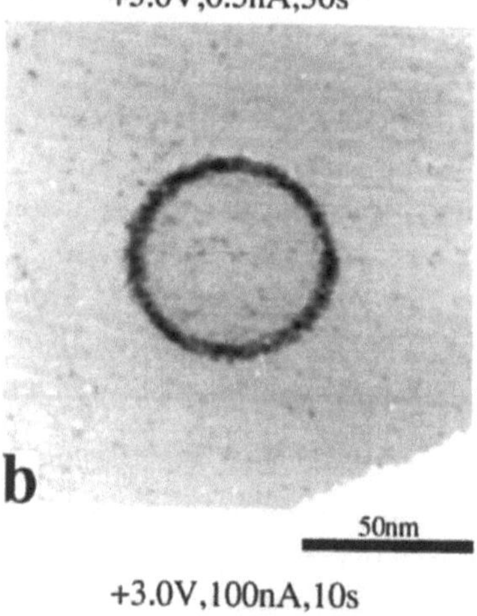

Fig. 4.1. Typical filled-state STM images of the Si(111) surface with (**a**) line-, (**b**) circle- and (**c**) spiral-shaped trench patterns fabricated by moving the tip on the surface with given conditions. Images are observed at a tip bias of $V_t = +2.0$ V and a tunneling current of $I_t = 0.5$ nA

itself observed by STM showed no particular structure, such as adatom arrangement or rest site atoms and rather showed small random protrusions. For the surface manipulation using a high voltage [20] or a large current [21], field evaporation and/or chemical interactions have been reported as manipulation mechanisms. However, we shall not discuss the details here. We investigated the electronic structure of the trench itself by STS and observed a clear band gap of approximately 1.4 eV, which indicates that surface states originating from dangling bonds (DBs) are somehow passivated by other atoms, such as oxygen or hydrogen [22]. At the moment, however, the surface structure and passivation mechanism of the trench region are not clear.

Figure 4.2 shows bias dependent STM images and cross-sectional views of a circle trench fabricated on the Si(111)–7×7 surface. Note that there is an atomic step at the bottom of the STM images in Fig. 4.2. In Fig. 4.2a, we clearly observe that the height of the inside of the square trench is lower in the empty-state STM image than the rest of the surface [23]. Although the images in Fig. 4.2 do not have sufficient resolution to observe the 7×7 atomic structure, we obtained atomically resolved images of 7×7 structure both inside and outside the square trench when the scan range was reduced. The atomic structure inside the trench was essentially the same as outside the trench. Thus the observed height in the STM image in Fig. 4.2a corresponds to the average over the surface area larger than the 7×7 unit. In the filled-state image in Fig. 4.2c, the height of the inside of the trench is slightly lower than outside. The corresponding cross-sectional views in Fig. 4.2 show the height difference more clearly. The height difference is more significant for a negative tip bias and for a lower tip voltage. It is also more significant for a larger tunneling current, as is shown in Fig. 4.3.

The tendency indicates that the height difference cannot be attributed to a structural change but rather to a change in electronic effect. The measured height difference is also unlikely to be due to a structural change like surface damage because the 7×7 reconstructed surface is intact both inside and outside the trench. This clearly shows that the surface is not damaged by trench formation. Since the density of states near the Fermi level is observed by STM, the effect should be related to electronic effects caused when the square trench is formed. We tried to fabricate a line trench across the circle-trench pattern using the same method we used to make the trench. The STM tip was moved on the surface across the circle trench, while maintaining a constant tunneling current of 300 nA and a tip bias voltage of +3.0 V.

The speed of the tip movement in this case was chosen to be 5 nm/s, faster than during square-trench formation, in order to clarify surface property changes. The results showed another interesting property change due to the closed circular trench. While a line trench was clearly fabricated outside the trench, almost no change occurred inside the square trench and the difference in the extent of the modification between the inside and the outside of the trench was clear. We have tested trenches with different shape, depth and

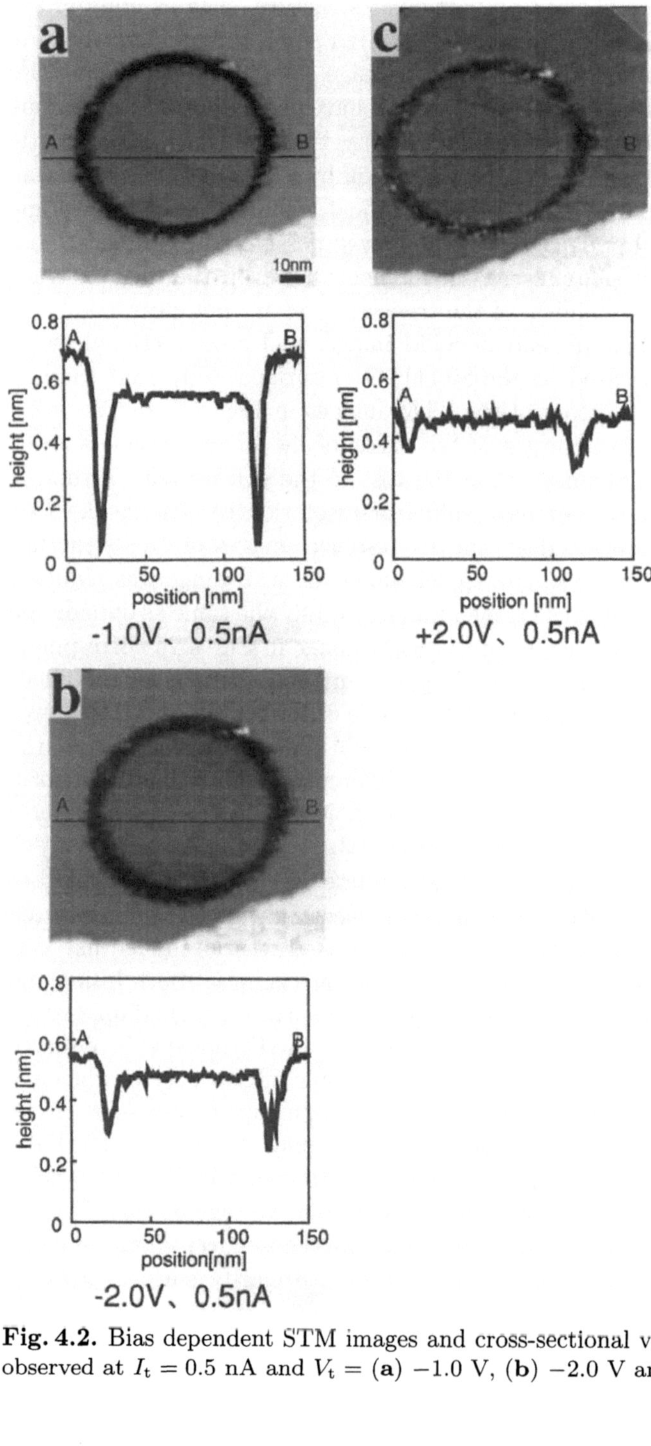

Fig. 4.2. Bias dependent STM images and cross-sectional views of a circle trench observed at $I_t = 0.5$ nA and $V_t =$ (**a**) -1.0 V, (**b**) -2.0 V and (**c**) $+2.0$ V

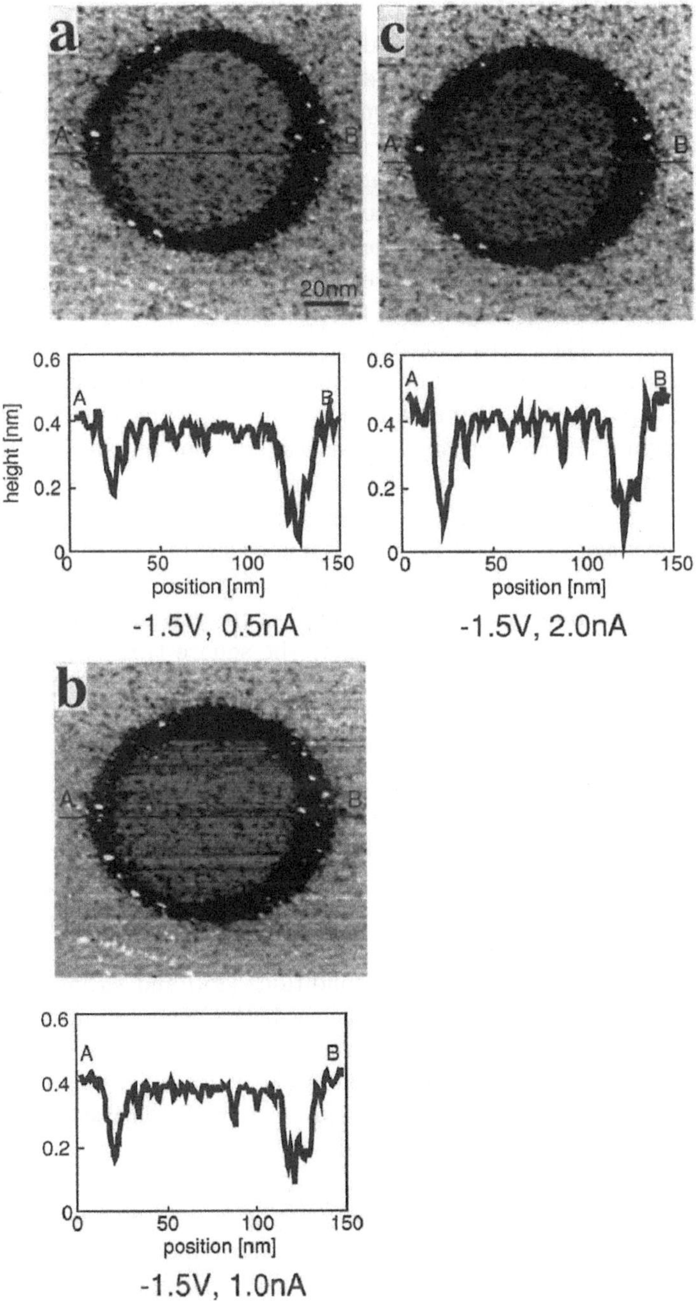

Fig. 4.3. Current dependent STM images and cross-sectional views of a circle trench observed at $V_t = -1.5$ V and $I_t =$ (**a**) 0.5 nA, (**b**) 1.0 nA and (**c**) 2.0 nA

width and find that the observed surface property changes are sensitive to the depth and width of the trench but do not depend on the trench shape when the trench is closed. For example, square and rectangular trenches give similar results to the circular trench. When the trench was not closed completely, we observed almost no surface property change [23]. We also tested the possibility of fabricating the line trench inside the open loop trench. We clearly observed a line-trench pattern written both inside and outside the open trench.

We shall suggest several possibilities for the mechanism underlying surface property changes in apparent surface height and line-trench fabricability, although this mechanism has not yet been fully elucidated.

One explanation for the trench fabricability change is possible if we assume that trench fabrication is based on electromigration. Electromigration was proposed to explain atom manipulation by the tunneling current between tip and sample [24]. For the Si(111) surface, adatoms on the 7×7 reconstructed layer are known to migrate on the surface by an electronic field and/or a large current. The current density generated by an STM current of 300 nA at a bias voltage of +3.0 V is of the same order, or one to two orders of magnitude larger, than that in the metal (Al, Cu, etc.) wires of ULSI technology. Thus electromigration may be induced in the present experimental conditions. When the surface is enclosed by the square trench, adatom migration on the surface becomes difficult because the trench constitutes a barrier, and the fabricability of the surface is thereby reduced, especially in the case of closed loop trenches.

It may also be that a surface stress is changing the electronic structure of the surface. By enclosing the surface with the trench, surface structures and surface electronic states may be changed by relaxation of surface stresses. Because STM is quite sensitive to surface electronic states, the inside and outside of the trench can be observed in a different way if electronic states are modified by surface stress relaxation. Furthermore, surface manipulation conditions depend on materials or surface conditions [25], and surface fabricability may change when the surface is enclosed. However, the apparent height change may not correspond to the surface structural change because the apparent height change depends on the STM bias voltage. The apparent height change is more evident as the bias decreases in both polarities, and is more noticeable for negative tip voltages than positive voltages.

Another possibility is electron confinement. If the conductance of the trenches is small enough, which may be the case judging from STS results, and the enclosed region is electronically isolated from the surrounding surface, the enclosed surface can be regarded as a quantum dot and the Coulomb blockade should occur when tunneling electrons come from, or go to, the tip [26]. Such an unpaired electron tunneling phenomenon decreases surface conductance and results in depressed images. Surface modification is often explained by field evaporation due to a high electric field between tip and sample. Since the surface is charged up during unpaired electron tunneling, the effective bias voltage is lowered. Consequently, the electric field on the surface right beneath the tip is also lowered, and field evaporation is prevented.

One further explanation, and perhaps the most plausible, is based on conduction through surface states. Several kinds of surface conductivity have been reported [27,28]. Hasegawa, S. et al. reported surface conductivity based on band bending. Electronic conduction through surface states was discussed by Hasegawa, Y. et al. in the light of their point contact experiment. If the resistance of the trench region is much higher than the surface reconstructed layer and is comparable to the tunneling resistance (~1 GΩ), a voltage drop occurs between the inside and outside of the trench. As a result, the effective bias voltage is lowered and surface modification is prevented because the electric field on the surface does not reach the threshold field for field evaporation. In addition, the distance between the tip and the sample decreases by a reduced bias voltage because the surface inside the trench is charged up. Hence, the STM image of the enclosed region is depressed compared with the normal surface.

Considering the bias dependence of apparent height change, the electron confinement and surface conductance models are more plausible because both are based on lowering of the effective bias voltage inside the trench. A detailed investigation to clarify the mechanism is discussed elsewhere [23].

4.4 Properties of Dangling Bonds on the Si(100)–2×1–H Surface

A typical STM image of the hydrogen-terminated Si(100)–2×1–H monohydride phase is shown in Fig. 4.4a. Although the surface states of the dangling bonds (DBs) of Si dimers are passivated by hydrogen adsorption, we observe cocoon-shaped Si dimers similar to those observed in the case of the clean Si(100)–2×1 surface. Missing hydrogen defects (DBs) are observed as white protrusions, since the DB states near the Fermi level are recovered [9,10]. In this particular sample-treatment condition, most of the protrusions we observe are unpaired DBs

An example of this is marked with a small arrow in Fig. 4.4a. A defect with two hydrogen atoms missing from a Si dimer (a DB pair) is imaged at the center of the dimer row in Fig. 4.4a (indicated by a larger arrow). Their ball-and-stick models are shown in Fig. 4.4b. The dark spots in Fig. 4.4a are missing Si-dimer defects. We also observed a local dihydride phase and a 3 × 1–H phase [10]. This surface can be routinely prepared over a wide range of atomic hydrogen flux (3×10^{-3}– 3×10^{-1} ML/s) and temperature (350–400°C), although DB density depends on conditions.

Bias dependent STM images of the DBs are shown in Figs. 4.5a and b. In Fig. 4.5a, several DBs are observed and are shown by arrows. Impurity adsorbates are also imaged as protrusions and look similar to the DBs in filled-state STM images. However, DBs are imaged differently in the relatively

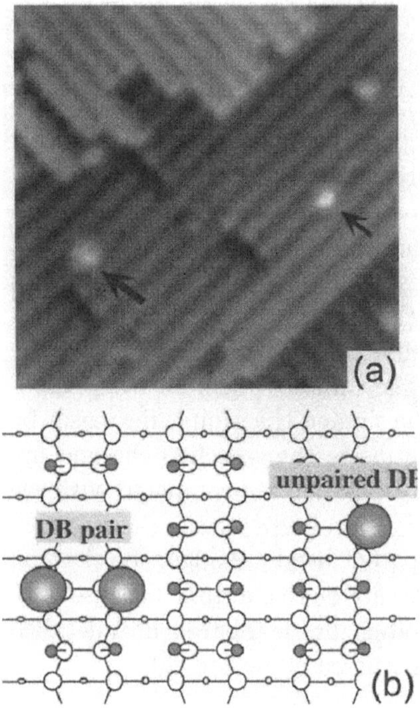

Fig. 4.4. (**a**) A typical filled-state gray-scale STM image of the Si(100)–2×1–H surface (12 nm×12 nm, $V_s = -2.0$ V, $I_t = 100$ pA). An unpaired DB (*small arrow*) and a DB pair (*larger arrow*) can be distinguished by their symmetry against the Si dimer row. (**b**) Schematics of an unpaired DB and a DB pair

low-biased empty-state images shown in Fig. 4.5a. Individual DBs are imaged as small protrusions surrounded by ring-shaped moats. This pattern can be understood as a Friedel oscillation of the density of states resulting from the charge localized at the dangling bond, similar to the case of Si dopants near the GaAs(110) surface recently reported by van der Wielen et al. [30].

An example of a DB wire fabricated by the STM tunneling current [11] is shown in Fig. 4.6. The DB wire in Fig. 4.6 is observed as a bright line running from top right to bottom left along a dimer row at a step edge. It is mainly made up of DB pairs in this case and is one dimer wide (0.8 nm) over most of the wire. We have found that tip-shape control by Field Ion Microscopy (FIM) [17,18] and/or the Needle Formation and Tip Imaging (NFTI) method [19] is particularly useful in controlling the width of DB wires fabricated by this method. Figure 4.7a shows an example of the square-shaped DB structure fabricated using a series of voltage pulses as described in Sect. 4.2 [31].

Using this method, we routinely obtained DB wires of width one to two dimers (0.8 to 1.6 nm). The wires forming the square in Fig. 4.7a are made of either unpaired dangling bonds or DB pairs. Since impurity adsorbates appear as protrusions in both filled- and empty-state images, they can be distinguish from DBs, which appear as small protrusions surrounded by ring-shaped moats in the empty-state STM image (Fig. 4.5) [29]. We confirmed that the DB structure in Fig. 4.7a does not contain Si or impurity atoms.

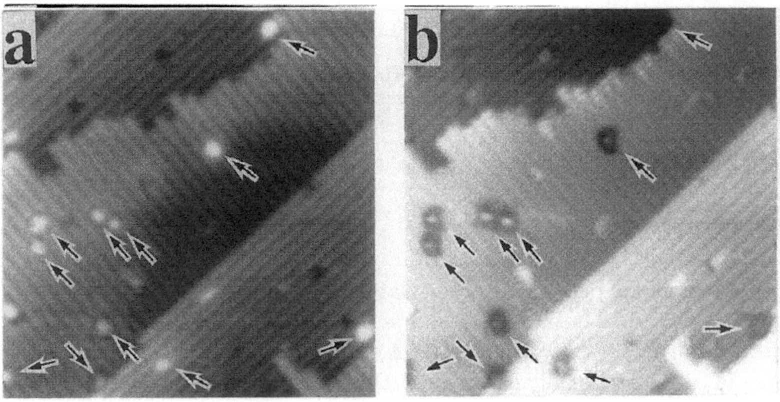

Fig. 4.5. Gray-scale STM images of the same surface area showing bias dependence of the DB image. Each DB, imaged as a round protrusion in the filled-state image (**a**) (*arrows*), is imaged as a small protrusion surrounded by a moat (*dark ring*) in the empty-state image (**b**). [25 nm × 25 nm, V_s = (**a**) −2.0 V and (**b**) +1.8 V, I_t = 20 pA]

Fig. 4.6. An example of a long DB wire fabricated using the STM tunneling current. (60 nm × 60 nm, V_s = −2.0 V, I_t = 20 pA)

When a pulsed voltage V_s over +3.0 V was used, the extraction of unpaired atoms became difficult and wide-area extraction and/or impurity adsorption from tip to surface occurred frequently.

When a pulsed voltage V_s less than +2.6 V was used, hydrogen extraction rarely occurred. In Fig. 4.7b, we show a schematic model of an ideal DB wire made only of unpaired DBs, parallel to the dimer-row direction. Part of the

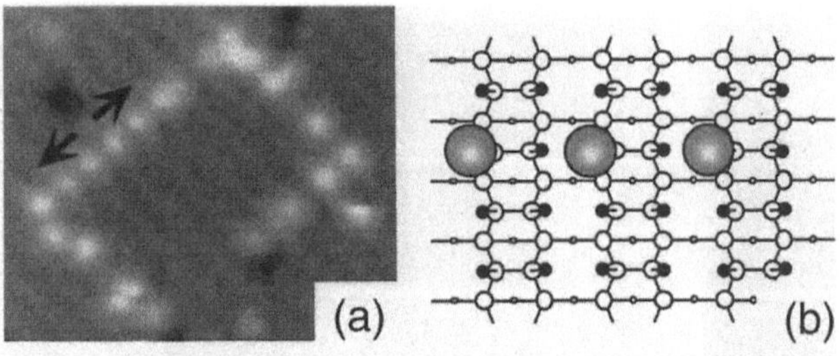

Fig. 4.7. (**a**) A square-shaped DB structure fabricated on the Si(100)–2×1–H surface that contains DB wires both parallel and perpendicular to the dimer-row direction (10 nm × 8 nm, $V_s = -2.0$ V, $I_t = 100$ pA). Part of the DB structure, marked by *two arrows* in (**a**), is made only of unpaired DBs and its schematic ball-and-stick model is shown in (**b**)

DB structure in Fig. 4.7a (marked by two arrows) contains the DB wire shown in Fig. 4.7b. In general, a DB wire may be made of DB pairs, unpaired DBs or a combination of both.

We later use selective adsorption of metal atoms on the DBs in order to fabricate atomic-scale metal structures on the silicon surface. For this purpose, we have tested several other ways of manipulating hydrogen atoms using STM. The making, erasing and moving of a DB were demonstrated. In order to make a DB, the sample bias was ramped down and kept at the manipulation bias for 1 s while the tunneling current was kept constant. This procedure was repeated while the manipulation bias was reduced step by step until a sudden tip-height change was observed with sample bias $V_s = -0.9$ V. Manipulation of a hydrogen atom was sometimes accompanied by a tip apex change and STM image degradation in that case (double tip image). We do not understand the exact mechanism of the hydrogen and tip apex manipulation. However, we speculate that it is due to strong chemical interaction between tip and surface and/or the field effect, since the tip–sample distance was reduced by 0.3 nm down to 0.4–0.6 nm during the procedure [32].

A DB was erased by placing the tip on top of a DB and applying a sample bias pulse of $V_s = -3.5$ V for 5 ms. We found that we could sense the local barrier height difference between the DB and the hydrogen-terminated area and we could apply the sample bias pulse until we observed that the manipulation was complete. The particular tip used for this experiment had been used for the DB patterning and NFTI method on the hydrogen-terminated surface, and presumably there were enough hydrogen atoms on the tip apex, one of which was manipulated by the bias pulse and transferred from tip to DB position.

We also observed that a DB was sometimes displaced by tip motion during STM imaging [34] and we used the tip–DB interaction to move a DB on the surface. We started with the surface observed by STM and placed the tip on top of a DB by observing the local barrier height. We then displaced the tip by a small amount and held the tip position (typically for a few seconds) until we observed that the tip was once more on a DB. During this procedure, we kept the sample bias and tunneling current at the imaging condition. We repeated this procedure several times and found that the DB was displaced by 3 times unit separation (0.384 nm) along the dimer-row direction and also was displaced to the next dimer row. We speculate that there is a repulsion interaction between the tip and the hydrogen atoms, which caused the DB to move in the opposite direction to the hydrogen atom. The interaction was effective for manipulating a surface atom at the place closest to the tip apex. This is consistent with theoretical calculations by Hirose and Tsukada [32].

Electronic structures of various kinds of DB wires were then examined by STS. We first show local electronic structures for a bare Si dimer on the clean Si(100)–2×1 surface, and a hydrogen-terminated dimer and an unpaired DB on the Si(100)–2×1–H surface in Figs. 4.8a through c.

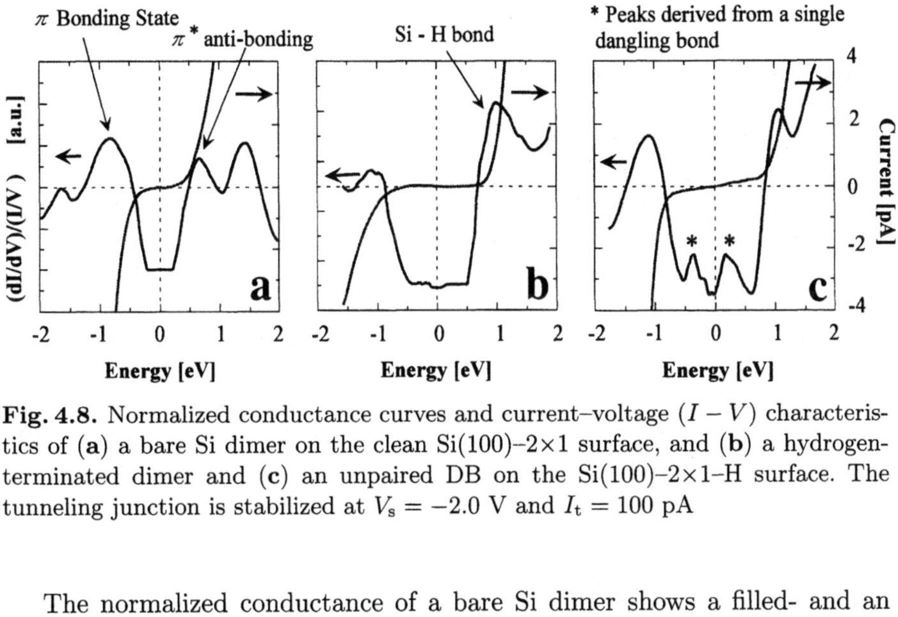

Fig. 4.8. Normalized conductance curves and current–voltage ($I-V$) characteristics of (**a**) a bare Si dimer on the clean Si(100)–2×1 surface, and (**b**) a hydrogen-terminated dimer and (**c**) an unpaired DB on the Si(100)–2×1–H surface. The tunneling junction is stabilized at $V_s = -2.0$ V and $I_t = 100$ pA

The normalized conductance of a bare Si dimer shows a filled- and an empty-state peak located at 0.9 eV below and at 0.5 eV above the Fermi energy E_F, respectively (Fig. 4.8a). Those peaks correspond to the local density of states resulting from π interaction between two DBs on a Si dimer. The splitting between bonding and antibonding states is responsible for the surface band gap observed on the normalized conductance and $I-V$ characteristics. For the hydrogen-terminated dimer, a wider band gap is observed and the peak of the normalized conductance near +1.1 eV is assigned to antibonding states of the Si–H chemical bond (Fig. 4.8b) [9,10]. An unpaired DB shows finite (non-zero) density of states near the Fermi level and shows two density-of-state peaks at −0.4 and +0.2 eV, that correspond to the density

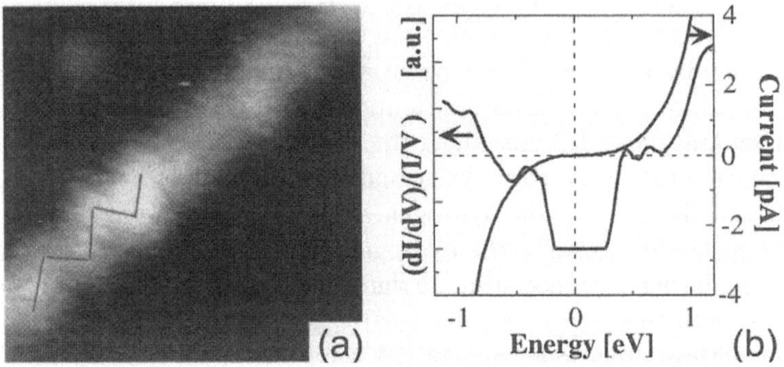

Fig. 4.9. (**a**) A zoom-up filled-state image of a DB wire parallel to the dimer-row direction which is mainly made of DB pairs, similar to the case in Fig. 4.6. (5 nm× 5 nm, $V_s = -2.0$ V, $I_t = 20$ pA.) (**b**) $I - V$ characteristics and a normalized conductance curve recorded on the DB wire shown in (**a**). The tunneling junction is stabilized at $V_s = -2.0$ V and $I_t = 100$ pA

of states resulting from unpaired DB states (Fig. 4.8c). These results agree well with those reported by Boland [9,10].

Figure 4.9a depicts part of a long DB wire fabricated by the STM tunneling current, similar to the one shown in Fig. 4.6.

It interesting to note that the dimers in this DB wire are buckled and the wire is observed as a zigzag line similar to the clean dimer-row structure stabilized by a defect or an adsorbate atom [18,33]. Because of the buckling feature and also because the DB wire is located at the center of a dimer row, it is reasonable to assume that the DB wire is made up of paired DBs. The tunneling spectra in Fig. 4.9b have a band gap of 0.5 eV. First-principles calculations by Watanabe et al. [13] predicted that DB wires made of DB pairs should have a band gap similar to a clean Si(100)–2 × 1 surface, for DB wires both parallel and perpendicular to the dimer-row direction. The gap is caused by splitting between bonding and antibonding states derived from the interaction between DBs of Si atoms on a dimer. The DB wire in Fig. 4.9a is mainly composed of DB pairs and has band gap 0.5 eV, in good agreement with theoretical calculations [13]. The buckled dimers are only stabilized near defects on the clean Si(100)–2 × 1 surface.

In the present case, buckling is stabilized either by the one-dimensional characteristics of the wire structure (Peierls distortion) or else by the side effects (both sides of the wire are somewhat disordered and this may stabilize the buckling). The local electronic structure was examined for a DB wire fabricated parallel to the dimer-row direction shown in Fig. 4.10a. The observed tunneling spectra are shown in Fig. 4.10b. The wire is in this case made of both unpaired DBs and DB pairs in a dimer row, as well as a couple of DBs in the next dimer row. The $I - V$ characteristics show a finite density of states at the Fermi energy, in contrast to the case of a DB wire composed

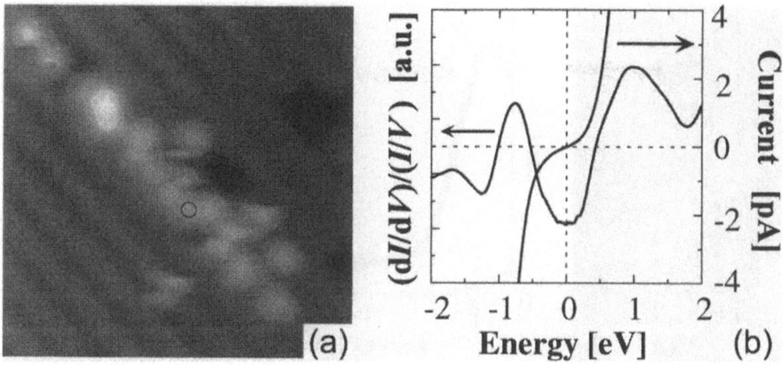

Fig. 4.10. (**a**) Filled-state STM image of a DB wire parallel to the dimer-row direction (7 nm× 7 nm, $V_s = -2.0$ V, $I_t = 100$ pA) and (**b**) $I - V$ characteristics and a normalized conductance curve recorded on the DB wire [marked with a *circle* in (**a**)]. The tunneling junction is stabilized at $V_s = -2.0$ V and $I_t = 100$ pA

only of DB pairs. Characteristic peaks at −0.8 and +1.0 eV are observed in the normalized conductance and an energy gap is not observed.

Figure 4.11a shows an example of the DB wire fabricated perpendicular to the dimer-row direction. The wire is made of three unpaired DBs and two DB pairs in a row as well as other DBs exhibiting a zigzag structure. There are also a couple of DBs located outside the DB wire. Figure 4.11b shows conductance curves ($dI - dV$) recorded on a hydrogen-terminated Si dimer (thin line) and on the dangling-bond wire marked with a circle in Fig. 4.11a (thick line).

Similar results are obtained for the tunneling spectra at the various positions of the DB wire in Fig. 4.11a. It is evident from the conductance curves in Fig. 4.11b that the DB wire in Fig. 4.11a has finite density of states at the Fermi energy, whereas no density of states is observed for the hydrogen-terminated dimer. Characteristic peaks at −1.6 eV, −0.9 eV and +0.8 eV are seen in the normalized conductance and an energy gap is not observed.

These STS results are now compared with the first-principles calculations by Watanabe et al. [13]. A surface-state band in the bulk band gap resulting from the DBs is predicted by theoretical calculations for the case of DB wires made up of unpaired DBs, formed either parallel or perpendicular to the dimer-row direction. When the Peierls instability is considered, the system is stabilized by only 14 meV, which is sufficiently small compared with the thermal energy at room temperature. We therefore expect a finite density of states at the Fermi energy for the DB wires in Figs. 4.10 and 4.11. The theoretical result also predicts that the DB wires composed of DB pairs should exhibit a band gap similar to the case of the clean Si(100)–2×1 surface, for DB wires both parallel and perpendicular to the dimer-row direction.

The DB wires in Figs. 4.10 and 4.11 are made of both unpaired DBs and DB pairs. Tunneling spectra indicate that there is a finite density of states

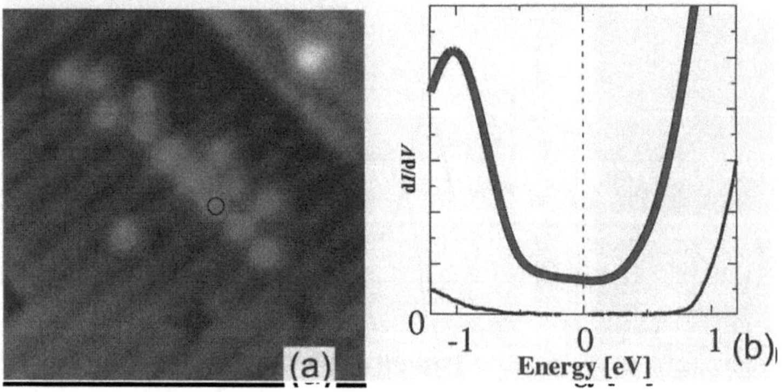

Fig. 4.11. (**a**) Filled-state STM image of a dangling-bond wire perpendicular to the dimer-row direction (7.5 nm× 7.5 nm, $V_s = -2.0$ V, $I_t = 100$ pA) and (**b**) conductance curves ($dI - dV$) recorded on a hydrogen-terminated Si dimer (*thin line*) and on the dangling-bond wire marked with a *circle* in (**a**) (*thick line*). The tunneling junction is stabilized at $V_s = -2.0$ V and $I_t = 100$ pA

at the Fermi energy, and it is clear that a semiconductive band gap does not exist for either case. We therefore propose that the density of states near the Fermi energy is due to the row of unpaired DBs in the DB wires. Two distinct peaks in the normalized conductance curve in Fig. 4.10 are located near the position of calculated bonding and antibonding band positions for the DB wires made of DB pairs.

Similar discussion is possible for the DB wire in Fig. 4.11. We temporarily conclude that the peaks in the normalized conductance curves at approximately −0.9 eV and approximately +0.9 eV correspond to the two characteristic peaks of the DB wires made of DB pairs. Following the discussion above, we propose that the electronic states of DB wires shown here are a mixture of the electronic states of both unpaired DBs and DB pairs. However, first-principles calculations by Watanabe et al. [13] deal with infinite perfect DB wires and do not fully treat the finite-length, disorder or electron correlation effects. Detailed theoretical and experimental investigations on these effects will be discussed in the future.

4.5 Interaction of Adsorbates with Dangling Bonds on Si(100)–2×1–H Surfaces and Atomic Wire Fabrication

It has been well documented that chemical reactions on semiconductor surfaces are determined, in the first approximation, by the nature of the surface dangling bonds (DBs). We therefore expect that selective adsorption of metal atoms on DBs can be used to fabricate atomic-scale metal structures on the silicon surface. We chose the gallium atom since it is one of the

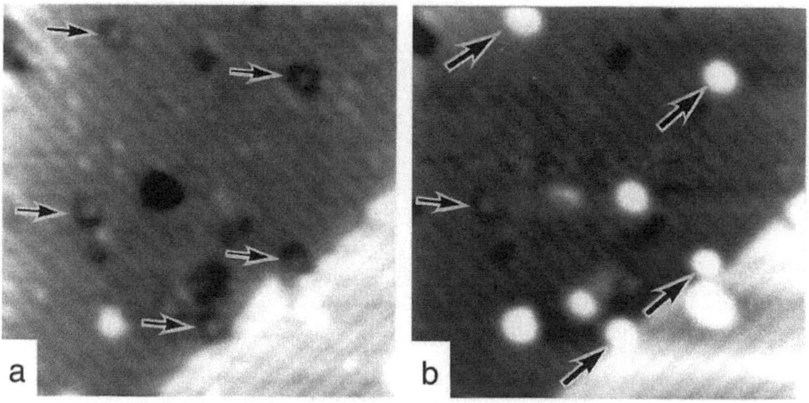

Fig. 4.12. Empty-state STM images of the hydrogen-terminated Si(100)–2×1–H surface (**a**) before and (**b**) after 0.005 ML of Ga is deposited. (17 nm×17 nm, $V_s = +1.8$ V, $I_t = 20$ pA.) Each DB is imaged as a small protrusion surrounded by a moat ring [*small arrows* in (**a**)]. Impurity adsorptions are also imaged as protrusions. In (**b**), Ga atoms are preferentially adsorbed onto the DBs (*large arrows*), except for a leftover DB (*small arrow*)

metal elements which shows surface adsorption even after annealing [15]. As mentioned above, each DB is imaged as a small protrusion surrounded by a ring-shaped moat in the relatively low-biased empty-state images, as shown by a small arrow in Fig. 4.12a. Impurity adsorbates are imaged as protrusions. Figure 4.12b shows the same area with thermally deposited 0.005 ML of Ga. It is evident from the STM image that Ga atoms preferentially adsorb onto the DBs (shown by large arrows in Figs. 4.12b). One DB in Fig. 4.12b (small arrow) remains without Ga adsorption. The images of some impurity adsorbates are larger in (b) than in (a) indicating adsorption of Ga atoms. Further Ga deposition results in the growth of Ga clusters, the density of which is comparable to that of initial DBs and impurity adsorbates. The Ga atoms also show stable adsorption onto the DBs. The first-principles calculation by Watanabe et al. [35] has shown that the barrier height of Ga atoms for surface migration is approximately 0.3 eV on the hydrogen-terminated area and 1.2 eV on the DB. We do not observe preferential adsorption at the step edges in Fig. 4.12b. Missing-dimer defects (seen in Fig. 4.12b as dark dots) are not reactive to Ga adsorption either. It is consistent with the model of hydrogen termination that the DBs at step edges and at missing-dimer defects should be passivated by hydrogen. Several metal atoms such as Au and Ni showed similar adsorption features to Ga atoms, although Ti atoms reacted with hydrogen-terminated Si dimers and showed random adsorption [36].

The tendency for preferential adsorption onto DBs is also observed for small molecules such as 1,4-bis[β-pyridyl-(2)-vinyl]benzene (P2VB) [37] and C_{60} [36]. An example for the case of the C_{60} molecule is shown in Fig. 4.13. Similar to Ga atoms, C_{60} molecules preferentially adsorb onto the DBs (shown

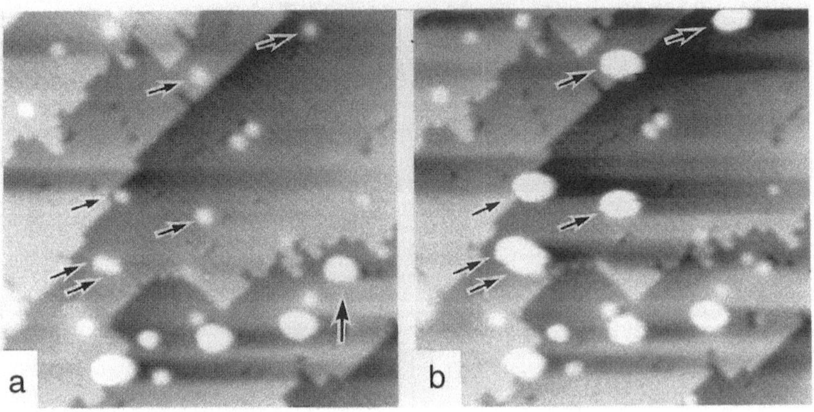

Fig. 4.13. Empty-state STM images of the hydrogen-terminated Si(100)– 2×1–H surface (**a**) before and (**b**) after 0.003 ML of C_{60} is deposited. (32 nm×32 nm, $V_s = -2.0$ V, $I_t = 20$ pA.) Each DB in (**a**) is covered with a C_{60} molecule in (**b**) (*small arrows*), while one C_{60} molecule is displaced by the tip–molecule interaction [*large arrow* in (**a**)]

by small arrows in Fig. 4.13b). Note that a small number of C_{60} molecules were already present in Fig. 4.13a. In the case of C_{60} molecules, in contrast to metal atoms such as Ga, Au and Ni, interactions among molecules are not strong enough to nucleate clusters starting from one-molecule adsorption, which predominates for the prepared Si(100)–2×1–H surface. This is because there are only a small number of randomly distributed, isolated DBs and adsorbed C_{60} molecules are thus also isolated from each other. As a result, only a very limited number of C_{60} molecular island formations was observed on the Si(100)–2×1–H surface (one island per square micron).

We have fabricated atomic-scale DB wires using the STM tunneling current (Fig. 4.14a) and thermally deposited Ga atoms in order to demonstrate a method for fabricating atomic-scale metal-atom structures on silicon surfaces (Fig. 4.14b). Since Ga atoms preferentially adsorb onto DBs, atomic-scale structures of Ga atoms have been realized on the silicon surface. However, the width of the Ga wire, is not constant, indicating that part of the wire is two or more Ga-atoms wide. In addition, Ga atoms are missing in some parts of the wire. Therefore, we are currently investigating the possibility of reshaping the Ga wire using individual DB manipulation and/or individual Ga atom manipulation.

Connecting atomic structures to bulk electrodes is one of the key issues for measuring properties of those structures and evaluating the performance of atomic devices. We have used a micron-order metal mask and formed metal wires together with bonding pads on the Si(100)–2×1–H surface by thermally depositing Au/Ti in situ. Figure 4.15 shows an example of the bulk electrodes with bonding pads formed on the sample surface and observed by scanning

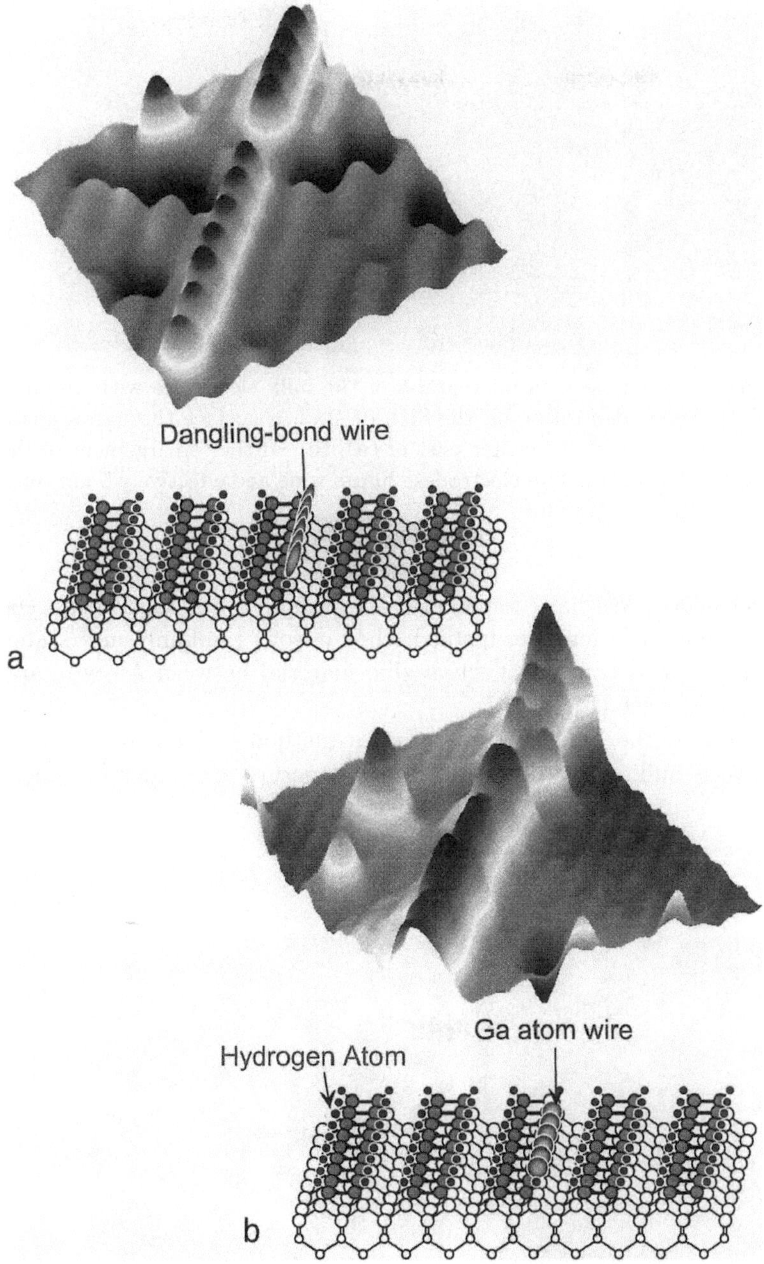

Fig. 4.14. Three-dimensional views of (**a**) an atomic-scale DB wire and (**b**) thermal deposition of Ga atoms on a DB wire, demonstrating a method of fabricating atomic-scale metal wires on the silicon surface. (6.5 nm× 6.5 nm, $V_s = -2.0$ V, $I_t = 20$ pA)

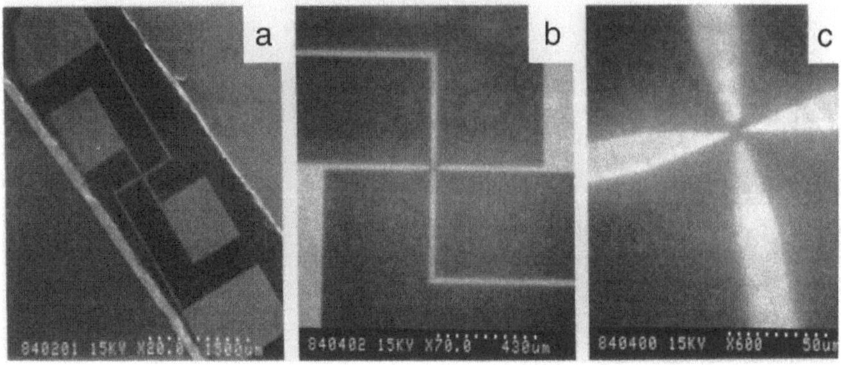

Fig. 4.15. (**a**) Scanning electron micrograph of the bulk electrodes with bonding pads formed by Au/Ti deposition on the Si(100)–2×1–H surface through a metal mask. (**b**) Enlarged view of the center part of (**a**). (**c**) Further enlargement of the center part of the Au/Ti thin film electrodes, 20 μm wide and effectively 5 nm thick

electron microscope. We have found that Ti atoms do not migrate on the hydrogen-terminated Si surface. Instead, they adsorb randomly and stably onto the surface. It is thus an effective glue material between Au wire and Si(100)–2×1–H surface (Fig. 4.16).

Detailed use of the bulk electrodes and a method of intermediate connection between bulk electrodes and atomic structure are under investigation.

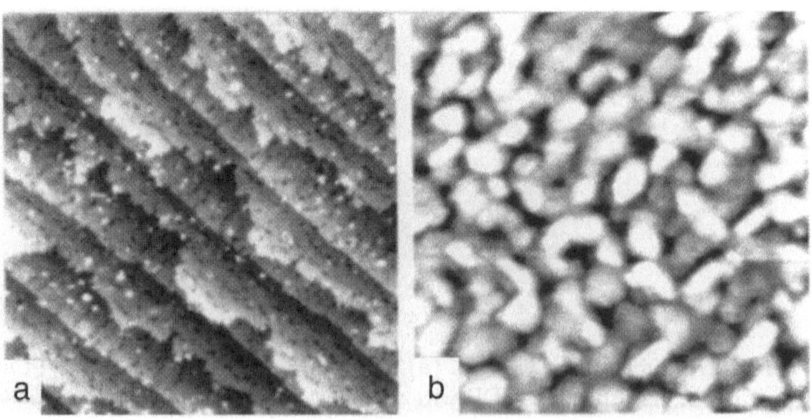

Fig. 4.16. Filled-state STM images of (**a**) an Au electrode formed by Au/Ti deposition on the Si(100)–2×1–H surface and (**b**) almost bare Si(100)–2×1–H surface at the intersection of four Au electrodes shown in Fig. 4.15c. (120 nm×120 nm, $V_s = -2.0$ V, $I_t = 20$ pA)

4.6 Conclusion

We have applied STM/STS and atom manipulation to fabricate and characterize nanoscale and atomic-scale structures on Si(111)–7×7 and hydrogen-terminated Si(100)–2×1–H surfaces. We observed that the surface properties of the Si(111)–7×7 surface were drastically modified by forming a closed trench made by applying a high voltage and/or a large tunneling current between the tip and the surface. The apparent height of the surface inside the closed trench measured by STM was lower than the outside surface. We discussed several possible mechanisms to explain the STM observations and proposed the most probable explanation based on conductance through surface states. We fabricated atomic-scale DB wires by extracting individual hydrogen atoms from the hydrogen-terminated Si(100)–2×1–H surface. Tunneling spectra of the DB wires showed a finite density of states at the Fermi energy. We discussed the origin of the density of states in the light of theoretical results derived by first-principles calculations. We pointed out the possibility of the mixed density of states resulting from both unpaired DBs and DB pairs. However, we do not rule out a possible explanation based on the strong electron correlation inherent in one-dimensional and/or atomic-scale structures. The Ga atoms on the monohydride phase are mobile and preferential adsorption of Ga on the DBs was observed. We demonstrated a method of fabricating atomic-scale Ga patterns on the silicon surface, using DB patterning by the STM current and thermal deposition of Ga on the DB patterns.

Acknowledgements

This work was carried out in collaboration with Drs. K. Nakajima, Y. Wada and M. I. Lutwyche, and Professors T. Nishi, S. Watanabe and T. Hasegawa.

References

1. R. Feynman, There's Plenty of Room at the Bottom, (invited talk at the Jet Propulsion Laboratory, February 1960). Reprinted in J. Microelectromechanical Systems **1**, 60 (1992).
2. G. Binnig, H. Rohrer, Ch. Gerber and E. Weibel, Phys. Rev. Lett. **49**, 57 (1982).
3. D.M. Eigler and E.K. Schweizer, Nature **344**, 524 (1990).
4. H. Uchida, D. Huang, J. Yoshinobu and M. Aono, Surf. Sci. **287/288**, 1036 (1993).
5. S. Hosoki, S. Hosaka and T. Hasegawa, Appl. Surf. Sci. **60/61**, 643 (1991).
6. Y. Wada, T. Uda, M. Lutwyche, S. Kondo and S. Heike, J. Appl. Phys. **74**, 7321 (1993).
7. M.C. Crommie, C.P. Lutz and D.M. Eigler, Science **262**, 218 (1993).
8. K. Takayanagi, Y. Tanishiro, M. Takahashi and S. Takahashi, J. Vac. Sci. & Technol. **A3**, 1502 (1985).

9. J.J. Boland, Phys. Rev. Lett. **65**, 3325 (1990).
10. J.J. Boland, Adv. Phys. **42**, 129 (1993).
11. J.W. Lyding, T.C. Shen, J.S. Hubacek, J.R. Tucker and G.C. Abeln, Appl. Phys. Lett. **64**, 2010 (1994).
12. T.-C. Shen, C. Wang, G.C. Abeln, J.R. Tucker, J.W. Lyding, P. Avouris, R.E. Walkup, Science**268**, 1590 (1995).
13. S. Watanabe, Y.A. Ono, T. Hashizume, Y. Wada, Phys. Rev. **B54**, 17308 (1996).
14. J. Nogami, Sang-il Park and C.F. Quate, Appl. Phys. Lett. **53**, 2086 (1988).
15. A.A. Baski, J. Nogami and C.F. Quate, J. Vac. Sci. Technol. **A8**, 245 (1990).
16. S. Watanabe, Y.A. Ono, T. Hashizume, Y. Wada, J. Yamaguchi and M. Tsukada, Phys. Rev. **B52**, 10768 (1995).
17. T. Sakurai, T. Hashizume, I. Kamiya, Y. Hasegawa, N. Sano, H.W. Pickering and A. Sakai, Prog. Surf. Sci. **33**, 3 (1990).
18. T. Hashizume, Y. Hasegawa, I. Kamiya, T. Ide, I. Sumita, S. Hyodo, T. Sakurai, H. Tochihara, M. Kubota and Y. Murata, J. Vac. Sci. Technol. **A8**, 233 (1990).
19. S. Heike, T. Hashizume and Y. Wada, Jpn. J. Appl. Phys. **34**, L1061 (1995).
20. A. Kobayashi, F. Grey, R.S. Wiliams and M. Aono, Science **259**, 1724 (1993).
21. Q.J. Gu, N. Liu, W.B. Zhao, Z.L. Ma, Z.Q. Xue and S.J. Pang, Appl. Phys. Lett. **66**, 1747 (1995).
22. S. Heike, S. Watanabe, Y. Wada and T. Hashizume, Phys. Rev. Lett. **81**, 890 (1998).
23. S. Heike, T. Hashizume, S. Watanabe, M.I. Lutwyche and Y. Wada, Jpn. J. Appl. Phys. **35**, L1367 (1996).
24. D.M. Eigler, C.P. Lutz and W.E. Rudge, Nature (London) **325**, 600 (1990).
25. S. Kondo, S. Heike, M. Lutwyche and Y. Wada, J. Appl. Phys. **78**, 155 (1995).
26. R. Wilkins, E. Ben-Jacob and R.C. Jaklevic, Phys. Rev. Lett. **63**, 801 (1989).
27. S. Hasegawa and S. Ino, Phys. Rev. Lett. **68**, 1192 (1992).
28. Y. Hasegawa, I.W. Lyo and P. Avouris, Appl. Surf. Sci. **76/77**, 347 (1994).
29. T. Hashizume, S. Heike, M.I. Lutwyche, S. Watanabe, K. Nakajima, T. Nishi and Y. Wada, Jpn. J. Appl. Phys. **35**, L1085 (1996).
30. M.C.M.M. van der Wielen, A. J. A. van Roiji and H. van Kempen, Phys. Rev. Lett. **76**, 1075 (1996).
31. T. Hitosugi, T. Hashizume, S. Heike, S. Watanabe, Y. Wada, T. Hasegawa and K. Kitazawa, Jpn. J. Appl. Phys. **36**, L361 (1997).
32. K. Hirose and M. Tsukada, Phys. Rev. **B51**, 5278 (1995).
33. R.M. Tromp, R.J. Hamers and J.E. Demuth, Phys. Rev. Lett.**55**, 1303 (1985).
34. F. Grey, C. Thirstrup and H. Busch, Large Clusters of Atoms and Molecules, T.P. Martin (Ed.) NATO ASI Series E, Vol. 313, p.463. Kluwer, Dordrecht (1996).
35. S. Watanabe, M. Ichimura, T. Onogi, Y.A. Ono, T. Hashizume and Y. Wada, Jpn. J. Appl. Phys. **36**, L929 (1997).
36. T. Hashizume et al. (unpublished).
37. K. Nakajima, T. Ikehara, T. Nishi, T. Hashizume, S. Heike, S. Watanabe and Y. Wada, J. Korean Phys. Soc. **31**, S39 (1997).

5 Theoretical Insights into Fullerenes Adsorbed on Surfaces: Comparison with STM Studies

Kaoru Ohno and Yoshiyuki Kawazoe

Summary. Theoretical investigations of electronic structures and other characteristics of fullerenes (C_{60}, etc.) using all-electron mixed basis approaches, which are superior to pseudopotential approaches with regard to core treatments, are reviewed in connection with Scanning Tunneling Microscopy (STM) measurements of fullerenes adsorbed on substrates. A universal feature of the Partial Density of States (PDOS) of the HOMO and LUMO levels is discussed and compared successfully with STM images measured by Hashizume et al. When electron charge transfer occurs from substrates, it is pointed out that anomalous intermolecular interactions may appear between fullerenes, due to induced dipoles. This possibility is discussed in detail using a simple analysis. On the other hand, when fullerene samples are irradiated by laser light etc., it is pointed out that, due to electronic excitations, special fullerene orientation may be favored at surfaces. This behavior is explained in terms of an ab initio result for low-temperature simple cubic phases of C_{60}. Some STM experimental evidence concerning these new features of fullerenes is briefly described and compared with theory.

5.1 Introduction

In the last decade, the development of Scanning Tunneling Microscopy (STM) [1] has brought considerable progress in surface physics with quite widespread applications [2]. It provides reliable real-space information on atomic alignments and more global atomic morphologies at surfaces. However, one should note that STM images do not necessarily reflect true atomic positions. This is because STM does not detect nuclei, but rather features related to electronic structures near the Fermi level. When the electrostatic potential at the sample surface is higher than the potential at the STM tip by an amount V, this V is called the (sample) bias voltage with the tip grounded.

If the electronic states of the tip are empty at the energy E and those of the sample surface are filled at the energy $E + eV$ (for the original system in the absence of the bias field), electrons can be transferred from the sample to the tip by electron tunneling. This may happen when V is negative. Conversely, if tip electronic states are filled at E and sample electronic states are empty at $E + eV$ (such a situation may happen when V is positive), electrons can be transferred from tip to sample. Generally, the tunneling current decreases exponentially when the distance between the tip and the sample surface increases. For a given distance, the tunneling current (from tip to sample) is roughly proportional to the total Density of States (DOS) which can contribute to electron transfer:

$$I \propto \frac{2\pi e}{\hbar} \int_{E_F - eV}^{E_F} D_{\text{tip}}(E) D_{\text{sample}}(E + eV) \mathrm{d}E \,. \tag{5.1}$$

In this equation, $D_{\text{tip}}(E)$ and $D_{\text{sample}}(E)$ mean the Local Density of States (LDOS) of electrons at the STM tip and at the sample surface where the tunneling current predominantly occurs. More explicit formulation including the dependence on the distance between the tip and the substrate was given as early as 1961 by Bardeen [3]. The most popular method used with the STM apparatus is to vary the distance between tip and surface in order to keep the tunneling current I constant. The image is made from the voltage applied to the z-piezo in order to control the tip–surface distance. Detailed work on the comparison between the calculated tunneling current and experimentally observed STM images has been carried out by Tersoff and Hamann [4] and by Tsukada et al. [5]. When the tip bias voltage is not so high, one may find that the STM images observed with minus and plus bias voltages correspond roughly to the Partial Density of States (PDOS) of the Highest Occupied Molecular Orbital (HOMO) levels and the Lowest Unoccupied Molecular Orbital (LUMO) levels.

It is with great success that STM images have so far been compared with the spatial distribution of the PDOS calculated using first-principles methods. One successful comparison between experiment and theory was made in a variety of systems involving C_{60} (see Fig. 5.1) and other fullerenes adsorbed on substrates. Samples of fullerenes adsorbed on substrates provide a lot of interesting phenomena due to charge transfer from the substrates, and photo-induced dimerization reactions, to be discussed later. Many experimental studies have been reported to date for systems of C_{60} adsorbed on GaAs(110) [6], Au(100) [7], Si(111) [8,9], Si(100) [10], Cu(111) [11], and others. The C_{60} molecule is not completely inert and C_{60} interacts with substrates and aligns commensurately on the GaAs, Au, Si and Cu substrates.

In the following, we review theoretical aspects of recent progress with electronic structure calculations which are closely related to STM studies on fullerenes. In particular, we pick up several interesting topics in a system of fullerenes adsorbed on substrates and present an analysis based on ab initio approaches. First of all, in order to determine a universal feature of the STM images of fullerenes adsorbed on substrates, we calculate the PDOS of two-dimensional or three-dimensional alignments of C_{60} (or C_{70}). The result compares well with the STM observations by Hashizume et al. made on C_{60} (or C_{60}–C_{70} mixtures) aligned on the Cu(111) substrate [11]. Second, we discuss the effect of electron charge transfer from substrate atoms to adsorbed C_{60} molecules. Here, we suggest that C_{60} thin films may possess ferroelectric ordering due to the induced dipole field, when the amount of

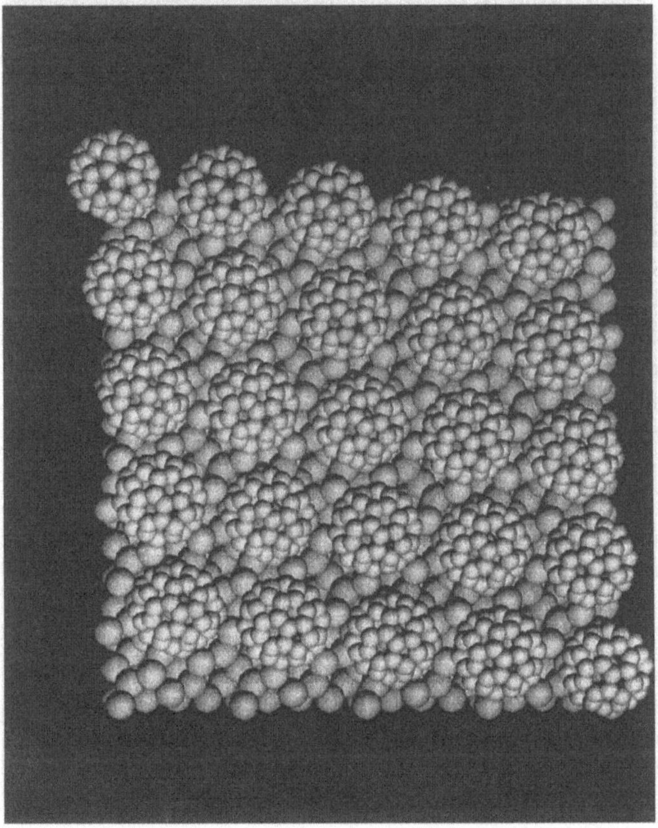

Fig. 5.1. Illustration of C_{60}'s adsorbed on a Si(100)–(2×1) surface. This adsorption geometry corresponds to the $c(4\times3)$ structure

charge transfer is large enough. One experimental possibility involves C_{60} layers adsorbed on GaAs (100) As-rich substrate, where the C_{60} lattice is expanded about 13% in overlayers, according STM measurements by Xue et al. [12]. Finally, we discuss the effect of laser-irradiation on a sample of fullerene thin films adsorbed onto substrates. The main interest here is to find out which kind of excitations of C_{60} molecules can appear at surfaces. We shall attempt to answer this question. The corresponding STM experiment has been performed by Kasuya et al. [13]. We also point out its possible relationship to the electronic structures of a bulk C_{60} simple cubic phase at low temperatures.

5.2 Fullerene Research Background

A fundamentally new class of carbon-cage molecules, C_{60}, discovered by Kroto et al. [14] in 1985, and other larger size carbon-cage molecules (C_{70},

C_{82}, C_{84}, etc.) have been extensively studied because of their interesting shape, made up of 6- and 5-membered rings, and their peculiar physical and chemical properties. Writing the number of facets, edges and vertices to be S, B and V, respectively, the Euler theorem on polyhedrons gives

$$S + V = B + 2\,. \tag{5.2}$$

Assuming the sp_2 bonding condition $B = 3V/2$ that three edges always start from one vertex, one may easily derive $S = V/2 + 2 = B/3 + 2$. Moreover, assuming that all the facets are only hexagons and pentagons, and putting these numbers as N_6 and N_5, respectively, one obtains $N_5 = 12$. That is, a closed cage composed only of six-membered and five-membered rings always has 12 five-membered rings. Carbon clusters having this structure are called fullerene, after Backminster Fuller who is famous for constructing buildings with the same architecture. On the other hand, Iijima et al. [15] found the 'carbon nanotube' which has a cylindrical structure, obtained by rolling a graphite sheet. Similarly, the carbon nanotube with one end closed by a fullerene half sphere, and the 'carbon onion' which is essentially a fullerene with multi-layer structure have also been found to be quite stable. Furthermore, the nanotube which has a seven-membered ring and spreads out like a funnel has also been observed. All of these structures may be promising as components for materials. A further important fact concerning fullerenes and nanotubes is that no structures with fused pentagons have been found. This rule is called the Isolated Pentagon Rule (IPR). If we assume the IPR, it is easily understood that the smallest cage is the C_{60} with a truncated icosahedral structure, possessing icosahedral point-group symmetry (I_h). It consists of 12 five-membered rings and 20 six-membered rings (see Fig. 5.1). This shape is identical to a soccer ball and has exceptional stability. The 30 bonds shared by two fused pentagons are double bonds with length 1.40 Å, while the 60 bonds constituting five-membered rings are the single bond of length 1.46 Å [16].

The success in mass production of C_{60} molecules by Krätschmer et al. [17] has advanced research on solid C_{60}. C_{60} molecules form a molecular crystal with a face-centered cubic (fcc) structure at room or higher temperatures. Saito and Oshiyama [18] calculated the electronic structure of an fcc C_{60} crystal and showed that it is a semiconductor with a direct band gap of 1.5 eV at the X point. Although C_{60} molecules are known to rotate (or ratchet) rapidly in the bulk crystal at room temperatures, knowledge of how rotations of C_{60} affect electronic structure is required. As will be discussed later, the C_{60} crystal shows orientational phase transitions. This problem was investigaed from first principles using the all-electron mixed basis approach [19,20]. There are two good intensive review articles [21,22] which review most important fullerene research appearing up until 1993 and 1995, respectively. Within recent memory, R.E. Smalley, H.W. Kroto and R.F. Curl were awarded the 1996 Nobel prize in Chemistry for their pioneering contributions to fullerene research.

C_{60} has given rise to a variety of interesting subjects. Relatively high-T_c superconductivity has been found in alkali-metal doped C_{60} crystals. K_3C_{60} exhibits superconductivity at 18 K [23,24]. Rb_3C_{60} [25,26], Cs_2RbC_{60} [27] and Cs_3C_{60} under pressure [28] exhibit superconductivity at 30 K, 33 K and 40 K, respectively. On the other hand, the doping of organic donor product such as tetrakis-dimethylamino-ethylene (TDAE) into C_{60} crystals shows a ferromagnetic transition at 16.1 K [29]. This behavior is interesting because the molecules only have valence *s*- and *p*-orbitals. Experimentally, endohedral metallofullerenes such as La [30,31], Y [32,33] and Sc [34,35] atoms encapsulated in C_{82} or C_{84} have been confirmed using arc-desorption or laser vaporization techniques. In addition, a posteriori insertion of atoms into already created C_{60} has been attempted, both experimentally and theoretically [36–40]. Photoconductivity [41,42] and nonlinear optics [43–45] also offer interesting subjects in the context of fullerene systems. The nucleation and growth of fullerene thin films and the adsorbed morphology of fullerenes on various substrates have also attracted much interest [46]. More recently, the possibility of polymerization of C_{60} has been investigated in several kinds of C_{60} sample. In alkali-metal doped C_{60} such as A_xC_{60} with $x = 1$ and A = K, Rb, Cs, the existence of a linear polymer network of C_{60} is reported by Stephens et al. [47]. Here, it is expected that each pair of facing double bonds will form an intermolecular four-membered ring called the 2+2 cycloaddition. In addition, by applying visible or UltraViolet (UV) light [48], an electron beam from the STM tip [49], or mechanical pressure [50,51], one may expect to find C_{60} polymerization.

5.3 Universal Features of C_{60} and C_{70} STM Images

In this section we discuss a universal feature of STM images of C_{60} (or C_{70}) theoretically. The result of a theoretical calculation via an ab initio method using the all-electron mixed basis approach will be compared with an experimental observation by STM images.

5.3.1 Ab Initio Calculation

Maruyama et al. [52] carried out ab initio band structure calculations for two-dimensional arrays of C_{60} and of C_{60}–C_{70} mixture using the all-electron mixed basis approach explained in the Appendix. [The calculational formalism is based on the Local Density Approximation (LDA) used in density functional theory.] This approach enables one to describe localized electron orbitals whilst considerably limiting the number of basis functions. It may be appropriate for comparing the resulting PDOS distribution with STM images, because it can also describe detailed structures in the vicinity of the atomic center. Good agreement was obtained between the resulting HOMO- and LUMO-PDOS and the STM images measured by Hashizume et al. [11]

Fig. 5.2. Unit cells assumed for ab initio calculations. The systems are two dimensional arrays of (**a**) only C_{60} molecules (case A), and (**b**) C_{60}–C_{70} mixture (case B). After Maruyama et al. [52]

with minus and plus bias voltages for the corresponding fullerene samples adsorbed on a Cu(111) surface. As for intramolecular bond lengths, all 6/6 bonds are set at 1.40 Å and all bonds belonging to five-membered rings are set at 1.46 Å. Figures 5.2a and b show, respectively, the input structures of C_{60} and of the C_{60}–C_{70} mixture, used for the calculation. The areas enclosed by solid lines represent unit cells whose shapes and sizes are chosen to be the same.

Case A. The first one is the ideal triangular lattice of C_{60} with a lattice constant of 10.2 Å. One of the six-membered rings in C_{60} is located at the top and parallel to the crystal plane. (The opposite six-membered ring is located at the bottom.) Three five-membered rings around the top six-membered ring form a small regular triangle, whose vertices face the adjacent C_{60} molecules.

In this orientation of C_{60}, the lateral five-membered ring faces the six-membered ring of the neighboring C_{60} and vice versa. The bonds constituting five-membered rings are all single bonds and therefore electron poor, while the double bonds constituting six-membered rings are electron rich. Therefore, this orientation is energetically favorable due to the electrostatic interaction.

Calculated band-energy dispersion curves are shown in Fig. 5.3a, which contain 100 levels around the HOMO–LUMO gap. The band gap is calculated to be 1.2 eV, which is larger than the previous result of 1.1 eV for the case of the fcc ($Fm\bar{3}$) symmetry [19]. Other basic features are similar to the 3D crystal band structures. Globally, there are two types of band. One is an intramolecular σ band, which has no dispersion. The other is an intermolecular π band, which has relatively strong dispersion. Nevertheless, it is found that dispersion is not strong enough to mix up energy levels across forbidden gaps of the isolated molecule. This indicates that each band keeps its original molecular character, such as h_u (HOMO) and t_{1u} (LUMO).

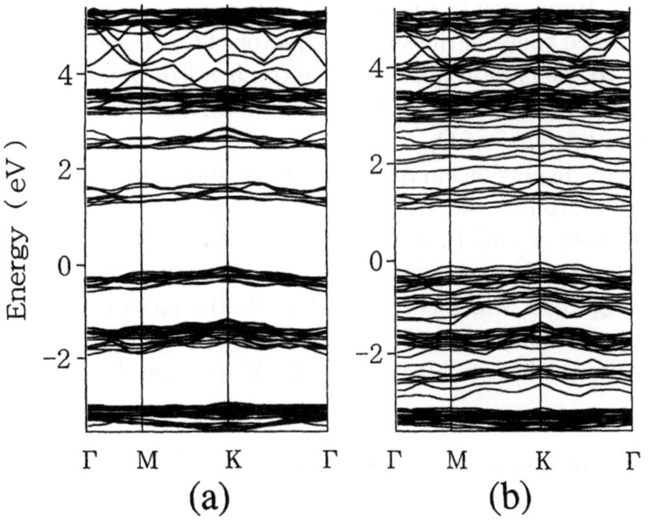

Fig. 5.3. Calculated band-energy dispersion curves for (**a**) the C_{60} system and (**b**) the C_{60}–C_{70} mixture. The unit used in the energy scale is [eV]. After Maruyama et al. [52]

Calculated spatial distributions of the Partial Density of States (PDOS) are drawn in Fig. 5.4. Here the distribution is only shown for the half region within the unit cell. The isosurface of density 0.005 Å^{-3} is drawn in the figure. Figure 5.4a shows the PDOS of the 10 HOMO bands, where the topmost six-membered ring is highlighted with a center hole just like a doughnut. Figure 5.4b shows the PDOS of the six LUMO bands, in which the three

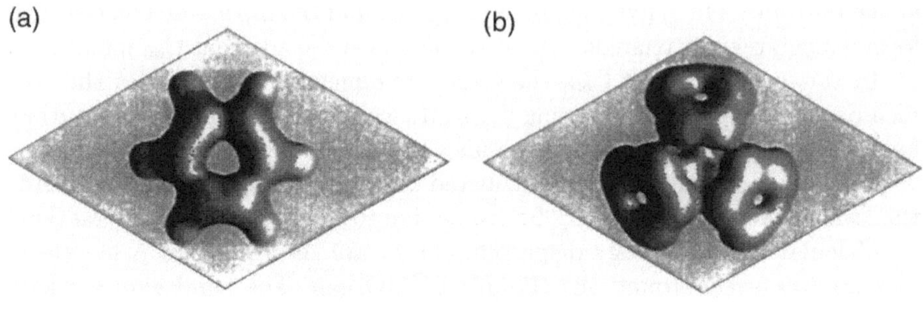

Fig. 5.4. Calculated spatial distribution of the Partial Density of States (PDOS) for the two-dimensional C_{60} system. (**a**) HOMO bands and (**b**) LUMO bands. After Maruyama et al. [52]

five-membered rings around the top six-membered ring are highlighted just like a three-leaf clover.

Case B. The second configuration is such that C_{70} and C_{60} are alternatively arranged in one direction. Here, each C_{60} is oriented as in case A. Concerning C_{70}, one of the six-membered rings on the shorter equator of C_{70} is located at the bottom and parallel to the crystal plane. At the top is a 6/6 bond which shares two fused six-membered rings. Four five-membered rings are located at the four corners in the top view of C_{70}.

Calculated band-energy dispersion curves in case B are shown in Fig. 5.3b, which contains 100 levels around the HOMO–LUMO gap. By comparing the band structures of cases A and B, it is found that many degeneracies disappear in case B, due to the lower symmetry of C_{70}. We may assume that the levels appearing in Fig. 5.3b, which correspond inside the forbidden gaps in Fig. 5.3a, are are mainly due to C_{70}. The HOMO–LUMO gap in case B is given by $E_g \sim 1.0$ eV, which is smaller than 1.2 eV for case A. This is because

- the intrinsic band gap of C_{70} is narrower than C_{60} [53–55],
- interatomic distances between C_{60} and C_{70} are slightly shorter than those in the pure C_{60} system.

Similarly to case A, dispersion is not strong enough to mix up bands with different characters, and forbidden gaps that exist in isolated molecules remain in the crystalline phase.

Calculated distributions of the PDOS are shown in Fig. 5.5, where the isosurface of density 0.005 $Å^{-3}$ is drawn. In Fig. 5.5a, which corresponds to 15 HOMO bands, the topmost six-membered rings are highlighted both in C_{60} and C_{70}. In Fig. 5.5b, which corresponds to six LUMO bands, three five-membered rings around the six-membered ring in C_{60} and four five-membered rings in C_{70} are highlighted.

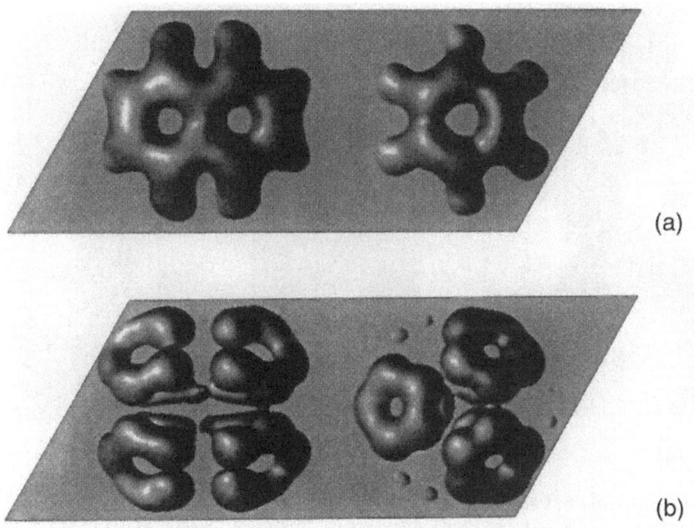

Fig. 5.5. Calculated spatial distribution of the Partial Density of States (PDOS) for the two-dimensional C_{60}–C_{70} mixture. (**a**) HOMO bands and (**b**) LUMO bands. After Maruyama et al. [52]

5.3.2 Comparison with Experimental Results Using STM

The result presented above indicates that the HOMO-PDOS or LUMO-PDOS of the two-dimensional array of fullerenes is not much changed from that of the isolated fullerenes, although a net dispersion appears in band dispersion curves. The similar HOMO- and LUMO-PDOS behavior has also been obtained for crystalline C_{60}'s [19,20].

The above result successfully reproduces the main features of Field-Ion Scanning Tunneling Microscopy (FI-STM) images observed by Hashizume et al. for first layer C_{60} (and C_{70}) molecules adsorbed on a Cu(111)–(1×1) surface.

Among many substrates, Cu(111)–(1×1) is suitable for C_{60} to grow epitaxially, because the lattice mismatch between the nearest neighbor distance (10.0 Å) of the bulk C_{60} crystal and 4 times the Cu–Cu nearest neighbor distance (10.2 Å) is small (2%). Hashizume et al. have reported clear intramolecular structures of the first layer C_{60} [11,56] (or C_{60}–C_{70} mixture [57]) on the Cu(111)–(1×1) substrate using the FI-STM technique. C_{60} (or C_{70}) molecules in these monolayer films show peculiar FI-STM images with a strong bias-voltage dependence. At the minus bias voltage −2.0V (sample bias voltage with respect to its Fermi level), individual C_{60} molecules appear in the STM image as a doughnut shape, which is almost round with a hole at the center but still maintains a weak threefold symmetry as shown in Fig. 5.6a. On the other hand, at the plus bias voltage of +2.0V, individual C_{60} molecules show a three-leaf clover shape as in Fig. 5.6b.

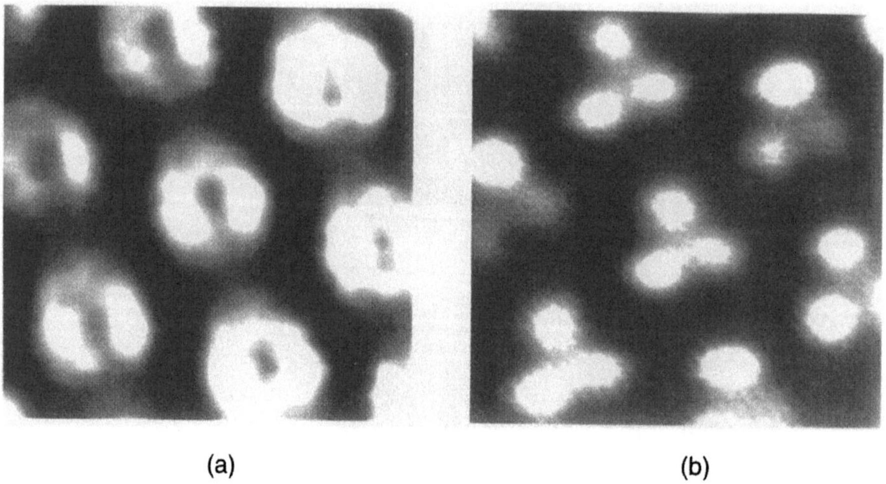

Fig. 5.6. FI-STM images observed for C_{60}'s adsorbed on a Cu(111) substrate. (a) Minus bias voltage and (b) plus bias voltage (tip grounded). After Hashizume et al. [11]

All molecules are well ordered on the surface with a special orientational preference, except for the rotating C_{60} molecules which appear only in a round spot irrespective of the bias voltage. Those rotating C_{60}'s are mainly located at low symmetry points such as step edges and point defects. The results suggest that most of the C_{60} molecules fix their orientation in the first layer even at room temperature, through C_{60}–C_{60} and substrate–C_{60} interactions.

It has been well understood that C_{60} molecules rapidly rotate or ratchet in a bulk medium at room temperature, and rotating C_{60} molecules have fixed orientation in some specific directions at 261 K [58]. This is a first order (orientational) phase transition between the high-temperature fcc and low-temperature simple cubic (sc) phases. At 90 K, a further orientational phase transition takes place [59]. Below this temperature, it is known that most of the fullerenes stop rotating [60]. Therefore, it is also interesting to study the orientational preference of C_{60} by analyzing peculiar bias-dependent STM images.

In the case of C_{70} coexisting with C_{60}, the situation is somewhat different [57]. In the STM image (with plus bias voltage) shown in Fig. 5.7, there are two types of C_{70} molecule: one is a completely round spot which has the same size as C_{60} but much brighter. This pattern can be identified to be a rotating C_{70} in a standing orientation, i.e., rotating about the long axis of a rugby ball C_{70} with this axis fixed vertically to the substrate. (Indeed, the sample made up of just C_{70} showed only these round spots, and their radius corresponds to the smaller equator of C_{70}.) The other pattern has a rectangular shape, which can be identified as a C_{70} lying down on the substrate. In this latter

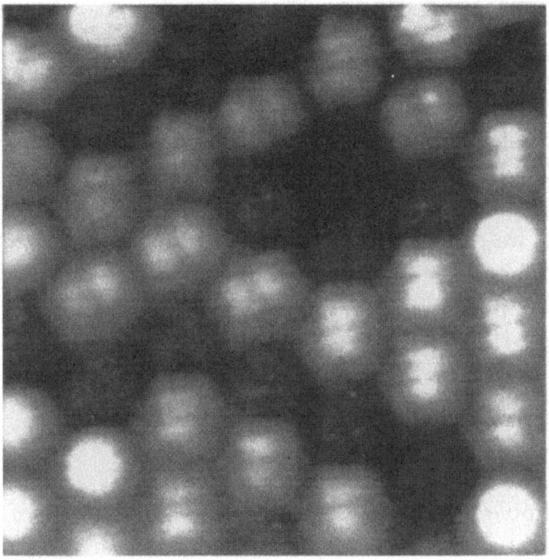

Fig. 5.7. FI-STM image with plus bias voltage (tip grounded) observed for C_{60}–C_{70} mixture adsorbed on a Cu(111) substrate. After Wang et al. [57]

pattern, it is found that there is a strong bias-voltage dependence. At negative bias voltages, each rectangle seems to be composed of two connected rings, while at positive bias voltages, each bright rectangle has an inscribed dark cross, as if there were four spots at every corner of the rectangle. Figure 5.7 corresponds to this latter case.

From this analysis, one may clarify the relationship between the fullerene configurations and electronic structures. We have identified the universal feature of STM images of C_{60} and C_{70}. That is, all single (double) bonds are brightened at plus (minus) bias voltages. With this knowledge, one may speculate on the orientation of fullerenes at surfaces from STM pictures. All these conclusions are valid only when the PDOS is not much different from that of nearly isolated fullerenes. This validity is guaranteed by the characteristics of fullerenes, which are almost inert and do not form strong bondings among themselves. However, in some samples, this may not be the case. For example, in the case of C_{60}'s adsorbed on Si(111), Hashizume et al. [9] reported a result in which the FI-STM showed three- or four-stripe patterns, which are somewhat different from the above-mentioned universal pattern. In this case, Kawazoe et al. [61] showed that a partial sum over some of the LUMO (or HOMO) bands may explain such striped patterns when C_{60}'s have a special orientation. For such a substrate, there is a strong bonding between fullerenes and substrates, which may considerably change the band structure.

It is instructive to note here that in the case of graphite and Si surfaces [62], for example, the PDOS of several bands from the Fermi level must be added correctly inside the first Brillouin zone, in order to compare theory with STM images.

5.4 Dipole Field Caused by Charge Transfer

A C_{60} molecule has electron affinity 2.65 eV [63,64]. Therefore, depending on substrates, there may be a net electron charge transfer from substrates to C_{60}'s. If the level of charge transfer is big enough, the interface dipole field causes big dipole moments in C_{60} molecules adsorbed in overlayers. Then the induced dipole moments of C_{60} may dramatically change their intermolecular interaction. In this section, we discuss this possibility. In particular, we consider here the case of C_{60}. One piece of experimental evidence for such a big change in the intermolecular interaction among C_{60} molecules has been reported recently by Xue et al. [12] for the case of an As-rich GaAs(100) substrate.

5.4.1 Effect of Induced Dipole Fields

We may deduce the induced dipole in C_{60} in overlayers from the amount of charge transfer from the substrate to the first layer C_{60} [65]. If there is an electron charge transfer from the substrate to C_{60}, a dipole layer is formed in this region. The width of this dipole layer, i.e., the distance between the transferred excess charge in the first layer C_{60} and substrate atoms, is of the order of the interatomic distance between carbon and substrate atoms, denoted by h. Then, the dipole moment per C_{60} amounts to

$$P_1 = Q_1 h \,, \tag{5.3}$$

where Q_1 is the amount of charge transferred from substrate atoms to one C_{60}. The electric field caused by this dipole layer is given by

$$\begin{aligned} E(\sigma, h, z) &= \frac{\sigma}{4\pi\varepsilon_0} \int \mathrm{d}x\mathrm{d}y \left[\frac{1}{x^2+y^2+z^2} - \frac{1}{x^2+y^2+(z+h)^2} \right] \\ &= \frac{2\pi\sigma}{4\pi\varepsilon_0} \int_0^\infty \mathrm{d}t \left[\frac{1}{t+(z+h)^2} - \frac{1}{t+z^2} \right] \\ &= \frac{\sigma}{2\varepsilon_0} \left[\log\left| \frac{t+(z+h)^2}{t+z^2} \right| \right]_0^\infty = \frac{\sigma}{\varepsilon_0} \log\left| \frac{z+h}{z} \right| , \end{aligned} \tag{5.4}$$

as a function of the two-dimensional charge density σ, the dipole distance h, and the distance z from the center of the first layer C_{60}. When z becomes large, this electric field behaves as $\sigma h/\varepsilon_0 z$. Here the charge density σ is estimated to be $Q_1\rho$ with ρ the two-dimensional number density of C_{60} in the first layer. This dipole field causes the dipole moment in C_{60} adsorbed in the second layer (see Fig. 5.8).

The induced dipole moment in second-layer C_{60} is given by

$$P_2 = \alpha E_2 \,, \tag{5.5}$$

$$E_2 = E(Q_1\rho, h, d) \,, \tag{5.6}$$

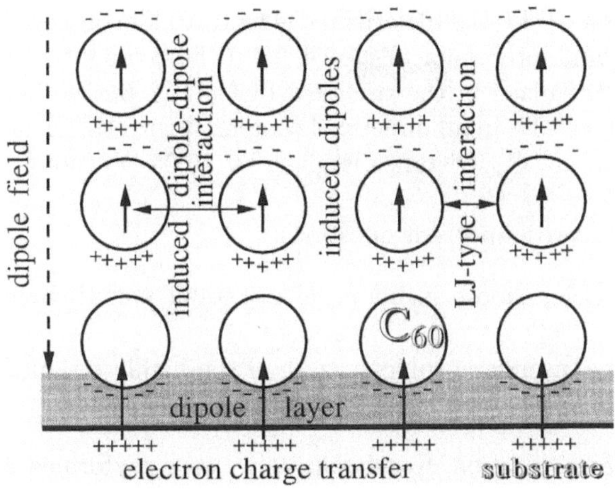

Fig. 5.8. Schematic illustration of the induction of dipoles in C_{60}, caused by the dipole field due to charge transfer from the substrate

where d is the interlayer distance of the C_{60} and α the dielectric polarizability of C_{60}. The induced dipole in the second-layer C_{60}'s as well as the first-layer dipole causes, in turn, the dipole in third-layer C_{60}, and so forth. The strength of the electric field at the third-layer C_{60} is given by

$$E_3 = E(Q_2\rho, D, d) + E(Q_1\rho, h, 2d) . \tag{5.7}$$

We can now estimate the induced dipole moment of the third layer as

$$P_3 = \alpha E_3 , \tag{5.8}$$

and in turn the induced charge $\pm Q_3$ in third-layer C_{60} as

$$Q_3 = P_3/D . \tag{5.9}$$

We can likewise estimate the induced dipole moment of C_{60} in the fourth layer, and so on. We will see later using these equations that the induced dipole moment in overlayers may become very large and it can affect intermolecular interactions between C_{60} molecules.

5.4.2 C_{60} Adsorbed on GaAs(001) As-Rich Surfacs

Recently, using STM, Xue et al. [12] observed a very interesting C_{60} multilayer structure adsorbed on the GaAs(001) As-rich 2×4 surface. According to their observations, the first layer C_{60} adsorbed on the deep (~ 2.8 Å) and wide (~ 8.0 Å) trough in between the As-dimer rows has a tendency to line up pairwise when coverage is increased. At saturation coverage, each valley is filled up with C_{60} pairs whose intermolecular distance is 10.54 ± 0.20 Å, evenly spaced by 24.0 Å (Fig. 5.9).

More strikingly, in the second and higher layers (at least up to 10 monolayers), C_{60} forms a highly 'strained' FCC(110)-oriented structure. The observed unit cell size, 16 Å×11.3 Å, and interplanar distance, 5.9 Å, have roughly the ideal axis ratio $\sqrt{2} : 1 : 1/2$ realized in the standard FCC(110), but with a lattice expansion of as much as 13% in all directions compared with the usual bulk FCC C_{60} in which the n.n. C_{60} distance is given by 10.05 Å (not by 11.3 Å).

Their observations raised two important questions:

- How does the first monolayer choose such a novel and stable paired-chain structure?
- What is the origin of the 'pseudomorphic' growth of the highly strained FCC(110) crystalline film?

In order to answer the first question, Kamiyama et al. have performed a molecular dynamics simulation [66] and one possible interpretation was discussed in the original paper by Xue et al. [12] using a simple analysis based on intermolecular interactions largely affected by the molecule–substrate interaction. We shall attempt to answer the second question in the rest of this section.

5.4.3 Ab Initio Calculation for C_{60} Adsorption

Li et al. [67,65] used a first-principle code (CASTEP), which is based on a plane wave expansion and a norm-conserving pseudopotential, to find the optimal adsorption height of one C_{60} molecule adsorbed at position S in Fig. 5.2, on the trough of the GaAs 2×4 surface. In a tetragonal unit cell, there are 40 As atoms and 44 Ga atoms (6 layers altogether), fixing all the positions to ideal ones at and near the surface. On the other hand, the ideal rigid structure is also assumed for C_{60} with an orientation in which the topmost and lowermost double bonds of C_{60} are aligned parallel to the As dimer row. The plane-wave cutoff energy used in this calculation is 300 eV.

By changing the height of the C_{60} center from the center of the As-atom on top of the trough, we estimated the total energy of the whole C_{60} and substrate system. The result is tabulated in Table 5.1.

From this data, we find that the optimal adsorption height, which is defined as the height of the center of the first-layer C_{60} measured from the substrate As atom, is about $h = 5.7$ Å. This value is comparable to the value

Table 5.1. Relative total energy values estimated by the present calculations for various adsorption heights

Adsorption height h [Å]	4.6	4.8	5.5	5.7	5.9	7.3
Relative energy [eV]	7.92	4.08	0.07	0.00	0.30	0.97

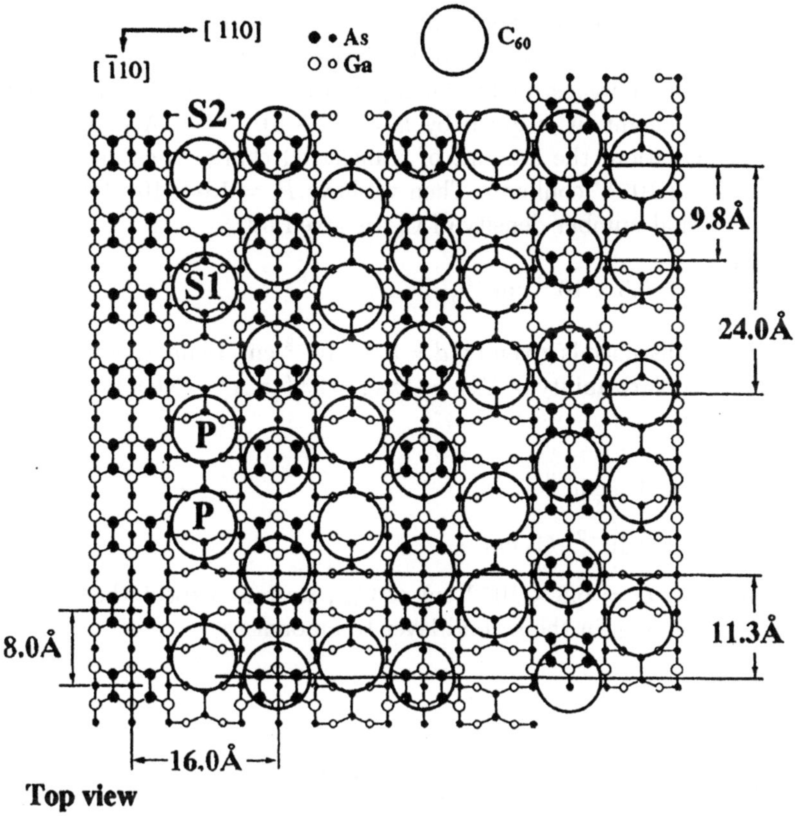

Fig. 5.9. Geometrical structure of the GaAs(001) As-rich 2×4 surface and possible adsorption configurations of C_{60}. After Xue et al. [12]

6.2 ± 0.6 Å, which can be obtained from STM data [12] as follows: starting from the height difference 3.8 ± 0.6 Å between the topmost carbon atoms in the first-layer C_{60} and the topmost As atoms, we add the distance 5.7 Å between the substrate As atoms under C_{60} and top-layer As atoms and then subtract the C_{60} radius, $D/2 = 3.3$ Å.

Then, at this optimal adsorption height, Li et al. [67,65] calculated the charge density distribution. From this distribution we can estimate how many electrons transfer from As atoms located on the bottom of the first-layer C_{60} molecules to these C_{60}'s. The estimated amount of charge transferred is about 1.76 electrons per C_{60}.

5.4.4 A Mechanism for FCC Lattice Expansion

Now we propose a mechanism for the 13% lattice expansion of the C_{60} fcc structure observed in the overlayers up to at least 10 monolayers. The main idea is to imagine also for C_{60} molecules in overlayers a dipole moment successively induced by the dipole field which is itself originally caused by the strong charge transfer from substrate to first-layer C_{60}'s.

For this system, if we estimate the dipole distance h to be the same as the adsorption height, it is given by $h = 5.7 \times 10^{-8}$ cm. According to the ab initio calculation, the amount of transferred charge is estimated as $Q_1 = 1.76e = 2.8 \times 10^{-19}$ coulomb. Then we have $P_1 = 1.6 \times 10^{-26}$ C · cm. The two-dimensional number density of C_{60} is given by

$$\rho = \frac{10^{16}}{16 \times 11.3} = 5.5 \times 10^{13}\ \mathrm{cm}^{-2} ,$$

and the interlayer distance is given by $d = 5.9 \times 10^{-8}$ cm. Using these values, together with the polarizability of C_{60}, $\alpha/4\pi\varepsilon_0 = 83/4\pi$ Å$^3 = 6.6 \times 10^{-24}$ cm^3 in [69–72], we find

$$E_2 = \frac{1.6 \times 10^{-4}}{4\pi\varepsilon_0}\ \mathrm{C \cdot cm^{-2}} ,$$

$$P_2 = \alpha E_2 = 1.1 \times 10^{-26}\ \mathrm{C \cdot cm} . \tag{5.10}$$

Using the diameter $D = 6.5 \times 10^{-8}$ cm of C_{60}, we may expect the charge $\pm Q_2$ induced in one C_{60} in the second layer to amount to

$$Q_2 = P_2/D . \tag{5.11}$$

Similarly, from (5.7), we obtain

$$E_3 = E(Q_2\rho, D, d) + E(Q_1\rho, h, 2d) = \frac{1.6 \times 10^{-4}}{4\pi\varepsilon_0}\ \mathrm{C \cdot cm^{-2}} . \tag{5.12}$$

We can then estimate the induced dipole moment of the third layer as

$$P_3 = \alpha E_3 = 1.3 \times 10^{-26}\ \mathrm{C \cdot cm} , \tag{5.13}$$

and in turn the induced charge $\pm Q_3$ in the third-layer C_{60} as

$$Q_3 = P_3/D = 2.0 \times 10^{-19}\ \mathrm{C} . \tag{5.14}$$

In a similar way, we can estimate the induced dipole moment of C_{60} in the fourth layer as $P_4 = 1.7 \times 10^{-26}$ C cm.

This is a very striking result because the induced dipole moment does not decrease at all, but rather increases gradually. Strictly speaking, the above recursive method for determining the induced dipole moment in each layer is not correct, since the induced dipole moments in all the layers should be determined self-consistently by taking into account influences from all existing C_{60} layers. At least from the present simple analysis, it is clear that the induced dipole moments of C_{60} in overlayers does not decrease or only slowly decreases with height above the interface.

We can estimate the magnitude of the resulting dipole–dipole interactions among overlayer C_{60}'s:

$$V_{\boldsymbol{P},\boldsymbol{P}} = \frac{\boldsymbol{P}^2}{4\pi\varepsilon_0 R^3} , \tag{5.15}$$

where $\varepsilon_0 = 8.854 \times 10^{-21}$ $C^2/\text{dyne}\,\text{cm}^3$ is the vacuum dielectric constant. The 13% lattice expansion observed by Xue et al. makes all the C_{60} layers exactly commensurate with the substrate. Such deformation in overlayers is possible only if the energy decrease due to the dipole–dipole interaction energy given by (5.15) exceeds the energy increase due to the Lennard–Jones (LJ)-type intermolecular interaction energy $\Phi(R)$ caused by lattice expansion (see Fig. 5.8). This criterion is expressed as

$$V_{\boldsymbol{P},\boldsymbol{P}}(R_1) \geq \Phi(R_1) - \Phi(R_0)\ , \tag{5.16}$$

where R_0 and R_1 are, respectively, the equilibrium nearest neighbor intermolecular distances in the pure C_{60} crystal and that in the overlayer C_{60}'s adsorbed on the As-rich substrate. The intermolecular potential energy $\Phi(R)$ in (5.16) is given, for example, by the Girifalco potential [68],

$$\begin{aligned} \Phi(\mathbf{R}) &\sim -\frac{300A}{r_0^2}\left[\frac{1}{R(R-r_0)^3} + \frac{1}{R(R+r_0)^3} - \frac{2}{R^4}\right] \\ &\sim -\frac{300A}{r_0^2}\frac{1}{R(R-r_0)^3}\ , \end{aligned} \tag{5.17}$$

where r_0 is the C_{60} hard-sphere diameter 7.1 Å and $A = 3.2 \times 10^{-59}$ erg cm^6. From knowledge of the ideal C_{60} crystal we have $R_0 = 10.05$ Å, while from the STM data [12] we have $R_1 = 11.3$ Å. Using these values and $P \sim 1 \times 10^{-26}$ C·cm, we can derive a rough estimate for the left-hand side of (5.16) as 6.2×10^{-13} erg, while for the right-hand side we derive $\Phi(R_1) = -2.3 \times 10^{-13}$ erg and $\Phi(R_0) = -7.0 \times 10^{-13}$ erg. Hence, (5.16) is certainly fulfilled. This fact guarantees that lattice expansion is caused by induced dipole moments of C_{60}'s adsorbed in overlayers.

5.5 Photo-Induced Excited States

STM measurements give information not only about ground states but also about excited states of fullerenes. First of all, the electronic excited states can be measured without difficulty by setting positive bias voltages. Moreover, excited states seen in atomic alignments can be made by laser-irradiating or heating the fullerene sample, and they can then be measured by the STM. Here, it is interesting to note that photo-induced excited states may favor some special fullerene orientations, in the sense that fullerenes may form dimers, trimers, and so on, or even polymers in some cases, although no one has yet detected the direct evidence of shortened intermolecular distances in the laser-irradiated samples as far as we know. If the laser power increases [73] or the sample is heated up to around 1000°C [74], the deformation of atomic alignments of C_{84} occurs more drastically and involves chemical reactions such as C_2 loss, dissociation or coalescence of fullerenes. In fact, STM can give a lot of valuable information on the surface structures of such samples.

The STM apparatus can sometimes be used to process the surface, i.e., to change and control atomic alignments of the sample surface, with an atomic manipulation [75–77]. Of course, a high level of competence may be required to make suitable surface alignments. There has been much success in this direction, although we shall not go into details here.

We begin now by discussing the use of electronic excited states to select a special intermolecular orientation of fullerenes. In order to consider this possibility from the theoretical point of view, we review the results of a theoretical calculation concerning the low-temperature C_{60} crystal, which has a simple cubic space group and reflects intermolecular interactions among C_{60}'s.

5.5.1 Insights from Ab Initio Calculations for Crystalline C_{60} Phases

Several crystalline phases of C_{60}, including the high-temperature rotating fcc phase ($Fm\bar{3}$ space group) and hypothetical low-temperature simple cubic phases ($Pn\bar{3}$ and $Pa\bar{3}$ space groups), with each molecule orientationally frozen, show all different electronic structures according to ab initio calculations [20] based on the all-electron mixed basis approach described in the Appendix. The degeneracies at special k-points, magnitudes of energy gaps, and widths of bands have been found to be sensitive to the orientation of fullerenes in the lattice. For the high-temperature fcc phase, good agreement is obtained for the angle-resolved DOS along the Γ-K symmetry axis when the calculation [78] is compared with photoemission data by Weaver et al. [79]. For the low-temperature phase, the ab initio result [20] suggests that the structure with $Pa\bar{3}$ space group and $-22°$ rotation around [111] directions, proposed by David et al. [60], is more stable compared to the structure with $Pn\bar{3}$ space group and the same rotation (see Fig. 5.10 which illustrates the mutually facing bonds). In the case of $Pn\bar{3}$, either pentagons or double bonds are facing each other, while in the case of $Pa\bar{3}$ a double bond faces the center of a pentagonal ring of the adjacent C_{60}. In the former case, excess charge around double bonds overlaps significantly. In order to realize less overlap of occupied charge density distributions of adjacent C_{60}'s, and in turn less electrostatic repulsive energy, the latter configuration is favorable. It might also be intuitively understandable that making the energy gap wider generally gains energy, although Madelung energy should also be taken into account. $Pa\bar{3}$ has wider gap and thus presumably more cohesive energy. All these considerations are consistent with our conclusion that the structure with $Pa\bar{3}$ space group and 22° rotation around [111] directions is energetically favored.

Isosurfaces of the spatial distribution of the PDOS calculated in [20] for the $Pn\bar{3}$ structure are shown in Fig. 5.11a for the 12 HOMO (h_u) bands and in Fig. 5.11b for the 12 LUMO (t_{1u}) bands. In this figure, only three of the four C_{60} molecules inside the simple cubic unit cell are shown. In the HOMO-band PDOS, all double bonds are brighter (higher in density) than

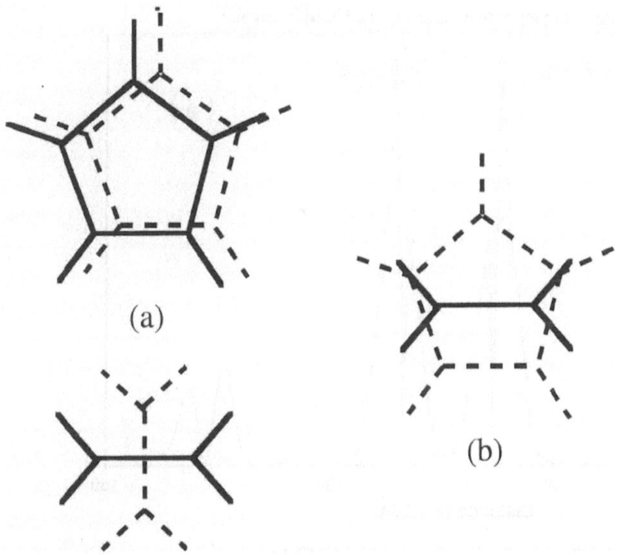

Fig. 5.10. Illustration of the stereographic location of mutually facing bonds of adjacent C_{60}'s. (**a**) In the $Pn\bar{3}$ phase with $-22°$ rotation, either single bonds (hexagonal rings) or double bonds are facing each other. (**b**) In the $Pa\bar{3}$ phase with $-22°$ rotation, a double bond faces the center of a pentagonal ring

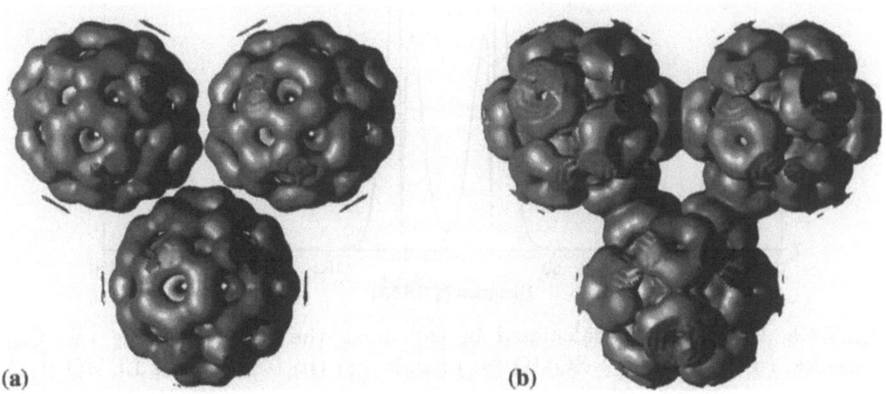

Fig. 5.11. Isosurfaces of the PDOS distribution [20] of (**a**) the 12 HOMO (h_u) bands and (**b**) the 12 LUMO (t_{1u}) bands

the rest, whereas in the LUMO-band PDOS, all single bonds constituting the five-membered rings are highlighted. These features are very similar to the previous result on the fcc ($Fm\bar{3}$ and also a less symmetrical case) C_{60} crystal [19].

Figure 5.12 plots the PDOS of the $Pn\bar{3}$ structure along the line connecting two C_{60} molecules for the same structure. As is clearly seen from this figure, the density at the intermolecular region of this structure is higher in the

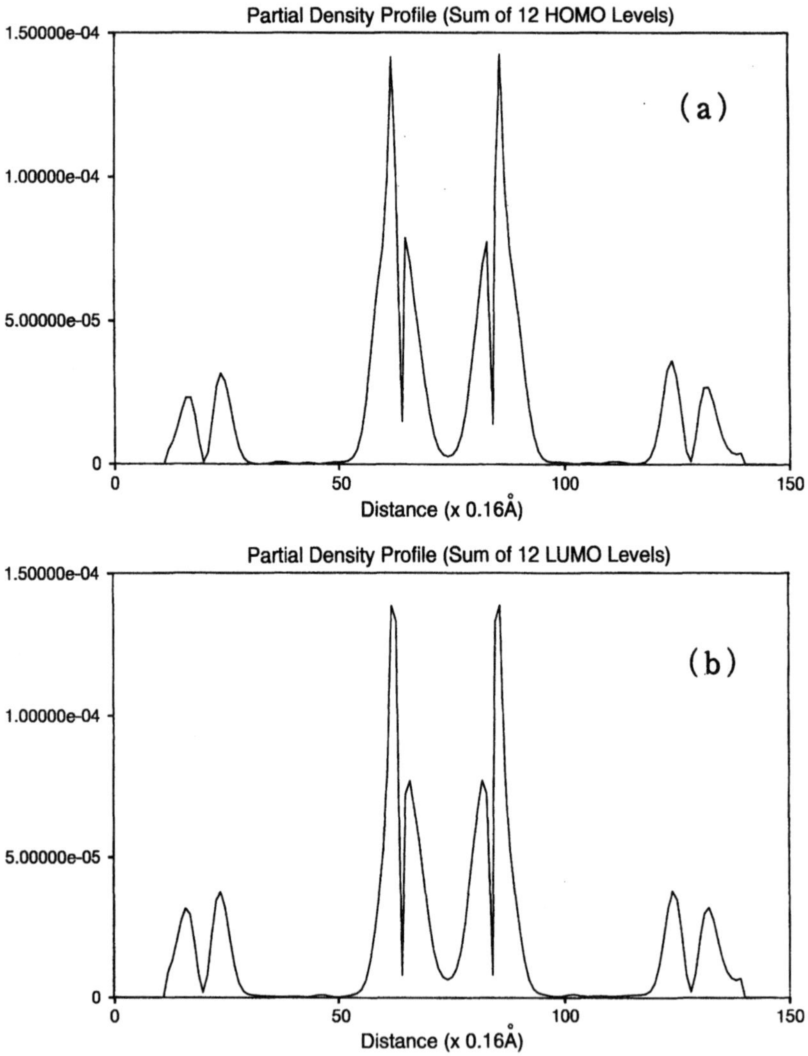

Fig. 5.12. PDOS profile calculated in [20] along the line connecting two C_{60} molecules. (**a**) is for the 12 HOMO (h_u) bands and (**b**) is for the 12 LUMO (t_{1u}) bands

LUMO bands than in the HOMO bands, because five-membered rings are facing each other among these molecules. Generally, the more the overlap in the intermolecular bonding orbitals, the lower the energy. Therefore, this structure with two five-membered rings facing each other can more easily be realized in the excited states, i.e., in the situation where electrons are excited from HOMO bands to LUMO bands. This gives a good starting point for discussing C_{60} geometries in excited states. In a similar way, for the $Pa\bar{3}$ structure, the overlap between the wavefunction around the double bond of one C_{60} and the wavefunction around the five-membered ring of the adjacent C_{60} increases in the excited states. Therefore, this orientation is also more stabilized in the electronic excited states.

5.5.2 Comparison with Experimental Work

The experiment we discuss here is due to Kasuya et al. [13], who used STM to observe the topmost layer of C_{60} or C_{84} films grown on a Si substrate, before and after Ar-laser irradiation. Of course, no anomalous behavior is observed before irradiation. Since every molecule is rotating, no internal structure is observed by STM in this sample. But after irradiation, some molecules which constitute a cluster of three or more molecules seem to have fixed rotation, and clear internal structure is observed. This may be evidence for the orientational ordering caused by the electronic excited states discussed above.

More strikingly, a very bright spot is detected in some of the C_{60} molecules when the sample bias voltage is minus. As the bias voltage becomes positive, each of these special spots becomes a very big bright star-shaped pattern (like an ancient picture of the Sun), which involves more than 7 molecules and has a six-fold symmetry. Such a big star-shaped pattern observed after irradiation indicates that the electronic structure of the center C_{60} molecule of this pattern became quite different from that of other C_{60}'s. There is a high probability that this molecule is deformed by laser power or is trapping an external object like an oxygen molecule. In order to check the appearance of such a big pattern, we performed an ab initio calculation based on the all-electron mixed basis approach (described in the Appendix) to calculate the PDOS of this system. In the calculation, the supercell is divided into $256 \times 256 \times 256$ meshes where one mesh corresponds to 0.116 Å in the XY plane and 0.053 Å in the Z direction.

To begin with, we assumed a two-dimensional array composed of seven C_{60} molecules: one C_{60} was located at the center and surrounded by the others, forming a big hexagonal shape. In this case, however, we could not obtain any similar pattern. We then changed the center molecule from C_{60} to C_{58} (a C_2-lost C_{60} molecule). All the surrounding molecules were the same C_{60}'s, that is, we assumed a hexagonal two-dimensional array composed of $C_{58}+6C_{60}$ clusters. In this case, we could obtain relatively extensive patterns which might correspond to the STM pattern observed by Kasuya et al. [13]. Such a big change in the electronic states may have a practical application, e.g., as a nanoscale memory device.

The total charge density distribution is shown in Fig. 5.13 (by isosurface plots). As can be seen from this figure, the center C_{58} molecule has two missing carbon atoms which were located at the topmost double bond. The spatial distributions of the PDOS corresponding to the 21 HOMO and LUMO levels are, respectively, shown in Figs. 5.14a and b, also by isosurface plots. The HOMO-PDOS (Fig. 5.14a) shows four strong spots at the center C_{58} molecule, due to dangling bonds. This pattern is quite different from the

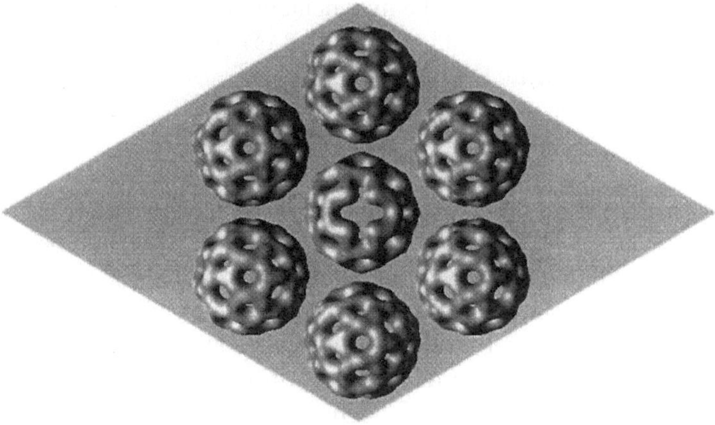

Fig. 5.13. Isosurfaces of the total charge density of the C_{58}+6C_{60} system

(a)

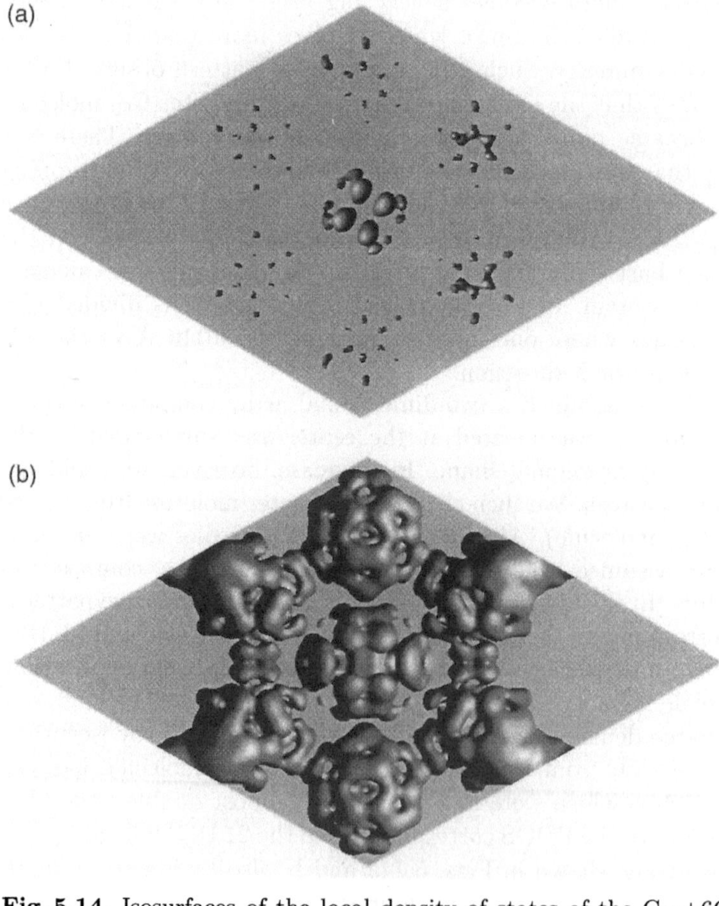

(b)

Fig. 5.14. Isosurfaces of the local density of states of the C_{58}+6C_{60} system. (**a**) corresponds to the 21 HOMO levels and (**b**) corresponds to the 21 LUMO levels ordinary pattern in which high densities occur mainly at double bonds. On the other hand, the LUMO-PDOS has a pattern similar to the ordinary one in which the density is high mainly at single bonds, i.e., at the five-membered rings, when the density used for drawing the isosurface is relatively high. However, if the isosurface density becomes low, a small but net density appears over four of the surrounding six C_{60} molecules and extends outwards from there, as shown in Fig. 5.14b. Thus, we certainly find a big star-shaped pattern, although lacking the six-fold symmetry which has been observed experimentally. Unfortunately, we cannot say that this hypothetical system is identical to the one observed experimentally, because there is no other evidence except for the similarity between the patterns. At this stage of the investigation, we cannot rule out other possibilities such as the adorption of an external molecule. Further study is desirable for this subject.

5.6 Conclusion

In this article, we have reviewed ab initio calculations of the Partial Density of States (PDOS) and other electronic properties of various fullerene systems in connection with STM research.

First we discussed two-dimensional alignments of C_{60} (or of C_{60}–C_{70} mixture) by means of the all-electron mixed basis approach [52]. The result was successfully compared with STM observations of C_{60} (or of C_{60}–C_{70} mixture) aligned on the Cu(111) substrate by Hashizume et al. [11]. A universal feature of STM images of C_{60} and C_{70} was discussed together with a possibility for identifying the orientation of fullerenes at surfaces on the basis of STM images.

The second topic was the effect of charge transfer from the substrate to fullerenes adsorbed in the first layer. We found that if the amount of charge transfer is not small, a strong dipole field caused by this dipole layer at the interface induces dipole moments in the C_{60}'s adsorbed on overlayers. The resulting dipole–dipole interaction among overlayer C_{60} molecules may become comparable to the original C_{60}–C_{60} interaction. An example of such behavior may have been found in the experiment performed by Xue et al. [12] for C_{60} adsorbed on GaAs (100) As-rich substrate. In this system, the amount of charge transferred from As atoms to the first-layer C_{60} was estimated to be 1.76 by a first-principles pseudopotential method [67,65]. Our theory suggested that the origin of the 13% lattice expansion of overlayer C_{60} is the induced dipole–dipole interaction.

The last subject concerned the use of electronic excited states in fullerene research. The difference between the electronic structures of HOMO and LUMO bands in a hypothetical simple cubic phase ($Pn\bar{3}$) may cause a change in the orientations of C_{60} molecules when an electron is excited from the HOMO band to the LUMO band. Such behavior, i.e., photo-induced rotational fixing, was observed by Kasuya et al. (using STM) in laser-irradiated

C_{60} samples [13]. After laser irradiation, STM images may reveal many drastic changes in the C_{60} samples. This use of electronic excited states might cause some 'damage' to samples. However, such experiments give the possibility of realizing some special atomic alignments which could not occur in nature. The study of electronic excited states in surface physics is just beginning and may lead to a variety of new subjects.

Acknowledgements

The authors thank Prof. T. Sakurai, Dr. T. Hashizume, Dr. Q.-K. Xue, Prof. A. Kasuya, Dr. W. Fu, Dr. Y. Maruyama and Dr. H. Kamiyama, for their valuable and helpful discussions. It is a pleasure to observe that most of the work described in this article is based on collaboration with them. The authors are also indebted to the Computer Science Group at the Institute for Materials Research, Tohoku University, for access to their HITAC S3800/380 supercomputing facilities.

Appendix: All-Electron Mixed Basis Approach

Here we describe the computational scheme in an all-electron mixed basis approach (which uses the Slater-type analytic atomic orbitals).

Historically, the mixed-basis approach was neatly formulated by Louie et al. [80] in order to provide an efficient treatment for spatial locality and asymmetry with a limited number of basis functions. The technique involves expanding wave functions in terms of both plane waves (PW) and atomic orbitals (AO). Louie et al. applied the mixed basis to expand the d-symmetry pseudo-wave functions of transition metal elements, which are given by using a pseudopotential. The pseudopotential scheme itself is a very powerful tool for investigating electronic structure. This scheme allows one to replace a strong electron–ion Coulomb potential with a soft potential effectively felt by valence electrons outside the core region. As a result, the true wave functions which come from calculation using the true atom are replaced by pseudo-wave functions which come from the pseudopotential, and what is called a pseudo-charge appears in the core region.

However, even though a pseudopotential is used, it is generally difficult to expand wave functions only in terms of PW's, because wave functions like $2p$ orbitals in the second-row elements and $3d$ orbitals in transition metal elements are strongly localized near the core region. In this all-electron mixed basis approach, PW's and Slater-type atomic orbitals (STO) are used to expand the total wavefunctions including the core. Numerical results show a drastic improvement with regard to numerical complexity: although 110 000 plane waves are required in the standard band structure calculation [81] in order to achieve a convergence on fcc C_{60}, we have obtained similar results using only 3000 plane waves with 240 atomic orbitals [19,78].

The all-electron mixed basis approach for band structure calculations has been successfully applied to similar experimental results of aligned C_{60} on Cu(111) [52] and to fcc and sc C_{60} [19,20,78].

In the mixed-basis formalism, the basis $|l\rangle$ denotes either PW or AO. Suppose that we are using N_{PW} plane waves and $4N_{\mathrm{A}}$ atomic orbitals accompanying the N_{A} atoms. Then for $l = 1, \ldots, N_{\mathrm{PW}}$, the basis $|l\rangle$ denotes the lth PW:

$$\phi_l^{\mathrm{PW}}(\boldsymbol{r}) = \langle \boldsymbol{r}|l\rangle = \frac{1}{\sqrt{\Omega}} \mathrm{e}^{-\mathrm{i}\boldsymbol{G}_l \cdot \boldsymbol{r}} . \tag{5.18}$$

For $l = N_{\mathrm{PW}} + 1, \ldots, N_{\mathrm{PW}} + N_{\mathrm{A}}$ it denotes the $1s$ atomic orbital (spin is neglected) centered at the mth atom, i.e.,

$$\phi_m^{1s}(\boldsymbol{r}) = \langle \boldsymbol{r}|l\rangle = \sqrt{\frac{\alpha_m^3}{\pi}}\, \mathrm{e}^{-\alpha_m|\boldsymbol{r}-\boldsymbol{R}_m|} , \tag{5.19}$$

and for $l = N_{\mathrm{PW}} + N_{\mathrm{A}} + 1, \ldots, N_{\mathrm{PW}} + 4N_{\mathrm{A}}$ it denotes one of the $2p_x$, $2p_y$, $2p_z$ atomic orbitals:

$$\phi_m^{2p_x}(\boldsymbol{r}) = \langle \boldsymbol{r}|l\rangle = \sqrt{\frac{\beta_m^5}{\pi}}(x - X_m)\, \mathrm{e}^{-\beta_m|\boldsymbol{r}-\boldsymbol{R}_m|} , \tag{5.20a}$$

$$\phi_m^{2p_y}(\boldsymbol{r}) = \langle \boldsymbol{r}|l\rangle = \sqrt{\frac{\beta_m^5}{\pi}}(y - Y_m)\, \mathrm{e}^{-\beta_m|\boldsymbol{r}-\boldsymbol{R}_m|} , \tag{5.20b}$$

$$\phi_m^{2p_z}(\boldsymbol{r}) = \langle \boldsymbol{r}|l\rangle = \sqrt{\frac{\beta_m^5}{\pi}}(z - Z_m)\, \mathrm{e}^{-\beta_m|\boldsymbol{r}-\boldsymbol{R}_m|} . \tag{5.20c}$$

Because these atomic bases are not orthogonal to the PW's, we start from the modified equation which guarantees orthogonality,

$$H\Psi_i = \varepsilon_i S\Psi_i , \tag{5.21}$$

where $\Psi_i = (\langle\Psi_i|1\rangle, \langle\Psi_i|2\rangle, \langle\Psi_i|3\rangle, \ldots)^\dagger$, and other time dependent variables α_m and β_m, while M_m represents the real nucleus mass of the mth atom, $S = \langle l|k\rangle$ denotes the overlap matrix of the wave function, and $H = \langle l|H|k\rangle$ and E denote the Hamiltonian of the electrons and the total energy of the system, respectively. Hereafter we use subscripts ij for electronic levels, lk for basis functions, and mn for atoms.

A feature distinguishing the present equations with the mixed basis from those of the earlier PW approach is the presence of the overlap matrix S in (5.21). This is due to the fact that the bases are not mutually orthogonal. Introducing the lower half triangular matrix U, which satisfies $S = UU^\dagger$ and is constructed in the Choleski decomposition, and writing $U^\dagger\Psi_i = \Phi_i$ and $H' = U^{-1}HU^{\dagger-1}$, we finally obtain:

$$\mu\dot{\Phi}_i = -\left(H' - \Phi_i^\dagger H'\Phi_i\right)\Phi_i . \tag{5.22}$$

Once one adopts this representation, the main algorithm for updating the wave function Φ_i is the same as in the original PW approach. To evaluate the charge density one needs to trace the wave functions to those in the original nondiagonal frame via the relation $\Psi_i = U^{\dagger-1}\Phi_i$. If the atomic wave function has extra parameters like exponential damping rates α_m and β_m, these parameters can also be treated as time dependent variables, as in (5.22).

The effective one-electron Hamiltonian reads

$$H = T + V\,, \quad T = -\frac{1}{2}\nabla^2\,, \tag{5.23a}$$

$$V(\boldsymbol{r}) = -\sum_m \frac{Z_m}{|\boldsymbol{r}-\boldsymbol{R}_m|} + \int \mathrm{d}\boldsymbol{r}' \frac{\rho(\boldsymbol{r}')}{|\boldsymbol{r}-\boldsymbol{r}'|} + V^{\mathrm{xc}}(\boldsymbol{r})\,, \tag{5.23b}$$

where we have used the atomic unit (a.u.), $\hbar = m_\mathrm{e} = e = 1$. In (5.23b), Z_m and $\boldsymbol{R}_m$ denote, respectively, the atomic number and position of the mth atom, $\rho(\boldsymbol{r})$ is the total electron density and $V^{\mathrm{xc}}(\boldsymbol{r})$ the exchange-correlation potential which is evaluated in real space under the local density approximation. In this formalism, the charge density $\rho(\boldsymbol{r})$ is directly evaluated from the expansion coefficients in real space, which is divided up as mesh $\times$ mesh $\times$ mesh with mesh = 64 (or 128, for example, for systems which have more than four C_{60} molecules in the unit cell). The Fast Fourier Transformation (FFT) is used many times in this algorithm. To calculate the electron–atom and electron–electron Coulomb potentials, i.e., the first two terms in (5.23b), it is convenient to evaluate their sum in Fourier space as

$$\bar{V}(\boldsymbol{G}) = 4\pi \frac{\bar{\rho}(\boldsymbol{G}) - \sum_m Z_m e^{-i\boldsymbol{G}\cdot\boldsymbol{R}_m}/\Omega}{G^2}\,, \tag{5.24}$$

provided that the charge density $\bar{\rho}(\boldsymbol{G})$ in Fourier space is known. In (5.24), Ω denotes the volume of the unit cell. In order to increase accuracy in the integration of potential matrix elements over real space around each atom, the local space is defined so as to contain each atom at its center. That is, both the real space charge density $\rho_m(\boldsymbol{r})$ and potential $V_m(\boldsymbol{r})$ are evaluated in the local space centered at the mth atom (for $m = 1 \sim N_\mathrm{A}$). This is the most time-consuming part in the present computer program.

Once the potential function is determined, the Hamiltonian matrix elements, $\langle l|H|k\rangle = \langle l|T|k\rangle + \langle l|V|k\rangle$, have to be evaluated. In the potential matrix elements $\langle l|V|k\rangle$, there are three types of combinations: PW–PW, PW–AO and AO–AO. For PW–PW, the calculation in Fourier space works well, similar to the standard PW expansion combined with a pseudopotential approach. Note that these PW–PW matrix elements do not have to be registered in computer memory, because the PW–PW block does not change after the diagonal transformation $H' = U^{-1}HU^{\dagger-1}$. On the other hand, the other combinations PW–AO and AO–AO are more easily calculated in real space (being localized within a small area around each atomic nucleus), since the $1s$ and $2p$ AO's are assumed to be well localized around each atomic center. The damping parameters α and β are fixed in the calculation: $\alpha = 1/0.098$ Å^{-1} and $\beta = 1/0.127$ Å^{-1}. [Therefore, the atomic orbitals in (5.19) and (5.20a–c) do not describe the full wavefunctions but should be considered as just a part of them. For such choices of AO's, one may neglect all overlaps between different atomic orbitals located at adjacent atomic centers, and thereby avoid the problem of overcompleteness.] Moreover, the real-space potential $V_m(\boldsymbol{r})$ centered at each atom separates into two parts: one is the spherically symmetric and analytic part $V_m^a(r) = -6\mathrm{e}^{-\zeta_m r}/r$ and the other is the numerical part $V_m^b(\boldsymbol{r}) = V_m(\boldsymbol{r}) - V_m^a(r)$. All relevant integrals involving $V_m^a(r)$ are carried out analytically.

For the matrix elements of the kinetic energy $\langle k|T|k\rangle$, one has three analogous combinations for PW–PW, PW–$1s$ and PW–$2p_x$, which can be expressed by analytical forms, while, for $1s$–$1s$ and $2p_\alpha$–$2p_\alpha$ $(\alpha = x, y, z)$, one may use the hydrogen-like analytic result, $\langle N_{\mathrm{PW}} + n|T|N_{\mathrm{PW}} + m\rangle = \frac{1}{2}\alpha_n^2\delta_{nm}$ and $\frac{1}{2}\beta_n^2\delta_{nm}$.

The nucleus charge Z_m used in (5.24) is renormalized by subtracting the contribution of the symmetric charge density from the $1s$–$1s$ part (at the mth atom). At the same time, the renormalized charge $\rho'(\boldsymbol{r})$, obtained from $\rho(\boldsymbol{r})$ by subtracting the $1s$–$1s$ part, is used. Finally, we evaluate the last term in (5.24) analytically in the region near the center of the mth atom and numerically elsewhere.

The standard band structure calculation can be performed within the Local Density Approximation (LDA). Since we use all electrons explicitly, including core electrons, it is not necessary to work with the pseudopotential and the obtained energy eigenvalues are absolute.

References

1. G. Binnig and H. Rohrer, Helv. Phys. Acta **55** (1982) 726; Rev. Mod. Phys. **59** (1987) 615.
2. H. Neddermeyer, Rep. Prog. Phys. **5** (1996); J.A. Kubby and J.J. Boland, Surf. **26** (1996) 61; Q.-K. Xue, T. Hashizume and T. Sakurai, Progr. Surf. Sci. **56** (1997) 1.
3. J. Bardeen, Phys. Rev. Lett. **6** (1961) 57.
4. J. Tersoff and D.R. Hamann, Phys. Rev. Lett. **50** (1983) 1998; Phys. Rev. B **31** (1985) 805.
5. M. Tsukada, K. Kobayashi, N, Isshiki and H. Kageshima, Surface Sci. Rep. **13** (1991) 265
6. Y.Z. Li, J.C. Patrin, M. Chander, J.H. Weaver, L.P.F. Chibante and R.E. Smalley, Science **252** (1991) 547; Y.Z. Li, M. Chander, J.C. Patrin, J.H. Weaver, L.P.F. Chibante and R.E. Smalley, Science **253** (1991) 429.
7. J.L. Wragg, J.E. Chamberlain, H.W. White, W. Krätschmer and D.R. Huffman, Nature **348** (1990) 623.
8. X.-D. Wang, V. Yu Yurov, T. Hashizume, H. Shinohara, Y. Saito, Y. Nishina and T. Sakurai, Jpn. J. Appl. Phys. **31** (1992) 450.
9. Y.Z. Li, M. Chander, J.C. Patrin, J.H. Weaver, L.P.F. Chibante and R.E. Smalley, Phys. Rev. B **45** (1992) 13837.
10. T. Hashizume, X.-D. Wang, Y. Nishina, H. Shinohara, Y. Saito, Y. Kuk and T. Sakurai, Jpn. J. Appl. Phys. **31** (1992) L880.

11. T. Hashizume, K. Motai, X.-D. Wang, N. Shinohara, Y. Saito. Y. Maruyama, K. Ohno, Y. Kawazoe, Y. Nishina, H.W. Pickering, Y. Kuk and T. Sakurai, Phys. Rev. Lett. **71** (1993) 2959.
12. Q.-K. Xue, T. Hashizume, Y. Hasegawa, H. Kamiyama, K. Ohno, Y. Kawazoe, H. Shinohara and T. Sakurai, Epitaxial Growth of a Strained C_{60} Overlayer on the GaAs(001)-As-rich 2×4 Surface (1996) to be published; T. Hashizume and T. Sakurai, Sci. Rep. RITU (Tohoku University) A **44** No.1 (1997) 17 [IMR Report No.2076].
13. A. Kasuya et al., in submission.
14. H.W. Kroto, J.R. Heath, S.C. O'Brien, R.F. Curl and R.E. Smalley, Nature **318** (1985) 162.
15. S. Iijima, I. Ichihashi and Y. Ando, Nature **356** (1992) 776.
16. R.D. Johnson, G. Meijer and D.S. Bethune, J. Am. Chem. Soc. **112** (1990) 8983.
17. W. Krätschmer, L.D. Lamb, K. Fostiropoulos and D.R. Huffman, Nature **347** (1990) 354.
18. S. Saito and A. Oshiyama, Phys. Rev. Lett. **66** (1991) 2637.
19. B.-L. Gu, Y. Maruyama, J.-Z. Yu, K. Ohno and Y. Kawazoe, Phys. Rev. B **49** (1994) 16202.
20. K. Ohno et al., Surf. Rev. Lett. **3** (1996) 735.
21. J.H. Weaver and D.M. Poirier, in Solid State Physics Vol. 48, 1. Academic Press, New York (1994).
22. M.S. Dresselhaus, G. Dresselhaus and P.C. Eklund, Science of Fullerenes and Carbon Nanotubes. Academic Press, New York (1996).
23. R.C. Haddon, A.F. Hebard, M.J. Rosseinsky, D.W. Murphy, S.J. Duclos, K.B. Lyons, B. Miller, J.M. Rosamilia, R.M. Fleming, A.R. Kortan, S.H. Glarum, A.V. Makhija, A.J. Muller, R.H. Eick, S.M. Zahurak, R. Tycho, G. Dabbagh and F.A. Thiel, Nature **350** (1991) 320.
24. A.F. Hebard, M.J. Rosseinsky, R.C. Haddon, D.W. Murphy, S.H. Glarum, T.T.M. Palstra, A.P. Ramirez and A.R. Kortan, Nature **350** (1991) 600.
25. K. Holczer, O. Klein, G. Grüner, S.-M. Huang, R.B. Kaner, K.-J. Fu, R.L. Whetten and F. Diederich, Science **252** (1991) 1154.
26. M.J. Rosseinsky, A.P. Ramirez, S.H. Glarum, D.W. Murphy, R.C. Haddon, A.F. Hebard, T.T.M. Palstra, A.R. Kortan, S.M. Zahurak and A.V. Makhija, Phys. Rev. Lett. **66** (1991) 2380.
27. K. Tanigaki, T.W. Ebbesen, S. Saito, J. Mizuki, J.S. Tsai, Y. Kubo and S.Kuroshima, Nature **352** (1991) 222.
28. T.T.M. Palstra, O. Zhou, Y. Iwasa, P. Sulewski, R. Fleming and B. Zegarski, Solid State Commun. **93** (1995) 327.
29. P.-M. Allemand, K.C. Khemani, A. Koch, F. Wudl, K. Holczer, S. Donovan, G. Grüner and J.D. Thompson, Science **253** (1991) 301.
30. Y. Chai, T. Guo, C. Jin, R.E. Haufler, L.P.F. Chibante, J. Fune, L. Wang, J.M. Alford and R.E. Smalley, Phys. Chem. **95** (1991) 7564.
31. R.D. Johnson, M.S. de Vries, J. Salem, D.S. Bethune and C. Yannoni, Nature **355** (1992) 239.
32. J.H. Weaver, Y. Chai, G.H. Kroll, C. Jin, T.R. Ohno, R.E. Haufler, T. Guo, J.M. Alford, J. Conceicao, L.P.F. Chibante, A. Jain, G. Palmer and R.E. Smalley, Chem. Phys. Lett. **190** (1992) 460.
33. H. Shinohara, H. Sato, Y. Saito, M. Ohkohchi and Y. Ando, J. Phys. Chem. **96** (1992) 3571.

34. H. Shinohara, H. Sato, M. Ohkohchi, Y. Ando, T. Kodama, T. Shida, T. Kato and Y. Saito, Nature **357** (1992) 52.
35. M. Takata, B. Umeda, E. Nishibori, M. Sakata, Y. Saito, M. Ohno and H. Shinohara, Nature **377** (1995) 46 (1995).
36. Z. Wan, J.F. Christian and S.L. Anderson, Phys. Rev. Lett. **69** (1992) 1352.
37. K. Ohno, Y. Maruyama, K. Esfarjani, Y. Kawazoe, N. Sato, R. Hatakeyama, T. Hirata and M. Niwano, Phys. Rev. Lett. **76** (1996) 3590.
38. T. Ohtsuki, K. Masumoto, K. Ohno, Y. Maruyama, Y. Kawazoe, K. Sueki and K. Kikuchi, Phys. Rev. Lett. **77** (1996) 3522.
39. M. Saunders, R.J. Cross, H.A. Jimenez-Vazquez, R. Shimshi and A. Khong, Science **271** (1996) 1693.
40. T. Ohtsuki, K. Ohno, K. Shiga, Y. Kawazoe, Y. Maruyama and K. Masumoto, Phys. Rev. Lett. **81** (1998) 385.
41. J. Mort, R. Ziolo, M. Machonkin, D.R. Guffman and M.I. Ferguson, Chem. Phys. Lett. **186** (1991) 284.
42. N. Minami and M. Sato, Synth. Met. **56** (1993) 3092.
43. H. Hoshi, N. Nakamura, Y. Maruyama, T. Nagagaa, S. Suzuki, H. Shiromaru and Y. Achiba, Jpn. J. Appl. Phys. **30** (1991) L1397.
44. Y. Wang and L.T. Cheng, J. Phys. Chem. **96** (1992) 1491.
45. N. Tang, J.P. Partanen, R.W. Hellwarth and R.J. Knize, Phys. Rev. B **48** (1993) 8404. Jpn. J. Appl. Phys. **30** (1991) L1397.
46. M. Haluška, H. Kuzmany, M. Vybornov, P. Rogl and P. Fejdi, Appl. Phys. A **56** (1993) 161.
47. P.W. Stephens, G. Bortel, G. Faigel, M. Tegze, A. Janossy, S. Pekker, G. Oszlany and L. Forro, Nature **370** (1994) 636.
48. A.M. Rao, P. Zhou, K. Wang, G.T. Hager, J.M. Holden, Y. Wang, W. Lee, X. Bi, P.C. Eklund, D.S. Cornett, M.A. Duncan and I. J. Amster, Science **259** (1993) 955.
49. Y.B. Zhao, D.M. Poirier, R.J. Pechman and J.H. Weaver, Appl. Phys. Lett. **64** (1994) 577.
50. Y. Iwasa, T. Arima, R.M. Fleming, T. Siegrist, O. Zhou, R.C. Haddon, L.J. Rothberg, K.B. Lyons, H.L. Carter Jr., A.F. Hebard, R. Tycho, G. Dabbagh, J.J. Krajewski, G.A. Thomas and T. Yagi, Science **264** (1994) 1570.
51. M. Nunez-Regueiro, L. Marques, J.-L. Hodeau, O. Bethoux and M. Perrous, Phys. Rev. Lett. **74** (1995) 278.
52. Y. Maruyama, K. Ohno and Y. Kawazoe, Phys. Rev. B **52** (1995) 2070; see also [11].
53. M.B. Jost, P.J. Benning, D.M. Poirier, J.H. Weaver, L.P.F. Chibante and R.E. Smalley, Chem. Phys. Lett. **184** (1991) 423.
54. J.H. Weaver, J. Chem. Solid **53** (1992) 1433.
55. K. Ohno, J.-Z. Yu, Y. Maruyama, Y. Kawazoe and T. Takahashi, Chem. Phys. Lett. **255** (1996) 367.
56. K. Motai, T. Hashizume, H. Shinohara, Y.S.H.W. Pickering, Y. Nishina and T. Sakurai, Jpn. J. Appl. Phys. **32** (1993) 450.
57. X.-D. Wang, V. Yu Yurov, T. Hashizume, H. Shinohara and T. Sakurai, Phys. Rev. B **49** (1994) 14746.
58. P.A. Heiney, J.E. Fischer, A.R. McGhie, W.J. Romanow, A.M. Denenstein, J.P. McCauley, Jr., A.B. Smith III and D.E. Cox, Phys. Rev. Lett. **66** (1991) 2911.

59. J.R.D. Copley, D.A. Neumann, R.L. Cappelletti, E. Prince, W.A. Kamitakahara, N. Coustel, J.P. McCauley, Jr., N.C. Maliszewskyj, J.E. Fischer, A.B. Smith III, K.M. Creegan and D.M. Cox, Physica B **180/181** (1992) 706.
60. W.I.F. David, R.M. Ibberson, J.C. Matthewman, K. Prassides, T.J.S. Dennis, J.P. Hare, H.W. Kroto, R. Taylor and D.R.M. Walton, Nature **353** (1991) 147; W.I.F. David, R.M. Ibberson, T.J.S. Dennis, J.P. Hare and K. Prassides, Europhys. Lett. **18** (1992) 219, 735.
61. Y. Kawazoe, H. Kamiyama, Y. Maruyama and K. Ohno, Jpn. J. Appl. Phys. **32** (1993) 1433.
62. N. Isshiki, K. Kobayashi and M. Tsukada, Surf. Sci. **238** (1990) L439; H. Kageshima and M. Tsukada, Phys. Rev. B **46** (1992) 6928.
63. S.H. Yang, C.L. Pettiette, J. Conceicao, O. Cheshnovsky and R.E. Smalley, Chem. Phys. Lett. **139** (1987) 233.
64. R.L. Hettich, R.N. Compton and R.H. Ritchie, Phys. Rev. Lett. **67** (1991) 1242.
65. K. Ohno, Z.Q. Li, H. Kamiyama, Q.-K. Xue, T. Hashizume, H. Shinohara and T. Sakurai, Sci. Rep. RITU (Tohoku University) A **43** No.1 (1997) 61 [IMR Report No.2061].
66. H. Kamiyama, K. Ohno and Y. Kawazoe, Sci. Rep. RITU (Tohoku University) A **41** No.2 (1996) 187 [IMR Report No.2005].
67. Z.-Q. Li, K. Ohno and Y. Kawazoe, unpublished.
68. L.A. Girifalco, J. Phys. Chem. **96** (1992) 858.
69. A.F. Hebard et al., Appl. Phys. Lett. **59** (1991) 2109.
70. P. Ecklund, Bull. Am. Phys. Soc. **37** (1992) 191.
71. A.A. Quong and M.R. Pederson, Phys. Rev. **B46** (1992) 12906.
72. P. Giannozzi and S. Baroni, J. Chem. Phys. (Let.) **100** (1994) 8537.
73. A. Adachi, MSc. Thesis in Tohoku University (1997); A. Adachi, W. Fu and A. Kasuya, in preparation.
74. X.-D. Wang, T. Hashizume, H. Shinohara, Y. Saito, Y. Nishina and T. Sakurai, Phys. Rev. B **47** (1993) 15923.
75. D.M. Eigler et al., Nature **34** (1990) 524.
76. P. Molinàs-Mata, A.J. Mayne and G. Dujardin, Phys. Rev. Lett. **80** (1998) 3101.
77. G. Dujardin, A. Mayne, O. Robert, F. Rose, C. Joachim and H. Tang, Phys. Rev. Lett. **80** (1998) 3085.
78. B.-L. Gu, Y. Maruyama, J.-Z. Yu, K. Ohno and Y. Kawazoe, MRS Symposium Proceedings Vol. 349: Novel Forms of Carbon II (1994) 289.
79. J.H. Weaver, unpublished; E. Sohmoen, J. Fink and W. Krtchmer, Z. Phys. B **86** (1992) 87; J.M. Themlin, S. Bouzidi, F. Coletti, J.M. Debever, G. Gensterblum, L.M. Yu, J.J. Pireaux and P.A. Thiry, Phys. Rev. B **46** (1992) 15602; G. Gensterblum, J.J. Pireaux, P.A. Thiry, R. Caudano, T. Buslaps, R.L. Johnson, G.L. Lay, V. Aristov, R. Günther, A. Taleb-Ibrahimi, G. Indlekofer and Y. Petroff, Phys. Rev. B **48** (1993) 14756.
80. S.G. Louie, K.M. Ho and M.L. Cohen, Phys. Rev. B **19** (1979) 1774.
81. N. Troullier and J.L. Martins, Phys. Rev. B **46** (1992) 1754.

6 Apparent Barrier Height and Barrier-Height Imaging of Surfaces

Akira Sakai

Summary. Tunneling barrier height is one of the most fundamental parameters in STM. It is this barrier height that leads to the exponential gap dependence of the tunneling current and hence enables STM to probe surfaces with atomic resolution. In a simple one-dimensional model of electron tunneling in STM, the barrier height has a definite meaning and its relation to the tunneling current and to the work functions of the tip and sample are both firmly established. In real STM, however, the barrier height derived experimentally from the gap dependence of the tunneling current cannot be simply interpreted as a potential barrier height since it contains complicated contributions from the electronic and atomic structures of the tip and sample. Because of this difficulty in analyzing barrier-height data, STM barrier-height imaging has long remained an auxiliary imaging technique, although it was invented in the early days of STM. However, in recent experiments, barrier-height imaging has been revived to procure unique information which cannot be obtained by other imaging methods. In these experiments, the sensitivity of the barrier height to the sample work function has been utilized for elemental identification on surfaces and for probing local changes in work function induced by adsorption and defect formation. In this article, we will present a brief review of experimental and theoretical studies of the barrier height in STM, together with some recent applications of barrier-height imaging to clean and adsorbed surfaces.

6.1 Introduction

The tunneling current in STM junctions varies exponentially with the tip–sample gap distance. This exponential gap distance is a characteristic of tunneling, and the tunneling current I can be generally written as $I \propto \exp(-2\kappa s)$, where s stands for the tip–sample gap distance. In the WKB model of tunneling, the coefficient κ is equal to $\sqrt{2m\phi}/\hbar$, where m is the electron mass and ϕ is the tunneling barrier height. Although the actual tunneling current in STM cannot be described by the simple WKB model, the exponential dependence of the STM current on gap distance allows us to define an 'apparent' tunneling barrier height by

$$\phi_{\mathrm{a}} = \frac{\hbar^2}{8m}\left(\frac{\mathrm{d}\ln I}{\mathrm{d}s}\right)^2 = \frac{\hbar^2}{8m}\left(\frac{1}{I}\frac{\mathrm{d}I}{\mathrm{d}s}\right)^2 . \tag{6.1}$$

When s is measured in units of [Å], the above definition becomes

$$\phi_a = 0.95 \left(\frac{d \ln I}{ds} \right)^2 = 0.95 \left(\frac{1}{I} \frac{dI}{ds} \right)^2 , \tag{6.2}$$

where ϕ_a is in units of [eV]. Although the tunneling current is the quantity one directly measures in STM, the intrinsic property of the STM junction is its tunneling conductance $G = I/V$, where V is the bias voltage. The apparent barrier height can be redefined in terms of G by

$$\phi_a = \frac{\hbar^2}{8m} \left(\frac{d \ln G}{ds} \right)^2 = \frac{\hbar^2}{8m} \left(\frac{d \ln I}{ds} - \frac{d \ln V}{ds} \right)^2 . \tag{6.3}$$

The previous definition (6.1) reduces to this definition when the bias voltage is constant. However, ϕ_a obtained with (6.1) becomes different from that of (6.3) when the bias voltage changes with gap distance. This will be discussed later in Sect. 6.2.1.

In the ideal case of one-dimensional tunneling between semi-infinite electrodes separated by a vacuum gap, ϕ_a agrees with the work function of the electrode, provided that the electron potential in the vacuum region is approximated by a rectangular barrier. In real STM junctions, however, the tunneling process is indeed far more complicated than this ideal tunneling. First, the electron tunneling is not one-dimensional: in spite of the current concentration at the tip apex, the current distribution still has nonzero spatial extension. Second, the tip and the sample in STM have no planar geometry. The tip has a conical shape with some atom-sized protrusions on its apex, and the sample surface, although appearing flat macroscopically, has a variety of microscopic corrugations. Furthermore, the tip and the sample have different work functions, and the potential in the gap region changes its profile with the gap distance. The bias voltage also modifies the potential profile. Because of these complexities, the apparent barrier height ϕ_a defined by (6.1) or (6.3) has no simple physical meaning and depends on details of the electron tunneling process in STM. It does not even necessarily agree with the actual height of the potential barrier between tip and sample.

In (6.1), changes in ϕ_a or in the gap distance s have the same effect on the magnitude of the tunneling current. In conventional STM image interpretation, one implicitly assumes that ϕ_a does not change appreciably over the sample surface. This assumption would be justified as long as one were investigating macroscopic features of the surface, where ϕ_a can be well approximated by the surface work function. On an atomic scale, however, ϕ_a is no longer constant and shows a spatial variation which is closely related to the local electronic structures of the surface. The two-dimensional mapping of ϕ_a, usually referred to as Barrier-Height (BH) imaging [1–7], therefore provides interesting information about the electronic structure of the sample surface, which in many cases complements spectral information obtained by Scanning Tunneling Spectroscopy (STS). On absorbed or defected surfaces, in particular, BH imaging becomes a unique tool for investigating site by site the charge transfer induced by adsorption or defect formation.

In this article, we present a brief review of experimental and theoretical studies of barrier height in STM. In the first four sections, Sects. 6.2.1–6.2.4, we describe how barrier height changes with gap distance, bias voltage, electrode material and tip contamination, respectively. After briefly describing experimental methods for measuring the barrier height in Sect. 6.3, BH imaging will be explained in Sect. 6.4. Applications of BH imaging to clean and adsorbed silicon surfaces will be presented in Sect. 6.5.

6.2 Properties of Barrier Height

6.2.1 Gap Dependence of Barrier Height

As the tip approaches the sample surface, greater overlap between tip and sample wave functions increases the tip–sample interaction and decreases the tunneling barrier height. In the proximity region with a gap distance of 0.5 nm or less, electron transport between tip and sample changes from tunneling to ballistic conduction. This process is known as a collapse of the potential barrier. Lang [8] has made a theoretical calculation on the potential profile between a flat jellium and a Na atom on a jellium representing the sample and the tip, respectively, and showed that the barrier height starts to decrease as the separation between jellium and Na atom becomes less than 0.5 nm. The barrier height vanishes entirely with further reduction in gap distance. Detailed clculations on the barrier height of Na–jellium junctions have also been performed by Hirose and Tsukada [9]. In particular, they report the gap dependence of the barrier height under nonzero biases.

For non-jellium metal surfaces, the potential barrier collapse has been investigated by Ciraci and coworkers [10]. They used an Al pyramidal tip and calculated the barrier profile for the Al tip above an Al(111) surface. Similar calculations were also carried out on the Ag-tip/Ag-surface junctions. These calculations showed that the barrier disappears for tip–sample distances less than 0.5 nm. For semiconductor surfaces, Kobayashi *et al.* [11] obtained the gap dependence of the barrier height between a pyramidal Al tip, consisting of 4 atoms in a fcc (111) stacking, and the un-reconstructed Si(111)–1×1 surface. Using the first-principle recursion matrix method, they showed that the barrier height starts to decrease even at the large tip–sample distance of 1 nm. They also studied the effects of bias voltage on barrier height and found an appreciable reduction in barrier height when the Si surface is biased to +2.0 V. At this bias, the barrier height vanishes at $\sim$ 0.5 nm. The results of this theoretical work are summarized in Fig. 6.1, where the calculated barrier height is plotted against the gap distance.

Since the collapse of the potential barrier is due to various local tip–sample interactions, it takes place quite locally just beneath the tip apex. At the tip–sample distance at which barrier collapse occurs, a small hole or a channel

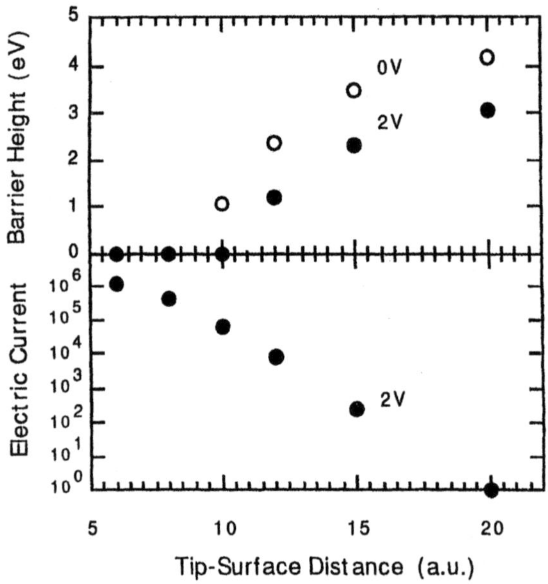

Fig. 6.1. Theoretical gap dependence of the barrier height between an Al tip and a Si(111) surface at zero bias and at +2 V sample bias (*upper panel*). Even after the collapse of the potential barrier, the tunneling current increases almost exponentially as the gap distance is reduced (*lower panel*). From [11]

opens in the barrier profile beneath the tip, and its channel width increases as the tip further approaches the surface [8–11]. Electron transport between tip and sample then becomes dominated by ballistic conduction through this channel. In spite of these drastic changes both in the electronic potential and in the conduction mechanism, the calculated STM current flowing through the junction shows little change at collapse of the potential barrier [11]. Even in the ballistic regime, the current retains its tunneling regime behavior and continues to increase exponentially as if there still existed the same potential barrier as at larger distances. This result indicates that the apparent barrier height ϕ_a defined by (6.2) remains almost unchanged until the tip finally makes strong interactions with the sample surface. This apparent discrepancy between the vanishing barrier height and the non-vanishing ϕ_a in the ballistic regime can be explained by the 'constriction effect' [10,11]. Since the channel width beneath the tip is of the order of 0.1 nm, confinement of electrons in this narrow channel costs an appreciable increase in their kinetic energy. This confinement energy acts as an effective potential barrier for electrons passing through the channel, and ϕ_a stays positive even when the real potential barrier becomes negative. Ciraci *et al.* [10] also showed that this constriction effect is site dependent.

The potential barrier is also lowered by the image potential which increases as $\alpha/(s - d_0)$, where d_0 is the position of the image plane relative to the sample surface, and α is a constant. The effect of image potential has been discussed before by several authors [12,13] who showed that the image potential contributes second-order effects in the effective barrier height ϕ_a and reduces ϕ_a by only a few percent, even though it causes an appreciable decrease in the real potential barrier height. Consequently, the presence of image potential has little effect on the gap dependence of ϕ_a.

When the tip comes very close to the surface, the STM current is greatly influenced by strong tip–sample interactions and becomes less gap dependent. As a result, the effective barrier height is reduced in the strong interaction regime. All theoretical work discussed above predicts such a decrease in ϕ_a at close proximity. However, the onset and the magnitude of the reduction appear to varying degrees to depend on the tip–sample system and the method of calculation. In Lang's calculation, the apparent barrier height starts to decrease at $s \sim 0.6$ nm and rapidly drops to 50% of its value at large s when $s \sim 0.4$ nm. On the other hand, calculations of the tunneling current through the Al-tip/Si-sample junctions [11] indicate almost no reduction in ϕ_a even at $s \sim 0.4$ nm.

Experimentally, the reduction in ϕ_a has been observed in various tip–sample systems [14–19], in agreement with the theoretical prediction. The gap dependence of ϕ_a measured by Lyo and Avouris [17] with a W tip and the Si(111)–7×7 surface is shown in Fig. 6.2. The apparent barrier height vanishes for tip displacements larger than 0.6 nm. However, in some experiments on clean metal surfaces [20,21], ϕ_a is found to stay constant until tip–sample contact. This means that ϕ_a remains unchanged even in the regime of strong tip–sample interaction. Explanations for these conflicting results on the gap

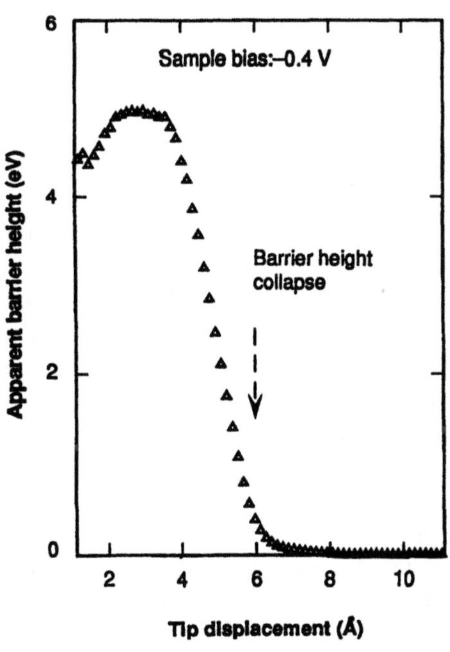

Fig. 6.2. Apparent barrier height as a function of the tip displacement for the Si(111)–7×7 W-tip system. From [17]

dependence of ϕ_a have been proposed recently by Olsen et al. [21], who have pointed out two factors which affect the change in ϕ_a with gap distance. One is the technical artefact which arises from the nonzero input impedance of the current amplifier used to measure the tunneling current. Under high current conditions, the input impedance of the current amplifier generates a non-negligible voltage drop which must be taken into account when obtaining the real bias voltage. Because of this voltage drop, the actual bias across the tip–sample junction is always less than the applied voltage, and this difference grows with increasing current or reduced gap distance. In such a case, the barrier height obtained with (6.1), assuming a constant bias, underestimates the intrinsic ϕ_a defined by (6.3), due to the gap dependence of V in (6.3). Olsen et al. showed that, while the ϕ_a derived from (6.1) decreases with gap distance, the ϕ_a from (6.3) stays almost constant down to tip–sample mechanical contact.

The another factor affecting the gap dependence of ϕ_a is tip deformation due to the interaction force acting between tip and sample. Usually the tip–sample force becomes stronger in magnitude as the tip approaches the sample. Therefore, a decrease in gap distance increases the tip–sample force, which in turn deforms the tip (or the sample) and changes the gap distance. As a result of this tip (sample) deformation, a change in the gap distance Δs becomes no longer equal to the tip displacement Δz. In barrier-height experiments, we always measure current changes with respect to the tip displacement z so that the barrier height is actually calculated as

$$\phi_a{}^{(m)} = \frac{\hbar^2}{8m}\left(\frac{\mathrm{d}\ln I}{\mathrm{d}z}\right)^2 .$$

When $\Delta s \neq \Delta z$, the barrier height thus obtained does not agree with ϕ_a defined by (6.1) and can be expressed as

$$\phi_a{}^{(m)} = \frac{\hbar^2}{8m}\left(\frac{\mathrm{d}\ln I}{\mathrm{d}z}\right)^2 = \phi_a\left(\frac{\mathrm{d}s}{\mathrm{d}z}\right)^2 . \tag{6.4}$$

Therefore one would obtain enhanced or reduced barrier height depending on the magnitude of $\mathrm{d}s/\mathrm{d}z$. When the interaction force is attractive, $|\mathrm{d}s/\mathrm{d}z| > 1$ and the measured barrier height becomes larger than ϕ_a, while it is reduced in the case of repulsive tip–sample force, where $|\mathrm{d}s/\mathrm{d}z| < 1$. Based on a simulation of the tip approach, Olsen *et al.* [21] concluded that the tip–sample force is attractive at close proximity and the resulting enhancement in the barrier height almost compensates the decrease in ϕ_a in the strong interaction regime. This explains the absence of gap dependence in their barrier-height data. The attractive tip–sample force is also indicated from the observed gap dependence of the STM corrugation height on Cu(001) [22].

Chen and Hamers [16], on the other hand, calculated $\mathrm{d}s/\mathrm{d}z$ for the W-tip/Si-sample system using a Morse-type interaction potential and showed that the interaction becomes repulsive, so that the barrier height is reduced, in the proximity regime. The gap dependence of the barrier height obtained from the theoretical $\mathrm{d}s/\mathrm{d}z$ is found to be in good agreement with the experimental barrier height on Si(111)–7×7. Note, however, that they attribute the entire gap dependence of the barrier height to that of $\mathrm{d}s/\mathrm{d}z$ and take into account neither the reduction in ϕ_a nor the effects of the gap-dependent bias described previously.

Summarizing the gap dependence of the apparent barrier height, all theoretical studies predict a reduction in ϕ_a in the strong interaction regime. On the other hand, two different types of behavior of ϕ_a have been observed experimentally: one is a collapse of ϕ_a like that shown in Fig. 6.2, and another is the gap-independent ϕ_a observed for some metal surfaces. The former behavior is consistent with the theoretical prediction but can also be explained as the result of $|\mathrm{d}s/\mathrm{d}z| < 1$ due to a *repulsive* tip–sample interaction. On the other hand, the latter behavior is apparently inconsistent with the predicted reduction in ϕ_a but can be reconciled to the theory by taking into consideration the effect of $|\mathrm{d}s/\mathrm{d}z| > 1$ due to the *attractive* tip–sample interaction. Therefore, the gap dependence of the barrier height and its interpretation appear rather controversial. The most critical issue is whether the tip–sample interaction is repulsive or attractive. The character of the tip–sample interaction is basically determined by the tip and sample materials, but the actual interaction may be much more sensitive to the tip condition, in particular, to tip contamination by foreign (sample) atoms. Part of the controversy in experimental results is very likely due to tip contamination. In order to correctly evaluate the gap dependence of ϕ_a and/or $\mathrm{d}s/\mathrm{d}z$, one thus has to carry out barrier-height and/or force measurements with atomically defined tip and sample. Since such an experiment has now become possible using AFM/STM combined with Field Ion Microscopy (FIM) [23], it is hopefully anticipated that the effect of $\mathrm{d}s/\mathrm{d}z$ on ϕ_a will be quantitatively clarified and that the intrinsic gap dependence of ϕ_a will soon be experimentally established not only on metal surfaces but also on semiconductor surfaces.

6.2.2 Bias Dependence of Barrier Height

Since the potential profile in the tunneling gap changes with the bias, the apparent barrier height should also be bias dependent. Unfortunately, there have been few experimental and theoretical studies directly treating the bias effects in ϕ_a, and the bias dependence of ϕ_a has not yet been well understood. Nevertheless, we can collect some fragmental information from various sources. First, the bias dependence of ϕ_a can be deduced from the Scanning Tunneling Spectroscopy (STS) data. As described in the STS literature [24], the bias dependence of the tunneling current comes from two factors: one is the combined electronic state densities of the tip and sample, and the other is the tunneling probability $T(s, V, E)$ for electrons of energy E. When the electronic state densities are assumed to be gap-independent, the apparent barrier height ϕ_a is related to $T(s, V, E)$ by

$$T(s,V,E) = \exp\left[-s\,\frac{2m}{\hbar^2}\,[\phi_a(V) + E_F - E]^{1/2}\right] , \tag{6.5}$$

so that the bias dependence of ϕ_a is a direct consequence of that of $T(s, V, E)$. In STS experiments, the contribution from T must be separated out from the measured $I-V$ characteristics to obtain electronic state densities. This problem has been discussed in detail by Stroscio and Feenstra [24]. The simplest approach is to employ an approximate formula for $T(s, V, E)$,

$$T(s,V,E) \approx \exp\left[-s\,\frac{2m}{\hbar^2}\,[\bar{\phi} - |eV/2| + E_F - E]^{1/2}\right] , \tag{6.6}$$

where $\bar{\phi}$ is the average work function of the tip and the sample. This formula approximates the WKB transmission probability for a trapezoidal potential barrier. From (6.5) and (6.6), we obtain

$$\phi_a(V) \approx \bar{\phi} - |eV/2| = (\phi_t + \phi_s)/2 - |eV/2| , \tag{6.7}$$

where ϕ_t and ϕ_s are the work function of the tip and the sample, respectively. (6.7) shows that ϕ_a decreases linearly with the bias in either polarity.

In their barrier-height imaging of metal surfaces, Hasegawa *et al.* [4] have investigated the bias dependence of ϕ_a and found that ϕ_a decreases linearly with the bias, in good agreement with (6.7). Also, the observed rate of decrease -0.2 eV/V has the same order of magnitude as -0.5 eV/V predicted from (6.7). Theoretically, the bias effect on barrier height has been calculated by Hirose and Tsukada [9] and by Kobayashi *et al.* [11], in their theoretical studies of high-bias effects in STM. Their calculations are made not on the apparent barrier height ϕ_a but on the real potential barrier height. According to their results at large tip–sample separations, the rate of decrease in the barrier height is -0.45 eV/V for the Na-tip/Na-sample junction, and -0.55 eV/V for the Al-tip/Si-sample junction, respectively. These values are also in good agreement with (6.7). Therefore, the simple trapezoidal approximation (6.7) for the barrier height gives a rather good account of the bias dependence of ϕ_a and of the real barrier height as well.

It should be noted that the bias dependence of ϕ_a has little influence on the tunneling current. In most STM experiments, the bias voltage is typically 2 V or less. Then according to (6.7), the bias-induced reduction in ϕ_a is 1 eV at the largest. When $\bar{\phi}$ = 4 eV and s =1 nm, this reduction in ϕ_a increases the tunneling current by a factor of 13. The same increase would be obtained if the gap distance changed by only 0.125 nm. Therefore, the usual STM/STS data does not reflect the bias dependence of ϕ_a. It is this minor role of the bias dependence of ϕ_a in STM/STS that simplifies the normalization of STS spectra. If ϕ_a had strong bias dependence, the STS data analysis would become quite complicated.

6.2.3 Material Dependence of Barrier Height

The apparent barrier height depends on the tip and sample materials primarily through their work functions. When the tip and the sample are made of the same metal, a reasonable approximation for ϕ_a should be the work function of that metal. In most STM experiments, however, the tip material is different from that of the sample. A standard recipe in such a case is to use the average work function $\bar{\phi} = (\phi_t + \phi_s)/2$ for ϕ_a, as in (6.7). However, the validity of this approximation has not been tested experimentally. A simple approach to this problem is to measure ϕ_a with tips of different ϕ_t. If ϕ_a varies linearly with ϕ_t, this would support the use of $\bar{\phi}$ as an approximate barrier height. Preliminary barrier-height measurements have been performed on Si(001)–2×1 using Al (ϕ_t = 4.1 eV), W (ϕ_t = 4.8 eV), and Pt (ϕ_t = 5.7 eV) tips. The dependence of ϕ_a on ϕ_t was, however, found to be ambiguous due to large fluctuations in ϕ_a, and the validity of the approximation $\phi_a \approx (\phi_t + \phi_s)/2$ has not been well established yet.

On certain semiconductor surfaces, the characteristic band structures of the surface states give rise to a significant effect on the apparent barrier height. This *band structure effect* comes into the barrier height through the electron energy E in (6.5). For electrons moving perpendicular to the barrier, E is the total kinetic energy of electrons. However, for electrons having a momentum $\boldsymbol{k}_{\parallel}$ parallel to the surface, E must be replaced by

$$E_z = E - \hbar^2 {k_{\parallel}}^2 / 2m^* , \tag{6.8}$$

where m^* represents the effective mass of electrons. Replacing E by E_z in (6.5) effectively increases the apparent barrier height according to

$$\phi_a = \phi_{a0} + \hbar^2 {k_{\parallel}}^2 / 2m^* , \tag{6.9}$$

where ϕ_{a0} is the apparent barrier height for $k_{\parallel} = 0$ electrons. Therefore the barrier height increases with $k_{\parallel}$ and becomes highest for electrons at the boundary of the first surface Brillouin zone. In metals, there are usually a number of states at and near the zone center, and electrons in these states contribute to the tunneling current. As a result, the band effect is not important on metal surfaces. On the other hand, certain semiconductor surfaces, such as Si(111)–2×1 and GaAs(110), have the zone-boundary surface states which also constitute HOMO and LUMO of the surface band. On such surfaces, the tunneling current at low biases flows in and out from these zone-boundary states, and the band effect in ϕ_a becomes quite significant. An example of the band-structure effect is shown in Fig. 6.3 where ϕ_a measured on the cleaved Si(111)–2×1 surface is plotted as a function of the sample bias voltage [25]. On this surface, the minimum surface band gap is located at $k_{\parallel}$ = 9.4 nm^{-1}, and the corresponding ϕ_a in (6.8) is 6 eV. This value is in good agreement with experimental ϕ_a at the band edge. As the bias voltage increases, electrons in states farther from the zone boundary start to contribute to tunneling, and ϕ_a decreases on both sides of the band gap.

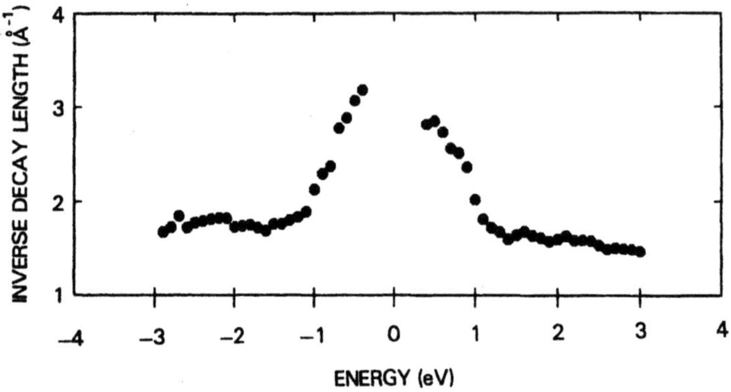

Fig. 6.3. Band effect in ϕ_a observed on the Si(111)–2×1 cleaved surface. Here ϕ_a is plotted as the inverse decay length κ which is related to ϕ_a through ϕ_a [eV] $= 3.8\kappa^2$ for κ in [Å^{-1}]. The energy is measured from the Fermi level. Adapted from [23]

At $|V|$ >1 V, ϕ_a is almost constant. The bias dependence of ϕ_a discussed in Sect. 6.2.2 is too small compared to a large band-structure effect on this surface. A similar band-structure effect is also reported on GaAs(110) [26,27].

6.2.4 Tip Contamination

Probably the most serious problem in barrier height measurements is tip contamination. In UHV-STM experiments, the tip apex can be contaminated by picking up sample atoms during tip approach and/or tip scanning. Recent atom-probe analysis [28,29] of the tip after scanning clearly demonstrates the presence of some sample atoms on the tip apex. The transfer of sample atoms to the tip is often indicated by a sudden change in STM contrast. But in many cases, tip contamination has no fatal effect on normal STM imaging and hence does not become a serious problem as long as the quality of STM images remains intact. On the other hand, the apparent barrier height is quite sensitive to tip contamination, and anomalously low or high barrier-height values are often obtained in measurements. Since the transfer of sample atoms is unpredictable, tip contamination often results in a large fluctuation in measured ϕ_a and makes it quite difficult to perform systematic studies of the barrier height in STM.

Unfortunately, little is known about the ϕ_a of contaminated tips. Since very low ϕ_a (less than 1 eV) is the most typical anomaly in the barrier height of Si surfaces, the effect of tip deformation described in Sect. 6.2.1 is generally believed to be responsible for this anomaly in ϕ_a. It is likely that the Si atoms picked up by the tip would be loosely bound to the tip surface and easily compressed by the repulsive tip–sample force, resulting in the low ϕ_a. However, it is not clear whether this explanation applies to all anomalous barrier heights. As discussed in Sect. 6.2.1, the tip–sample force is attractive on some metal surfaces, so that compression of the contamination layer on the tip is unlikely for these surfaces. This might be the reason why measured ϕ_as on metal surfaces exhibit relatively small fluctuations [4,7,21,22].

Besides the deformation of the contaminated layer, when foreign atoms are adsorbed on the tip they directly modify the tip electronic state and change the barrier height. Stievenard *et al.* [26,27] have recently calculated ϕ_a for a bare W tip and W tips having various single atoms (W, Ga and As) on their apices. According to their results, the adsorption of Ga and As atoms on the W tip leads to significant reduction (larger than 2 eV) in ϕ_a, particularly at small gap distances. They attribute the observed low ϕ_a on the GaAs(110) surface to such a contamination of the W tip by sample atoms. On the other hand, STM imaging of metal surfaces with oxygen-adsorbed tips [30] shows little influence of tip oxygen atoms on the measured barrier height. The effect of foreign tip atoms on ϕ_a is, therefore still quite unclear; neither their deformation nor their modification of the tip electronic state is well understood. Since this effect on ϕ_a is closely related to important topics such as anomalous corrugation in STM [31] and the discrimination of chemically different atoms using STM [30,32], further studies of this problem are strongly warranted. Measurements of ϕ_a with controlled tips artificially contaminated by specific foreign atoms on their apices will be of particular importance.

6.3 Measurements of Barrier Height

The apparent barrier height can be measured most directly by monitoring the tunneling current as the tip approaches (or retracts from) the sample surface under constant bias voltage. This measurement of the $I - z$ characteristics is referred to as the $z(I)$ or $I(z)$ method. The measured $\ln I$ is plotted against tip displacement z, and ϕ_a can be obtained from the slope of the plot. As mentioned in Sect. 6.2.1, one has to be careful about the difference between the real bias and the applied voltage, and also the difference between the tip displacement and the real change in gap distance. An example of the $z(I)$ plot measured on the Au(110)–2×1 surface [7] is shown in Fig. 6.4.

An alternative to the $z(I)$ method is to keep the tunneling current constant and measure the change in the bias voltage V as a function of z. This is the $z(V)$ method. In this method, the slope of the $\ln V - z$ plot at constant I gives the apparent barrier height, as indicated in (6.3). Figure 6.5 shows the $\ln V - z$ plot obtained on the same surface as in Fig. 6.4 [7]. In principle, the two methods should yield the same ϕ_a provided that the current varies linearly with the bias [7,13]. By comparing the data shown in Figs. 6.4 and 6.5, however, Olesen *et al.* [7] found that the barrier height obtained with the $z(V)$ method is 30% lower than that obtained with the $z(I)$ method. Since the $z(I)$ method gives results in good agreement with the barrier-height imaging

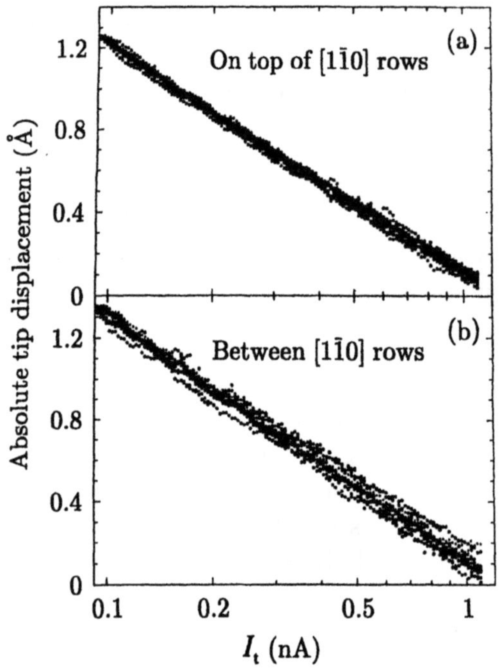

Fig. 6.4. $z-I$ characteristics observed on the Au(110)–2×1 surface. Two plots, obtained (**a**) on top of and (**b**) between the close-packed [1$\bar{1}$0] rows, are shown in the figure. From [7]

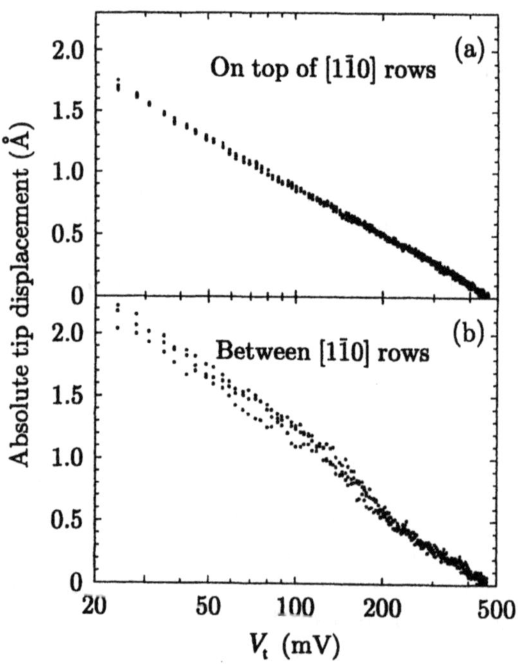

Fig. 6.5. The $z - \ln V$ data measured on the Au(110)–2×1 surface. Note that the voltage is plotted on a logarithmic scale. As in Fig. 6.4, (**a**) and (**b**) represent the data obtained on top of and between the [1$\bar{1}$0] rows, respectively. From [7]

data described later, they suggested that the low ϕ_a values from the $z(V)$ method are due to some nonlinearity in the $I-V$ characteristics.

Comparative study of the $z(I)$ and the $z(V)$ methods has also been carried out by Stievenard *et al.* [26,27] on GaAs(110). Contrary to the experiment by Olesen *et al.* [7], they found that the $z(V)$ method gives higher barrier height than that of the $z(I)$ method. Their results, however, appear inconclusive considering the band structure effect on this surface and the possible tip contamination with Ga and As atoms as mentioned in Sect. 6.2.4.

6.4 Barrier-Height Imaging

6.4.1 Experimental Method of Barrier-Height Imaging

Both the $z(I)$ and the $z(V)$ methods measure the barrier height at a couple of selected points on the surface. To obtain a two-dimensional map of the barrier height, one has to scan the tip over the surface and acquire the barrier height data while simultaneously taking a topographic image of the sample surface. This method of imaging the distribution of ϕ_a is referred to as Barrier-Height (BH) imaging, or ϕ_a-mapping. In BH imaging, one has to measure ϕ_a, or equivalently the derivative $\mathrm{d}I/\mathrm{d}s$, at each point on the surface. A modulation method is usually employed to obtain $\mathrm{d}I/\mathrm{d}s$. In this method, a small AC signal is superposed on the driving voltage of the z-piezo so that the gap distance is modulated. To first order in the modulation amplitude, the modulated component of the tunneling current is proportional to $\mathrm{d}I/\mathrm{d}s$, which can be detected by a lock-in amplifier. Then the apparent barrier height can be calculated using (6.2). The logarithmic derivative $\mathrm{d}\ln I/\mathrm{d}s$ can also be directly obtained if one detects the modulated signal from the logarithmic amplifier in the STM feedback circuit. Typically, the modulation frequency is 1–3 kHz, and the modulation amplitude is 0.02–0.03 nm. By employing a modulation frequency higher than the cutoff frequency of the STM feedback loop, the measurement of ϕ_a can be made without disturbing the conventional STM imaging, and the ϕ_a map can be obtained simultaneously with the corresponding constant-current topograph. The barrier height is usually measured at every pixel in the STM image. This inevitably leads to long scanning time, particularly when data averaging at each pixel is required to reduce noise in ϕ_a. Therefore, one has to make a compromise between fast scanning and good S/N ratio in ϕ_a.

An alternative method of BH imaging [7] is to interrupt the feedback control and measure the change in the tunneling current ΔI induced by a preset tip displacement Δs. After the measurement, the tip is brought back to the original position, and the feedback control starts again. The barrier height is calculated from (6.2), approximating $\mathrm{d}I/\mathrm{d}s$ by $\Delta I/\Delta s$. This BH imaging is therefore an analog of the CITS method in STS.

6.4.2 Image Contrast

Barrier-height images exhibit both macroscopic and microscopic (atomic) contrasts which must be considered separately [33]. Concerning the macroscopic contrast, there is little ambiguity in its interpretation as representing the corresponding macroscopic-scale changes in the surface work function. Therefore, in this section, we focus our attention on the atomic-scale contrast in the barrier height, which results in atom-resolved BH images of some metal and semiconductor surfaces.

When discussing the spatial variation in ϕ_a, it is convenient and customary to use the inverse decay length κ instead of ϕ_a, which is defined by

$$\kappa = 2m\phi_a{}^{1/2}/\hbar\ . \tag{6.10}$$

With this κ, the gap dependence of the tunneling current can be expressed as

$$I \propto \exp(-2\kappa s)\ . \tag{6.11}$$

According to the Tersoff–Hamann theory [34] of STM, the tunneling current is proportional to $|\psi|^2$ for the surface wave function ψ measured at the tip center. Then (6.11) indicates that the amplitude of the surface wave function outiside the surface decays as $|\psi| \propto \exp(-\kappa s)$. Thus κ represents the decay rate of the surface wave function into the vacuum region. For tunneling currents from surface filled states, 2κ is also the decay rate of the surface charge density outside the surface.

On flat surfaces, the surface wave function decays uniformly over the surface, and κ shows no lateral variation. On the other hand on corrugated surfaces, the decay rate of the surface wave function, or the charge density, is no longer uniform and becomes site dependent. The non-uniform decay of the surface wave function on highly corrugated surfaces can be understood by considering the contour map of the electronic charge density outside the surface schematically illustrated in Fig. 6.6. As seen in the figure, contour lines far from the surface are all straight lines parallel to the surface. However, near the surface, they become more wavy, reflecting the atom arrangement on the surface. On top of protruding sites, contour lines are more densely packed than on recessed sites. This means that the surface wave function decays

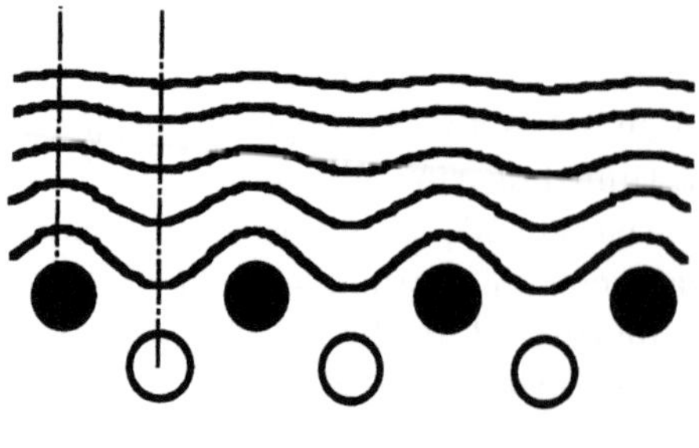

Fig. 6.6. Schematic of the charge density outside the surface. In the vacuum region, the charge density decays exponentially with increasing distance from the surface. The decay rate is, however, not uniform: the charge density decays faster on top of protruding atoms than on recessed sites

faster and hence ϕ_a becomes higher on these protruding sites. Because of this site dependent variation in ϕ_a, surface protrusions and depressions appear in BH images as bright and dark sites, respectively. A BH image, therefore, shows the same atomic-scale contrast as the topographic image of the surface.

The relation between the lateral variation in ϕ_a and the constant-current STM topograph can also be derived formally by decomposing the tunneling current into a uniform term $I_0(z)$ and a spatially varying term $\Delta I(x, y, z)$,

$$I = I_0(z) + \Delta I(x, y, z) . \tag{6.12}$$

The corresponding apparent barrier height then becomes

$$\phi_a(x, y) = \phi_{a0} + \left(\frac{\hbar\phi_{a0}}{2m}\right)^{1/2} \frac{\partial \Delta I(x, y, z)/I_0(z)}{\partial z} , \tag{6.13}$$

where ϕ_{a0} represents the uniform barrier height calculated from $I_0(z)$. The lateral variation in ϕ_a thus comes from $\Delta I(x, y, z)$ which is related to the constant-current topograph $z(x, y)$ through the following equation,

$$z(x, y) = -\frac{\Delta I(x, y, z)}{\mathrm{d}I_0(z)/\mathrm{d}z} . \tag{6.14}$$

It should be noted that the surface corrugation concerned here is that of the constant-current STM topograph, which does not necessarily agree with the actual surface geometry. For example, if an adsorbed atom is imaged as a depression in the STM image, it also appears as a dark site in the BH image, even though the adsorbed atom is protruding from the surface. The atomic contrast in BH images thus always agrees with that of corresponding STM images, as long as the apparent barrier height is determined by the decay rate of the surface wave function. However, this positive correlation between BH and STM images breaks down when there are local disturbances in the surface electronic structure. For example, step edges show dark BH contrast due to the depolarization effect at step edges [35]. Also electropositive alkali atoms on Si surfaces are imaged as dark sites in BH images while they appear as bright spots in STM images. It is these non-topographic features that make BH imaging a useful method for surface probing. Examples of these local BH contrasts will be shown and discussed in Sects. 6.5.1 and 6.5.2.

6.4.3 Artefacts in BH Imaging

In addition to the *intrinsic* BH contrasts discussed above, there appear some artificial contrasts in BH images. When the tip is located above a surface slope, for example, the tip displacement changes the gap distance by $\mathrm{d}s = \mathrm{d}z \cos\theta$, where θ is the tilting angle of the surface normal of the slope. Then from (6.4),

$$\phi_a = \phi_{a0} \cos^2\theta < \phi_{a0} , \tag{6.15}$$

where ϕ_{a0} represents the apparent barrier height on the flat part of the surface. In this case, ϕ_a becomes locally reduced on surface slopes. For the same reason, ϕ_a is reduced when the tip axis is tilted from the surface normal. This tip tilting, however, brings uniform reduction in ϕ_a and hence creates no additional lateral contrast in BH images.

A more serious problem of tip tilting is that it effectively generates tip dithering when the z-motion of the tip is modulated. When the tip axis is tilted by θ toward the x-direction, a z-modulation of amplitude Δz generates tip dithering of amplitude $\Delta z \sin\theta$ along the x-direction. Then a simple analysis shows that this dithering gives an apparent change in ϕ_a as $\Delta\phi_a/\phi_a = 2\theta(\partial z/\partial x)$, where $z(x,y)$ is a constant-current STM topograph. Hence, an artificial contrast which varies as $\partial z/\partial x$ appears in BH images. Since most surfaces have nearly sinusoidal corrugation profile, $\partial z/\partial x$ often coincides with $z(x,y)$, apart from a phase shift between them. Since the atomic-scale BH contrast agrees with $z(x,y)$, as described in Sect. 6.4.2, this artificial contrast proportional to $\partial z/\partial x$ appears quite similar to the intrinsic BH contrast. To discriminate this artefact, the BH image must be compared with the corresponding STM topographic image to check whether or not there is a phase shift between them. If there is an appreciable phase shift, the tip dithering effect becomes quite likely. It is therefore crucially important always to take BH images simultaneously with STM images. Note that tip tilting is not the only source of tip dithering. For example, the AC coupling between z and xy electrodes of the piezo scanner induces tip dithering when the z-modulation signal is applied to the z electrode. In this case, θ represents the strength of the coupling between scanner electrodes.

6.5 Applications of BH Imaging

6.5.1 Metal Surfaces

On low-index surfaces of metals, mobile conduction electrons smooth out the surface charge distribution to produce a flat surface. Therefore, the surface corrugation of most metal surfaces is very small compared to that of semiconductor surfaces. The STM surface corrugation of metals is typically 5–10 pm which is 1/20–1/30 that of Si(111)–7×7. The atomic BH contrast reflecting STM corrugation is also very small and hardly detectable. Consequently, BH imaging of metal surfaces only shows uniform contrast. The magnitude of ϕ_a on such a flat surface is expected to agree with the surface work function. Experimental values obtained on various metal surfaces [21] confirm this agreement, though relatively low values around 2 eV are found in some measurements [3,22]. Hasegawa *et al.* [4] have carried out BH imaging of Cu(111) and found that the measured ϕ_a (4.8 eV) agrees well with the work function of Cu(111) (4.93 eV). They also showed that the deposited Au layers on Cu do not have the same barrier height as the substrate. The BH image of the Au-deposited Cu surface is shown and compared with the

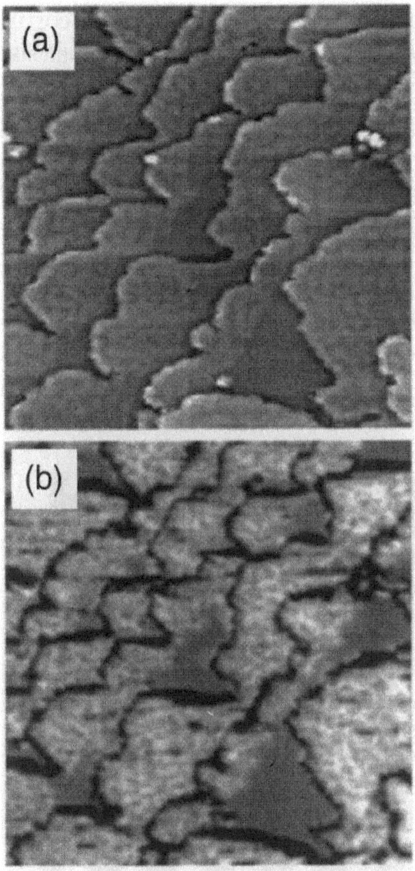

Fig. 6.7. (**a**) STM and (**b**) BH images of the Au-deposited Cu(111) surface. From [4]

STM image in Fig. 6.7. In the BH image, the bright area represents the Au monolayer on which ϕ_a increases by 7%. This difference reflects the fact that Au has higher work function than Cu. The BH image clearly shows that the Au layer grows from the step edge. In the corresponding STM image, however, the contrast difference between the Au layer and the Cu terrace is much weaker, and the Au layer can be identified only from the presence of a misfit-dislocation network. Similarly, Pd islands on Cu(111) can be easily identified via BH imaging [4]. These observations suggest the interesting possibility of using BH contrast to distinguish chemically different species on the surface.

In Fig. 6.7, the monoatomic steps are imaged as dark lines. This dark BH contrast is due to the depolarization effect of steps. On this surface, the reduction in ϕ_a at the step edge is 0.9 eV. Dark contrast at surface steps is also reported for other metal surfaces [3,35].

The Au(110)–2×1 surface is one of a few exceptional metal surfaces which have high corrugation amplitude comparable to that of semiconductor surfaces. The missing-row reconstruction of this surface results in alternating close-packed atomic rows and deep-trough atomic rows running along the

$[1\bar{1}0]$ direction. In the $[001]$ direction across rows and troughs, the corrugation amplitude is ~ 0.1 nm. Because of this large surface corrugation, the decay rate of surface wave functions on top of atomic rows differs greatly from that on troughs. This local variation in κ gives a BH contrast varying along the $[001]$ direction. Experimental BH images of Au(110)–2×1 clearly reveal such a contrast variation and they are in good agreement with the constant-current STM images of this surface. Figure 6.8 shows the BH image of Au(110)–2×1 [3]. Alternating bright and dark lines correspond to the$[1\bar{1}0]$ atomic rows and troughs, respectively. In this BH image, the atomic arrangement inside the close-packed atomic rows is resolved, showing even better resolution than STM images.

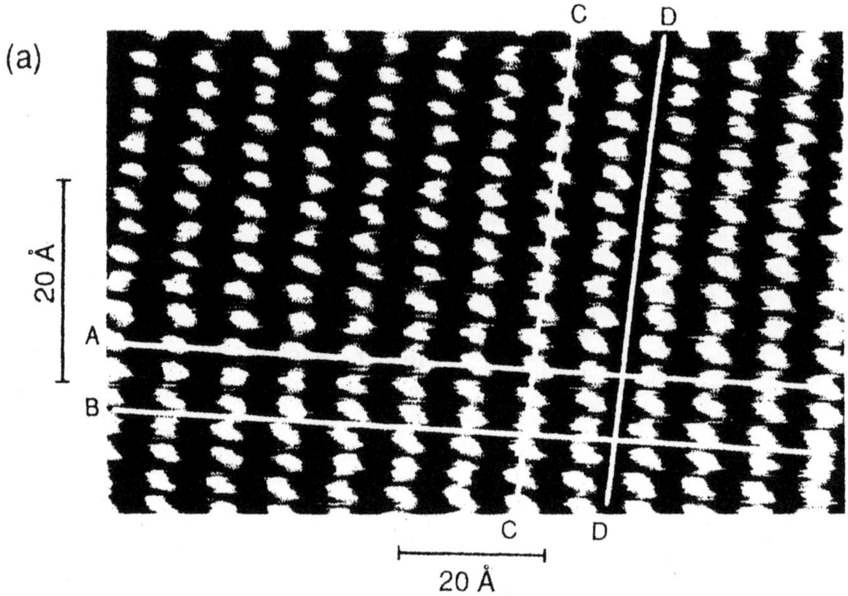

Fig. 6.8. Atom-resolved BH image of the Au(110)–2×1 surface. From [3]

On this surface, Schuster *et al.* [3] found an anomalously low barrier height $\phi_a = 0.1$ eV on troughs, which cannot be explained by the variation in κ. They suggested that the interpretation of BH image contrast in terms of κ cannot be valid on highly corrugated surfaces. Recent re-examination of the BH on Au(110)–2×1 by Olesen *et al.* [7], however, observed no such an anomaly in the barrier height. On atomic rows and troughs, the measured barrier heights are $\phi_a = 4.3$ eV and $\phi_a = 3.2$ eV, respectively. Their values are thus consistent with the variation in κ. This result re-establishes the validity of the decay rate interpretation of the atomic BH contrast. It is also worth noting that the corrugation-related variation in ϕ_a on the Au(110)–2×1 surface becomes as large as 1 eV.

6.5.2 Semiconductor Surfaces

Surface reconstruction of low-index semiconductor surfaces tends to increase the interatomic distance between surface atoms and enhance the surface corrugation. As in the case of Au(110)–2×1, the BH contrast of these corrugated surfaces is dominated by the lateral variation in κ. Consequently, BH images of semiconductor surfaces are essentially not much different from the corresponding STM images. The example of an atom-resolved BH image of Si(111)–7×7 [5] is shown in Fig. 6.9. The average barrier height in Fig. 6.9 is 3.1 eV, which is slightly lower than the average work function of Si and W but close to the reported value (3.5 eV) of the barrier height at large separations [16]. The 7×7 modulation in ϕ_a often amounts to 1 eV in amplitude. This spatial variation in ϕ_a is comparable to that observed on the Au(110)–2×1 surface and may not be exceptional on corrugated surfaces.

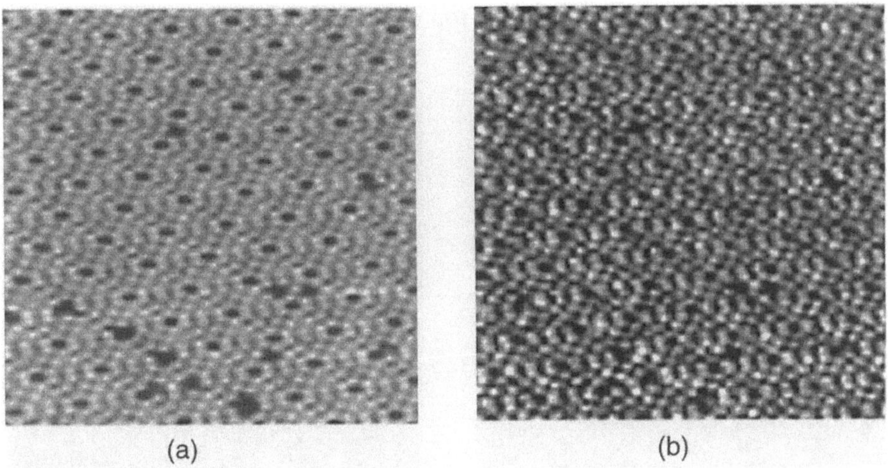

Fig. 6.9. (**a**) STM and (**b**) BH images of the clean Si(111)–7×7 surface. From [5]

On clean and defect-free surfaces, BH imaging gives no new information beyond that obtained with STM. This is partly the reason why BH imaging has so rarely been applied to semiconductor surfaces, where numerous STM studies have already been performed. BH imaging shows its unique capability when it is applied to surfaces which have defects and/or adsorbates. These surface defects or adsorbates generally induce local charge transfer to or from substrate atoms and generate surface dipoles at defect or adsorption sites. Such induced dipoles enhance or reduce the surface double-layer potential, depending on their orientation, and change the surface work function accordingly:

$$\Delta\phi = -ne\Delta qd/\varepsilon_0 \ , \tag{6.16}$$

where Δq is the amount of charge transfer, d is the length of the dipole and n is the average number of dipoles per unit area. Equation (6.16) describes the change in the work function averaged over the macroscopic surface area. However, it does not tell us how much change in the surface potential takes place just on the dipole sites and how far the change extends around these sites. By virtue of its atomic-scale spatial resolution, BH imaging is best suited to study these issues.

An example of BH imaging of an adsorbed surface is given in Fig. 6.10 which shows both the STM and the BH images of Si(111)–7×7 after exposure to 1 L of oxygen at room temperature [5]. It is well known [36,37] that a small amount of oxygen adsorbed on Si(111)–7×7 at room temperature produces both bright and dark sites. Bright sites are interpreted as Si adatoms with one oxygen atom inserted into the Si backbond. Further adsorption of oxygen on top of the Si adatom at a bright site passivates its dangling bond and turns the adsorption site into a dark site. Since adatoms at dark sites have no density of states near the Fermi level, they are imaged as depressions even though oxygen atoms are sitting on top of the surface Si adatoms.

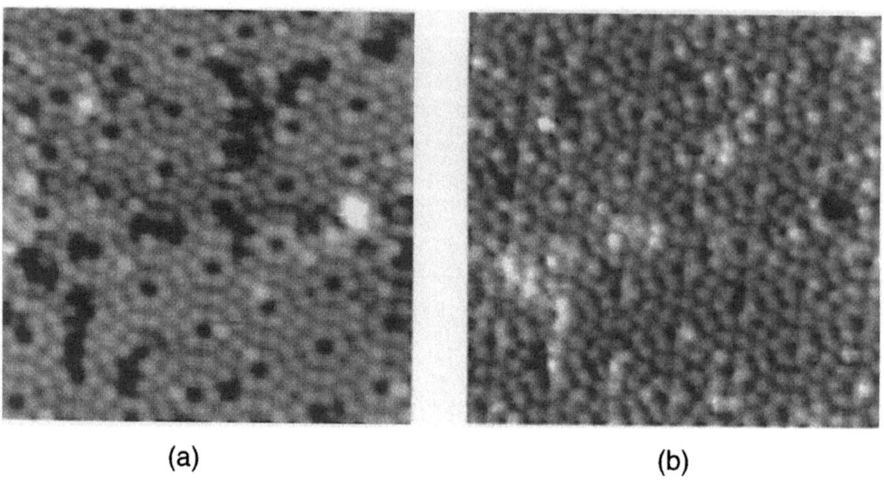

Fig. 6.10. (**a**) STM and (**b**) BH images of the Si(111)–7×7 surface after exposure to 1 L of oxygen at room temperature. Oxygen-reacted sites are imaged as higher barrier spots in the BH image. From [5]

In the BH image, on the other hand, both dark and bright sites show bright contrast and appear as high-barrier sites. Some dark spots remaining dark in the BH image are surface defects. Since the oxygen atom is more electronegative than Si, oxidation of surface Si atoms generates the inward-pointing surface dipole and increases the local surface potential according to (6.16). Observation of the high local barrier at O-reacted sites in Fig. 6.9 is thus consistent with the predicted charge transfer from Si to O. The relative

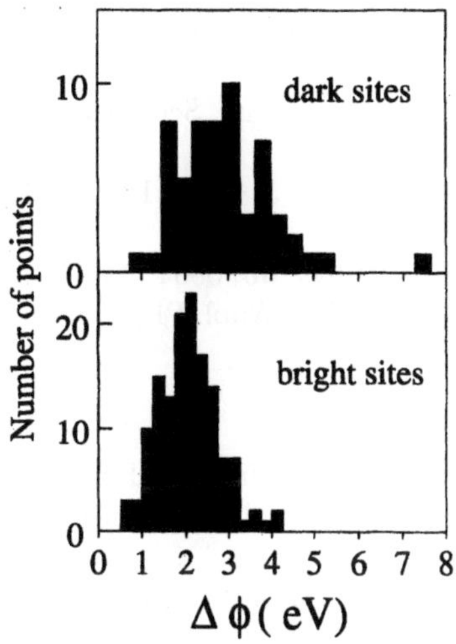

Fig. 6.11. Barrier-height increase $\Delta\phi$ measured at dark and bright sites on oxygen-adsorbed Si(111)–7×7 surfaces. From [5]

increase $\Delta\phi_a$ in the barrier height from the substrate value at dark and bright sites is shown in Fig. 6.11. Comparison of two distributions of $\Delta\phi_a$ clearly indicates that dark sites have slightly larger $\Delta\phi_a$ than bright sites. This result is also consistent with the adsorption configuration of the dark site where the oxidized Si bonds to two oxygen atoms and gives larger charge transfer than at the bright site, which involves only one oxygen atom [38].

As seen in this example, BH imaging of adsorbed and defected surfaces is quite useful for probing local charge transfer at defect/adsorption sites. On semiconductor surfaces, however, quantitative analysis of $\Delta\phi_a$ is obscured by the large corrugation-induced variation in ϕ_a which already exists on clean Si(111)–7×7. In the case of oxygen-adsorbed Si(111)–7×7, bright sites would show a positive $\Delta\phi_a$ even without any charge transfer, since these sites are imaged as protrusions in the constant-current STM topograph. The same argument shows that dark sites would have negative $\Delta\phi_a$ even if there were no dipole formation. Therefore, the oxygen-induced $\Delta\phi_a$ should assume higher values at dark sites and lower values at bright sites than is indicated in Fig. 6.10. Since the topographic contribution to $\Delta\phi_a$ cannot be measured from the adsorption-induced part of $\Delta\phi_a$, the quantitative study of $\Delta\phi_a$ cannot be made straightfoward on corrugated surfaces. Nevertheless, information about site-specific charge transfer cannot be easily obtained by other methods, and BH imaging will continue to be a useful and unique tool for probing local charge transfer on an atomic scale.

References

1. B. Marchon, P. Bernhardt, M.E. Bussell, G.A. Somorjai, M. Salmeron and W. Siekhaus, Phys. Rev. Lett. **60**, 1166 (1988).
2. J.S. Villarrubia and J.J. Boland, Phys. Rev. Lett. **63**, 306 (1989).

3. R. Schuster, J.V. Barth, J. Wintterlin, R.J. Behm and G. Ertl, Ultramicroscopy **42–44**, 533 (1992).
4. Y. Hasegawa, J.F. Jia, K. Inoue, A. Sakai and T. Sakurai, Surf. Sci. **386**, 328 (1997).
5. S. Kurokawa, M. Yuasa, A. Sakai and Y. Hasegawa, Jpn. J. Appl. Phys. **36**, 3860 (1997).
6. N. Horiguchi, K. Yonei and M. Miyao, Jpn. J. Appl. Phys. **36**, 3864 (1997).
7. L. Olesen, E. Lægsgaard, I. Stensgaard and F. Besenbacher, Appl. Phys. **A 66**, S157 (1998).
8. N.D. Lang, Phys. Rev. **B 37**, 10395 (1988).
9. K. Hirose and M. Tsukada, Phys. Rev. **B 51**, 5278 (1995).
10. S. Ciraci and E. Tekman, Phys. Rev. **B 40**, 11969 (1989); S, Ciraci, A. Baratoff, and I.P. Batra, Phys. Rev. **B 41**, 2763 (1990); E. Tekman and S. Ciraci, Phys. Rev. **B 42**, 1860 (1990).
11. N. Kobayashi, K. Hirose and M. Tsukada, Jpn. J. Appl. Phys. **35**, 3710 (1996).
12. G. Binnig, N. Garcia, H. Rohrer, J.M. Soler and F. Flores, Phys. Rev. **B 30**, 4816 (1984).
13. J.H. Coombs, M.E. Welland and J.B. Pethica, Surf. Sci. **198**, L353 (1988).
14. J.K. Gimzewski and R. Möller, Phys. Rev. **B 36**, 1284 (1987).
15. Y. Kuk and P.J. Silverman, J. Vac. Sci. Technol. **A 8**, 289 (1990).
16. C.J. Chen and R.J. Hamers, J. Vac. Sci. Technol. **A 9**, 230 (1991).
17. I.-W. Lyo and Ph. Avouris, Science **253**, 173 (1991).
18. R. Berndt, J.K. Gimzewski and R.R. Schlittler, Ultramicroscopy **42–44**, 528 (1992).
19. Ph. Avouris, I.-W. Lyo and Y. Hasegawa, J. Vac. Sci. Technol. **A 11**, 1725 (1993).
20. U. Dürig, O. Züger and D.W. Pohl, Phys. Rev. Lett. **65**, 349 (1990).
21. L. Olesen, M. Brandbyge, M.R. Sørensen, K.W. Jacobsen, E. Lægsgaard, I. Stensgaard and F. Besenbacher, Phys. Rev. Lett. **76**, 1485 (1996).
22. A.R.H. Clarke, J.B. Pethica, J.A. Nieminen, F. Besenbacher, E. Lægsgaard and I. Stensgaard, Phys. Rev. Lett. **76**, 1276 (1996).
23. G. Cross, A. Schirmeisen, A. Stalder, P. Grütter, M. Tschudy and U. Dürig, Phys. Rev. Lett. **80**, 4685 (1998).
24. J.A. Stroscio and R.M. Feenstra, Methods of Tunneling Spectroscopy. In Scanning Tunneling Microscopy, J.A. Stroscio and W.J. Kaiser (Eds.). Academic Press, San Diego (1993).
25. J.A. Stroscio, R.M. Feenstra and A.P. Fein, Phys. Rev. Lett. **57**, 2579 (1986).
26. B. Grandidier, J.P. Nys, D. Stievenard, X. de la Broïse, C. Delerue and M. Lannoo, Appl. Phys. **A 66**, S977 (1998).
27. D. Stievenard, B. Grandidier, J.P. Nys, X. de la Broïse, C. Delerue and M. Lannoo, Appl. Phys. Lett. **72**, 569 (1998).
28. J.C.H. Spence, U. Weierstall and W. Lo, J. Vac. Sci. Technol. **B 14**, 1587 (1996).
29. T. Shimizu, J.-T. Kim and H. Tokumoto, Appl. Phys. **A 66**, S771 (1998).
30. L. Ruan, F. Besenbacher, I. Stensgaard and E. Lægsgaard, Phys. Rev. Lett. **70**, 4079 (1993).
31. J. Winterlin, J. Wiechers, H. Brune, T. Gritsch, H. Hofer and R.J. Behm, Phys. Rev. Lett. **62**, 59 (1989).
32. M. Schmidt, H. Stadler and P. Varga, Phys. Rev. Lett. **70**, 1441 (1993).

33. R. Wiesendanger, Scanning Probe Microscopy and Spectroscopy, Sect. 1.12. Cambridge University Press, Cambridge (1994).
34. J. Tersoff and D.R. Hamann, Phys. Rev.**B 31**, 805 (1985).
35. J.F. Jia, K. Inoue, Y. Hasegawa, W.S. Yang and T. Sakurai, Phys. Rev.**B 58**, 1193 (1998).
36. Ph. Avouris, I.-W. Lyo and F. Bozso, J. Vac. Sci. Technol. **B 9**, 424 (1991).
37. R. Martel, Ph. Avouris and I.-W. Lyo, Science **272**, 385 (1996).
38. B. Schubert, Ph. Avouris and R. Hoffmann, J. Chem. Phys. **98**, 7593 (1993); ibid. **98**, 7606 (1993).

7 Mesoscopic Work Function Measurement by Scanning Tunneling Microscopy

Y. Hasegawa, J.F. Jia, T. Sakurai, Z.Q. Li, K. Ohno and Y. Kawazoe

Summary. In this paper, after a brief review of the mesoscopic (local) work function, we present our recent studies using Scanning Tunneling Microscopy. We discuss the thickness dependence of the work function for Au/Cu(111) and Pd/Cu(111), the potential profile on Au(111) and Cu(111) steps, and the work function on reconstructed S/Pt(111).

7.1 Introduction

The work function is one of the fundamental parameters describing the properties of material surfaces. It is defined as a minimum energy required for extracting one electron from the surface of a material. This parameter thus appears in explanations of many phenomena involving electronic emission and transfer through surfaces, including chemisorption, ion scattering, field emission and so on [1,2]. Since the beginning of this century, the work function has been discussed mainly in connection with thermionic emission, first observed by Edison in 1884. Research progress on the work function has been accelerated since the technical development of UHV and related surface characterization methods.

One of the difficulties with work function measurements arises from the fact that the work function is very sensitive to various atomic and nanometer scale features of the surface. For instance, it depends on the kind and amount of adsorbates on the surface. Alkali-metal adsorption is a typical example. Adsorbed in very small amounts, they greatly reduce the work function on metal and semiconductor surfaces. The work function also depends on how those adsorbates are arranged on the surface, i.e., whether they are randomly distributed or aggregated into clusters. Another relevant factor is the local and atomic structure and morphology of the surface, i.e., orientation of the surface, steps on the surface, and so on. Therefore, in order to understand the physics involved here, the work function must be measured locally on the nanometer/mesoscopic scale. In this paper, after a brief review of work function measurements, in particular, those with spatial resolution, we present our recent work function measurements with nanometer scale spatial resolution using Scanning Tunneling Microscopy (STM).

7.2 Work Function

7.2.1 How the Work Function is Related to Material Properties

There are electrons in materials, but they do not go out into the vacuum unless extra energy is supplied. At least the energy corresponding to the work function is required in order to allow one electron out of the material. The work function is thus often written as a difference in energy between the vacuum potential (V_{vac}) and the Fermi energy (E_F) of the material (Fig. 7.1). According to a simple model of the work function, there are two factors which attract electrons into materials. One is related to bulk properties of the materials, and the other to surface properties. Because of the latter, the work function itself depends on surface properties.

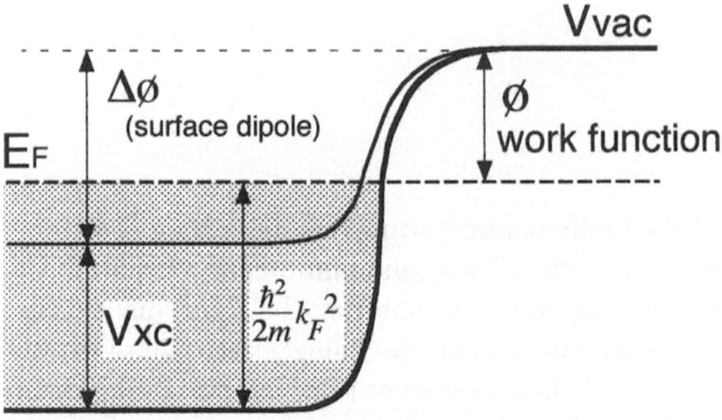

Fig. 7.1. Schematic of the electronic potential at a surface. V_{xc} is the exchange-correlation potential and $\Delta\phi$ is the potential due to an electrical dipole moment formed at the surface. The work function ϕ is described as a difference between the potential in vacuum (V_{vac}) and the Fermi energy (E_F)

The bulk component of the attractive potential is due to positive ion cores existing in the material. Almost all potentials due to ion cores are screened by other electrons, except for an area surrounding the electron. Because of Coulomb repulsion, other electrons are expelled from the site where the electron is located. In addition, electrons with the same spin are expelled further by the Pauli exclusion principle. As a result, the potential exerted on the electron is lowered by the remaining positive ion cores which are located at the area where other electrons are expelled (exchange-correlation potential V_{xc}). How electrons screen the potential or how electrons are expelled depends on bulk electronic properties of the materials. In the case of the simple model, it mainly depends on the electron density of the materials. Therefore, the work function also depends on those bulk properties.

On the other hand, another factor is related to surface properties. On the surface, electrons are spilt out of the area where positive ions are located,

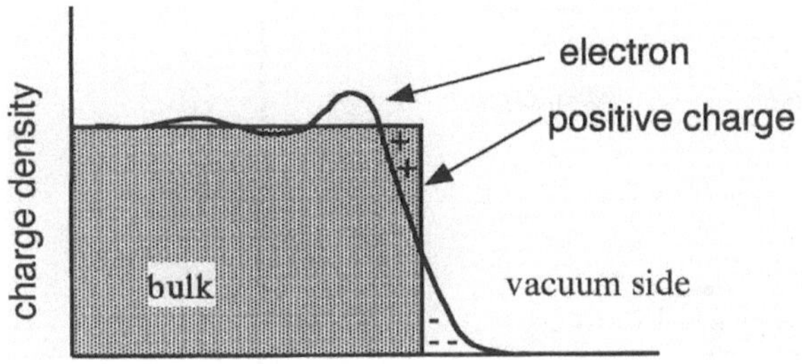

Fig. 7.2. Charge distribution at the surface. Electrons spill out into the vacuum to produce a dipole moment

and into the vacuum. This distribution of electrons forms a dipole moment with a positive charge on the bulk side and a negative charge on the vacuum side, as shown in Fig. 7.2.

The surface dipole produces a step-like potential which is low on the bulk side, contributing to the work function. According to the theory of Lang and Kohn [3,4], this component also depends on the electron density of the material. They demonstrated that calculated values of the work function, including the two components under the scheme of the jellium model (a model in which electrons are situated in a uniform background of positive charges) and the Local Density Approximation (LDA), fit very nicely with experimentally measured results (Fig. 7.3), except for its dependence on atomic scale properties of the surface.

The strength of the surface dipole changes slightly with surface properties. For instance, it depends on the atomic density of the surface. As can be seen from Fig. 7.4, when the surface is atomically corrugated, a dipole moment is formed in the opposite direction to the surface dipole. This is because the electron distribution over the surface is smeared in order to reduce the electron kinetic energy. This is known as the Smoluchowski effect [5]. As a result, lower work functions are expected on sparse surfaces than on closely packed surfaces. This is consistent with experimental results. In the case of Cu, an fcc metal, the closed-packed (111) surface has the highest work function, whilst (110) has the lowest among the three basic surface planes.

Based on the work function scheme explained above, we can understand why various features of materials affect values of the work function. Firstly it depends on the elements the material is composed of. This is due to the bulk contribution discussed above. In addition, it depends on the surface orientation at the point where the electron is emitted, also explained above. Besides these factors, adsorption on the surface changes its work function. For instance, deposition of alkali metals on surfaces reduces their work functions, whereas exposure to oxygen generally raises them. These effects are due to the formation of dipole moments on the surface with the opposite/same direction relative to the surface dipole. Finally, the work function depends on the flatness of the surface. For instance, surface steps reduce the work func-

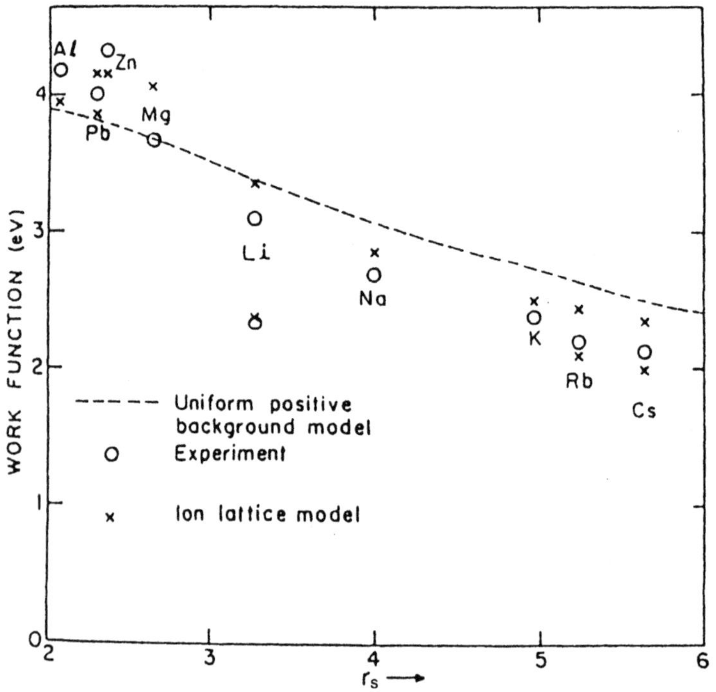

Fig. 7.3. Comparison of experimentally measured work function with that calculated by Lang and Kohn using the jellium model. r_s is a radius (in units of the Bohr radius) for the volume occupied by one electron in the material. Small r_s means high electron density. From [3]

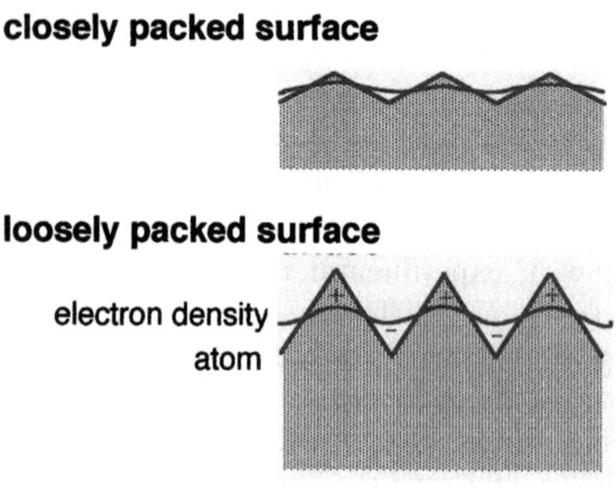

Fig. 7.4. Since the electron distribution is smeared in order to reduce its kinetic energy, a dipole moment is formed on corrugated surfaces whose direction opposes the surface dipole, thereby reducing the work function. This is called the Smoluchowski effect

tion. This is explained by the same smearing electron density used to explain the dependence on surface atomic density. Around step edges, the electron density is depleted on the upper side whilst accumulating on the lower side (Fig. 7.5). As a consequence, a dipole moment is formed with opposite polarity to the surface dipole, resulting in reduction of the work function. The lower work function on stepped surfaces was observed in experiments done by Wagner et al. using samples with vicinal surface (Fig. 7.6) [6–8].

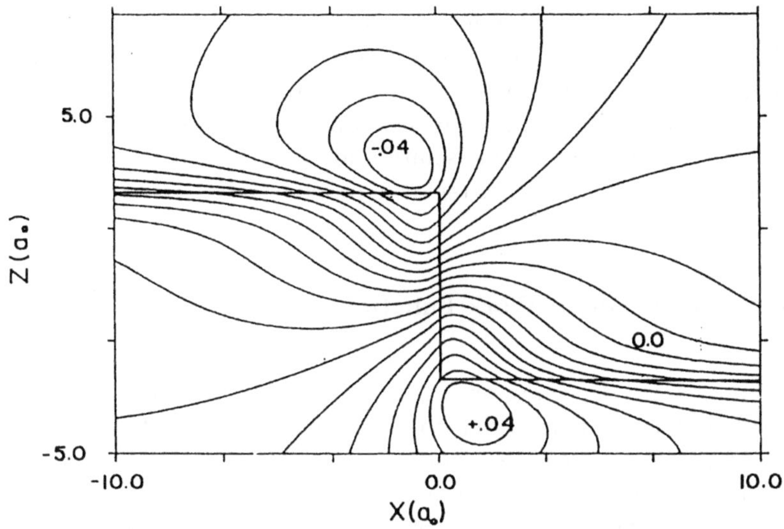

Fig. 7.5. Electron distribution around step edges calculated with a jellium model. From [38]

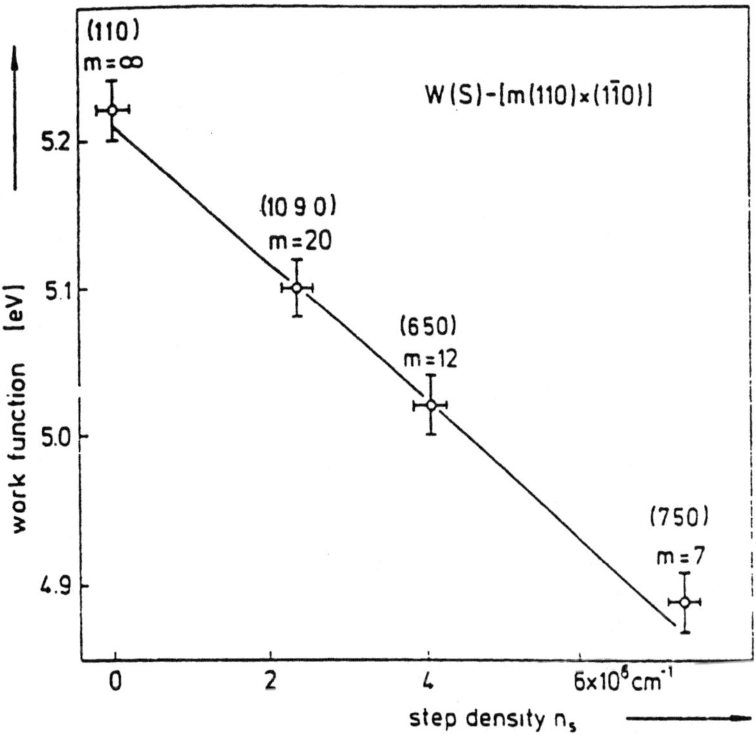

Fig. 7.6. Work function measured on a tungsten vicinal surface. Proportional decrease in work function with step density is clearly demonstrated [6]

It is thus clear that the work function is related to atomic scale surface features. However, it also depends on nanometer scale features, such as step bunching and clustering of adsorbates. Since it is often very difficult to characterize and control surface features on such scales, ambiguity still remains in work function measurements. One solution for avoiding such problems is to measure it locally using a nanometer scale probe.

7.2.2 Mesoscopic (Local) Work Function

When alkali atoms are deposited on a material surface, the work function of the surface decreases. This is due to dipole moment formation by the adsorbates. At initial coverage, as shown in Fig. 7.6a, the work function drops at a constant rate because each adsorbate has a constant amount of dipole moment and the number of adsorbates (dipole moments) increases with coverage. Around 0.7–0.8 ML the rate slows down and then the work function begins to reach saturation. The slowing down of the reduction rate is explained by interactions among the adsorbates. The work function variation mentioned above is measured using conventional work function measurement techniques, providing a value averaged over the surface. If the work function were measured locally, results would be completely different. They would depend on the site at which it is measured (Fig. 7.7b). If measured on the alkali adsorbates, the work function should be at its lowest value at initial coverage and increase when depolarization effects start to occur (curve A in Fig. 7.7b). On the other hand, the work function measured locally on the bare substrate should start with the same value as on the clean surface, then decrease as coverage increases and the nearest alkali atoms come closer to the measured site (curve B). The work function measured locally on a nanometer scale behaves quite differently to the averaged macroscopic work function.

As is mentioned before, the work function is defined as $V_{\mathrm{vac}} - E_{\mathrm{F}}$. Since V_{vac} is constant in vacuum and E_{F} is constant in the material, the work function should not depend on the site over the material surface. However, this is only the case when the work function is considered macroscopically. In fact, if the potential of some part of the surface is lowered, the effect of the lowered potential is smeared out with distance from the surface (Fig. 7.8) while the potential rises gradually as $1/s$ due to the image potential [2]. The value we focus on in this review is the work function defined as an energy difference between the vacuum potential, when we assume the whole surface to be covered by the area, and the Fermi energy E_{F} (ϕ'_1, ϕ'_2 and ϕ'_3 in Fig. 7.8). This is closely related to the surface potential, which varies spatially on the atomic scale. Since it relates the conventional macroscopic work function to an atomic scale surface potential, we call it the mesoscopic work function.

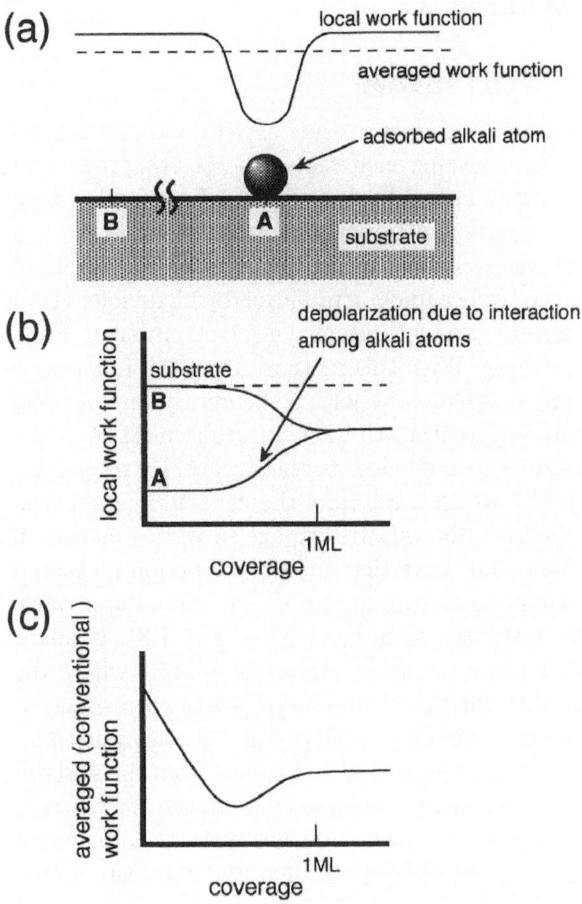

Fig. 7.7. Difference between macroscopic and local work functions, in the case of alkali metal adsorption. (**a**) The local work function drops only above adsorbates while remaining constant over the substrate. The macroscopic work function is an average of the adsorbate and substrate work functions. (**b**) Local work function with a coverage of deposited alkali metal: *curve* A: on the adsorbates and *curve* B on the substrate. (**c**) Averaged macroscopic work function as a function of coverage. Note the minimum although neither *curve* A nor *curve* B in (**b**) have minima

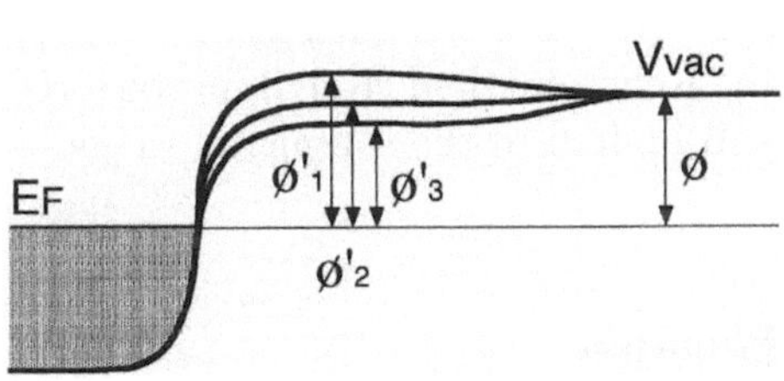

Fig. 7.8. Relation between mesoscopic and macroscopic work functions. Variation of the mesoscopic work function (ϕ_1', ϕ_2' and ϕ_3') is smeared out at a distance to become a macroscopic work function. In order to detect the mesoscopic work function, a probe has to be located close to the surface

7.3 Experimental Techniques

7.3.1 Conventional Macroscopic Techniques

Work functions have been measured using two kinds of method. One detects electrons emitted from the surface by some excitation, and the other measures the contact potential, which is due to the difference in work function between the sample surface and probe materials. A typical example of the former uses photoemission. Work function values can be measured from the threshold of electron energy emitted from the surface under irradiation by ultraviolet (UV) light. An electron beam can also be used for inducing electron emission. Field emission is included in this category. With this method, the work function is estimated by fitting the current vs. field curve with a theoretical one. Among those methods using the contact potential, the Kelvin probe method is the best known. The probe electrode is located close to the surface. By measuring the voltage applied on the probe which minimizes the capacitance between sample and probe, one can measure the relative change in work function of the sample surface. It should be noted here that the work function measured using photoemission is not equal to the minimum of the mesoscopic work function over the surface (for instance, ϕ'_3 in the case of Fig. 7.8). It might be thought that electrons first come out from materials at sites where the surface potential is lowest, so that the threshold energy would correspond to that value. This is not true, however. As illustrated in Fig. 7.8, the potential is averaged over the surface (with some weights) at a distance from the surface, and electrons start to emit when the energy becomes equal to the macroscopic work function ϕ, rather than ϕ'_3. More discussion on this matter can be found in [2]. Conventional techniques to measure work functions have no spatial resolution and thus provide only averaged information over the probed area of the surface. However, the field emission pattern gives us information about spatial variation over the surface of a very sharp needle. Attempts to measure the work function of a small area on a flat surface have been pursued using scanning techniques with focused local electron/UV probes [9] or imaging techniques such as PhotoEmission Electron Microscopy (PEEM) [10]. Spatial resolution of these methods goes down to the micrometer scale, insufficient to reveal work function properties related to atomic and nanometer scale features on surfaces. Recently, the tip of an Atomic Force Microscope (AFM) has been used as a probe electrode for the Kelvin method (Kelvin-probe force microscope) [11], but its resolution is still limited to the submicron range.

7.3.2 Experimental Techniques for the Mesoscopic Work Function

In order to measure the mesoscopic work function, we need to use probes which can be located very close ($\leq$ 10 Å) to the surface. [Not too close, however. It should be far enough away ($\geq$ 2 Å) to ensure that the potential due to the surface dipole reaches saturation.] A few techniques have been developed to measure it. One of the techniques uses adsorbed xenon atoms as probe. By measuring the energy level of Xe atoms (5*p* level) adsorbed on the surface using photoemission spectroscopy, one can measure the potential the Xe atoms feel on the surface. This technique is called PAX (Photoemission from Adsorbed Xenon) [2,12]. Image states, which are electronic states bound just above the surface by the image potential, are also used for estimating the mesoscopic work function. Using 2-Photon PhotoEmission spectroscopy (2PPE), which can be utilized to measure unoccupied states, energy levels of image states are measured. From an extrapolation of their energy levels, the potential in vacuum (V_{vac}) can be estimated [13,14]. Since the xenon interatomic distance is ~ 3 Å and image states are confined within a range of less than 10 Å from the surface, both provide the above-mentioned mesoscopic work function, although they do not have any spatial resolution.

7.3.3 Work Function Measurement by STM

Scanning Tunneling Microscopy also uses a probe tip which is located very close to the surface. It can thus be used to measure the mesoscopic work function. The height of the potential barrier between the tip and sample where electrons tunnel is closely related to the sample work function (see Fig. 7.9). According to tunneling theory, the height of the tunneling barrier or work function can be obtained by measuring how the tunneling current responds when the gap distance changes. If it is measured whilst scanning the tip over the surface, an image can be made. Compared with PAX and 2PPE, which have no spatial resolution in principle, STM enables us to image the spatial variation of the mesoscopic work function [15,17]. As will be described in detail later, the mesoscopic work function image can be made simultaneously with the corresponding topographic image. From comparison of these images,

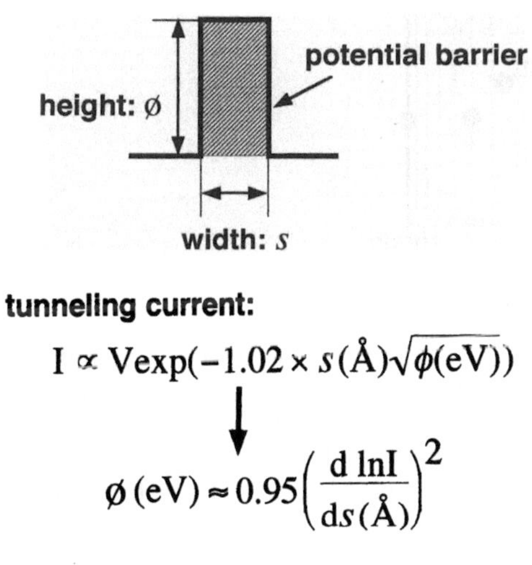

Fig. 7.9. Work function measurement using tunneling current

useful information can be directly obtained about the work function which would not be accessible with other techniques. A simple quantum theory of electron tunneling tells us that the tunneling current varies with gap distance according to $I \propto \exp(2\sqrt{2m\phi/\hbar})$, where ϕ is the tunneling barrier height, which corresponds to the work function, m is the electron mass, and $\hbar$ is Planck's constant. A value of the work function is then calculated simply using $\phi = 0.95(\Delta \ln I/\Delta s\ [\text{Å}])^2$ [15,16]. By applying a sinusoidal voltage on the z-axis piezo in order to modulate the tip–surface distance s and then detecting a tunneling current response at the same frequency with a lock-in amplifier, the value of $\Delta \ln I/\Delta s$ can be measured experimentally. The modulation frequency should be higher than the cutoff frequency of the feedback loop of the STM system ($\sim$ 1 kHz) but lower than the response frequency of the STM current amplifier. In our experiments, which will be described later, it was set at 2.0 kHz. Figure 7.10 is a plot of the output frequency dependence of a lock-in amplifier measured for a Cu(111) surface in UHV. At low frequency, the feedback circuit responds to the modulation, negating its effect. At high frequency, the output is reduced because of the limited response of the current amplifier. 2.0 kHz falls in a plateau range, indicating that it is a suitable value. At small distances from the surface ($\leq$ 10 Å), the image potential cannot be neglected. Since the STM tip is located around 5 Å from the surface, the barrier height in the gap is lowered relative to the surface work function. Because of the image potential, which varies inversely with distance, the potential at distance s from the surface becomes $\phi(s) = \phi_\infty - \alpha/s$, where $\alpha \sim 10$ eV Å and ϕ_∞ is the work function. However, STM does not measure $\phi(s)$ at the tip position or the maximum of

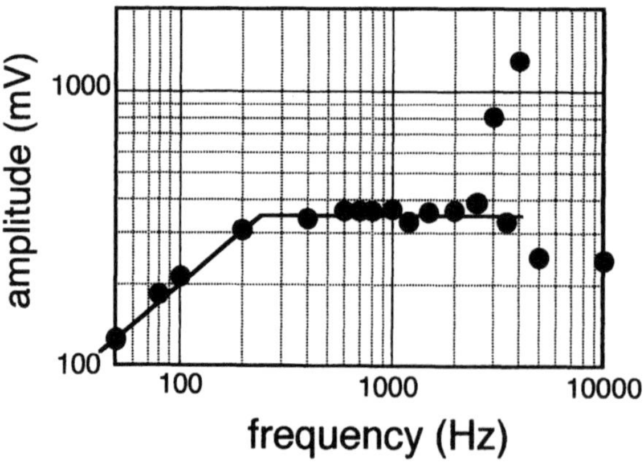

Fig. 7.10. Amplitude of detected tunneling current modulation when a modulated wave is applied on the z-axis piezo, measured as a function of modulation frequency. When the frequency is low, the feedback circuit responds to the modulation, negating its effect. At high frequency, a current amplifier used in this experiment cannot follow because of its limited response. A peak observed around 3–4 kHz is due to resonance of the current amplifier

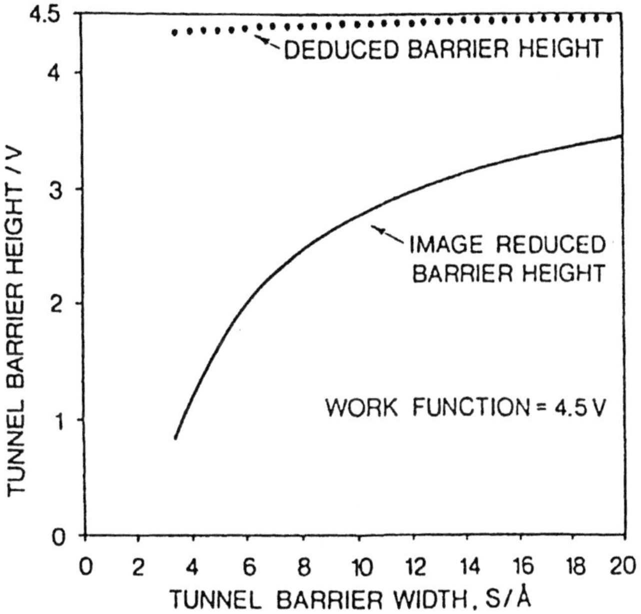

Fig. 7.11. Effect of image potential on the (calculated) STM work function measurement. *Solid line*: maximum height of potential barrier, which collapses at small gap distances. *Dotted line*: deduced barrier height or work function evaluated using the $\Delta(\ln I)/\Delta s$ method. The image potential is included in a calculation of the tunneling current I. Although the height of the tunneling barrier falls, in particular at short distances, the deduced barrier height is close to the work function of the surface

ϕ but rather the value calculated from the $\Delta(\ln I)/\Delta s$ measurement. Fortunately, as Binnig et al. have shown [18], at least in the range that concerns us here, the s dependence in the potential enters only to second order. The image potential lowers the barrier height, apparently reducing the response of the tunneling current with gap distance. The apparent reduced response is, however, compensated by the variation of barrier height with gap distance through the image potential. As a result, the work function obtained with the $\Delta(\ln I)/\Delta s$ measurement gives the surface work function accurately in the first order approximation (Fig. 7.11) [19]. Theoretical studies considering details related to the STM configuration, together with experimental studies, also support this result [19–22].

Another possible factor which might cause deviation is a bias voltage. In STM, a bias voltage has to be applied between the tip and sample in order to produce the tunneling current. The applied voltage also modifies the shape and height of the tunneling barrier and thus also the value measured using the $\Delta(\ln I)/\Delta s$ method. We measured the work function using this method for various values of bias voltage in order to check its effect. The result using a Cu(111) surface and (111) oriented tungsten tip is plotted in

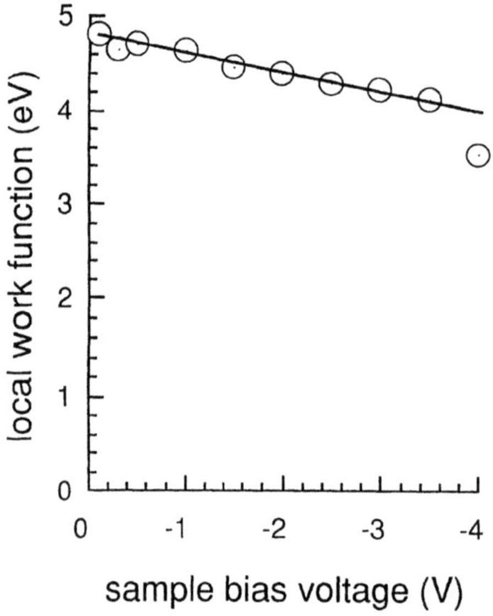

Fig. 7.12. Dependence of STM measured work function on a bias voltage applied to a Cu(111) sample

Fig. 7.12 [23]. From these measurements, it turns out that the measured work function gradually drops as the bias voltage increases to -3.5 V, with a ratio of ~ 0.2 eV/V. At the low bias voltage limit, it reaches around 4.8 eV, close to an average value for the work function of Cu(111) and W(111). The effect of the bias voltage can thus be compensated for by extrapolation.

In our experiment, a sinusoidal voltage is applied on the z-axis piezo in order to modulate the gap distance. When the surface is tilted from the z direction, the modulated amplitude does not completely contribute to the gap distance. As a result, the measured work function is apparently lowered by a factor of $\cos q$ (Fig. 7.13) [24]. The apparently lowered work function appears at areas where the surface is tilted locally, such as steps. This so-called topographic effect, however, can be estimated by measuring the tilting angle from the corresponding STM image and can subsequently be eliminated numerically.

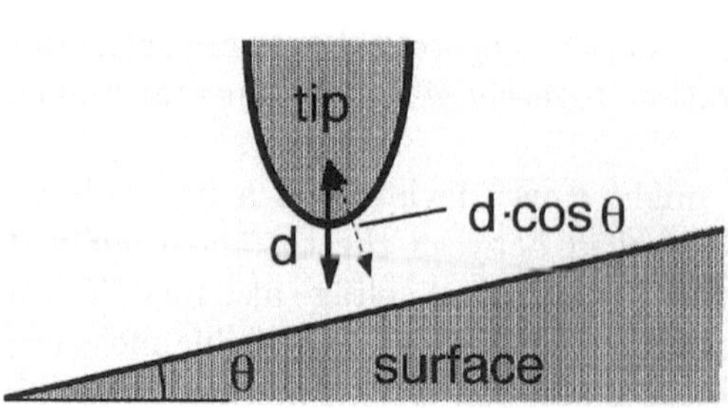

Fig. 7.13. Topographic effect in work function measurements using STM. When the surface is tilted with respect to the direction of the z-axis piezo, the modulation amplitude in the gap distance is reduced to $d \cos \theta$. As a result, the measured work function is reduced apparently by a factor of $\cos^2 \theta$

7.4 Results

7.4.1 Experimental Aspects

In this section, we review our recent experimental results on the mesoscopic work function using STM with Au/Cu(111), Pd/Cu(111) and S/Pt(111) surfaces. All experiments were carried out with our home-made UHV-STM system, using a base pressure $< 1.0 \times 10^{-10}$ torr [25]. The Cu(111) and Pt(111) samples were cleaned in situ by repeated cycles of Ar bombardment and annealing. Gold and palladium were evaporated onto the unheated Cu(111) surface from an evaporator consisting of a Au/Pd wire wrapped in Ta foil. Sulfur adsorption was achieved by exposing the clean Pt(111) substrate in an H_2S gas atmosphere (< 10 L). All STM images were obtained in constant-current mode (0.1–1.0 nA) at room temperature. A single crystalline (111)-oriented tungsten tip was used as STM probe. For the work function measurement [23,26], we applied a modulated voltage to the z-axis piezo with a frequency of 2 kHz. The amount of modulation in the gap distance is 0.23 Å much smaller than the gap distance 5.5–6.0 Å which is estimated from the tunneling current (0.1 nA) and bias voltage (−2.0 eV on the sample). When the gap distance is quite small, the measured work function becomes small because of collapse of the tunneling barrier due to the image potential [27,28]. The gap distance we used in this experiment is, however, much larger than those distances at which the current is saturated and thus the measured work function is reduced.

7.4.2 Thickness Dependence of the Work Function: Au and Pd on Cu(111)

Thin film formation is one method for creating unique materials. By just depositing a few monolayers of some element, the characteristics of a material can change. A question arises here: how many layers are needed to recover the same properties as in the bulk material? If something is deposited on a surface, the work function changes relative to the original surface. By further deposition, it should eventually reach the work function of the deposited material. How does the work function change from that of the substrate to that of the overlayer? Does it change gradually, abruptly or in an oscillatory manner?

We studied work functions of Au/Cu(111) and Pd/Cu(111) systems using STM in order to resolve these questions. Figure 7.14a shows an STM image taken on the Au/Cu(111) system, and Fig. 7.14b shows a work function image taken simultaneously with the STM image. Figure 7.15 is the magnified image of both topography and work function with a cross-section profile of the work function. The Au coverage is approximately 0.8 ML. It is known that alloy formation does not occur at room temperature [29]. The STM image shows that gold layers nucleate and grow from the lower step edges of the Cu

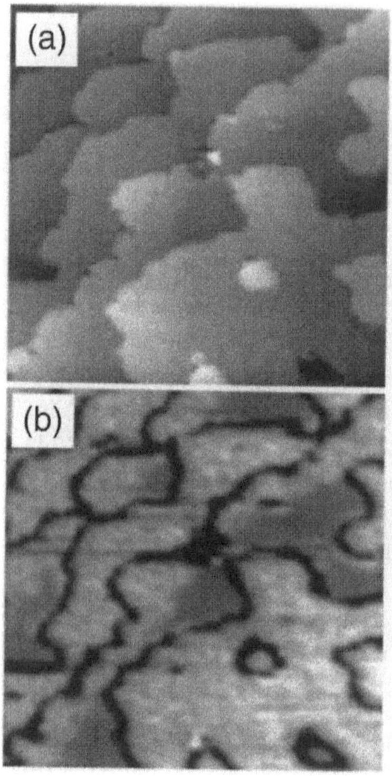

Fig. 7.14. (**a**) STM image and (**b**) work function image obtained on a Au/Cu(111) surface. Applied sample bias voltage is −2.0 V and tunneling current is 0.1 nA. The size of the observed area is ∼ 500 Å ×500 Å. The Au coverage is about 0.8 ML. The Au layer nucleates and grows at the lower side of step edges, making the (111)-oriented plane. The Au layer is imaged slightly brighter than the Cu substrate in the STM image (**a**) because of its larger lattice size. In the work function image (**b**), brighter imaging corresponds to a higher work function. It shows that the Au layer has the higher work function and that the step edges have lower work function

substrate, forming the (111)-oriented plane (Fig. 7.16). Because of the large lattice mismatch between the Au layer and Cu substrate [distances between atoms are Au(111): 2.88 Å, Cu(111): 2.55 Å], a network of misfit dislocations forms on the Au layers, demonstrated by the periodically arranged triangular dark lines in the STM image (see Fig. 7.15). The spacing of dislocation lines corresponds to 15 to 18 times the Cu(111) lattice, implying that the Au layer is compressed by 5–6% compared with bulk Au. The distinct contrast between the Au layer and Cu substrate is observed in the corresponding work function image (Fig. 7.14b), indicating that the work function is larger on the Au overlayer than on the Cu substrate. Step edges show dark contrast in the work function image, which is discussed later.

Similar measurements have been performed for the Pd/Cu(111) system. Figure 7.17 shows STM and work function images for the Pd/Cu(111) system. The growth of Pd on Cu(111) is similar to the Au case, although Pd overlayers are more dendritic. A similar network of misfit dislocations is observed on the Pd layer but it is not so ordered as in the case of the Au layer. From the periodic spacing of the misfit dislocation, the Pd overlayer is estimated to be compressed by only 2–3%. In the STM image of Fig. 7.17a, islands of the first Pd layer form on the wide Cu substrate terraces and several small islands of the second Pd layer form on the first Pd layer. In the corresponding

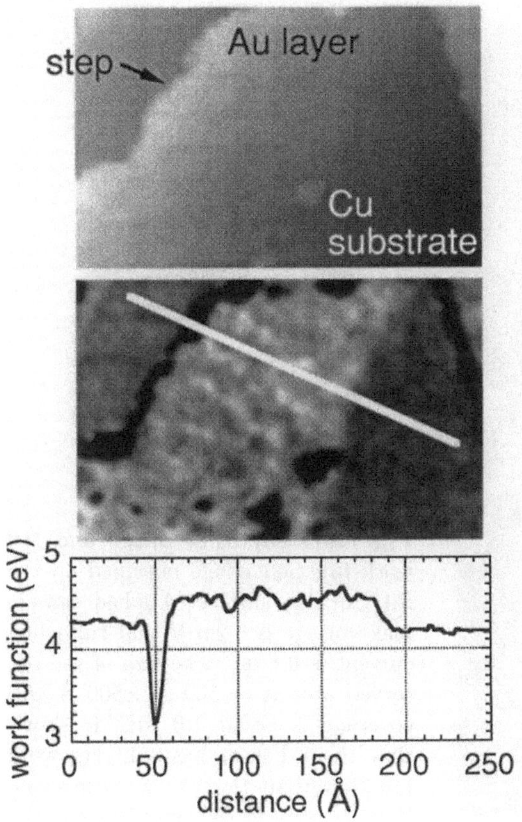

Fig. 7.15. Magnified STM image (*top*), work function image taken at the same time (*center*), and a plot of the work function measured along the *white line* drawn in the work function image

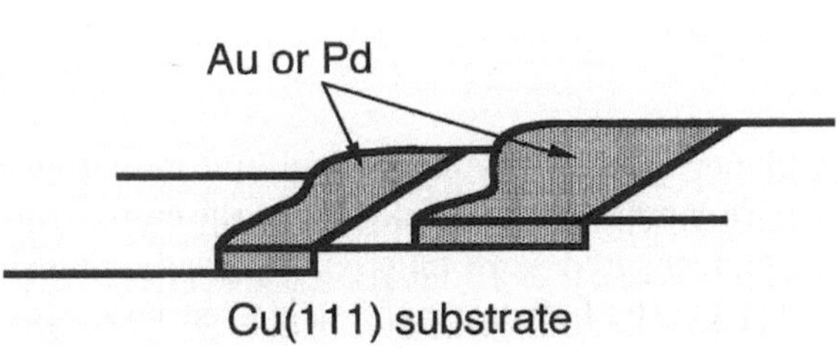

Fig. 7.16. Schematic of Au/Pd growth on Cu(111) substrate. The overlayer grows from the lower side of substrate step edges

work function image (Fig. 7.17b), islands of the second Pd layer are imaged brighter than the first Pd layer. For quantitative analysis, more data were collected from the multilayer Pd growth and the results are summarized in Fig. 7.18 together with the reported values for Cu(111), Au(111) and Pd(111) surfaces [14]. (All the measured data points plotted in Fig. 7.18 are shifted upward by 0.4 eV in order to compensate for the bias voltage effect.) For the Au/Cu(111) system, Wallauer et al. have already studied the thickness dependence of the work function using 2PPE [29]. These data are also included in the figure. Our data for the first and second Au layers are in agreement

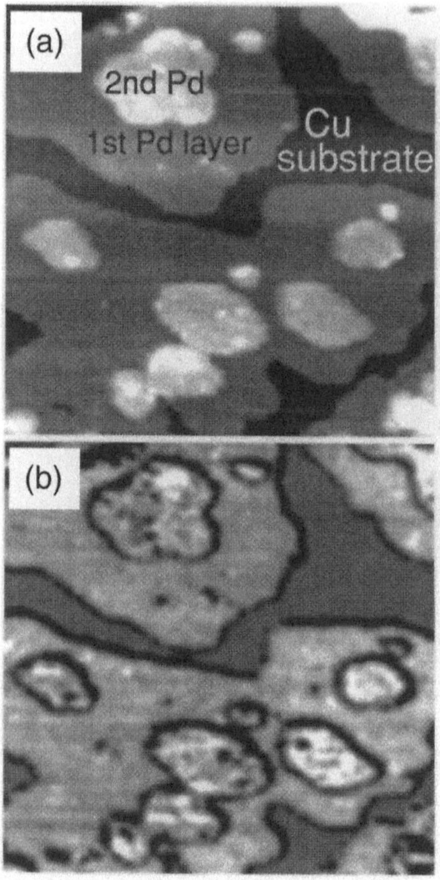

Fig. 7.17. (**a**) STM image and (**b**) work function image obtained on the Pd/Cu(111) surface. Applied sample bias voltage is −2.0 V and tunneling current is 0.1 nA. The size of the observed area is ∼ 500 Å ×500 Å. Au coverage is about 1.0 ML. It shows that the Pd layer has a higher work function than the Cu substrate and that the second Pd layer has an even higher work function than the first Pd layer

with the data obtained by Wallauer et al., showing gradual and monotonic increase toward the bulk value with increasing Au thickness. In the case of the Pd/Cu(111) system, however, the measured work function is already almost the same as that of bulk Pd(111) for the first layer and it is even larger for the second and third Pd layers before reaching the bulk value at the multiple (more than 10 ML) Pd layer.

We believe that this is the first experimental evidence that the work function of the metal adsorbate exhibits overshooting behavior. All previous work so far reported has concluded that the work function changes monotonically to reach the bulk value [Ag on Pd(111), Au(111) and Cu(111) substrates, and Au on Pd(111) and Cu(111) substrates] [13,14,29]. The present observations are distinctly different from those for Au and others. In order to understand our experimental observations we have performed a theoretical study using a first-principles total energy calculation [30] employing ultra-soft pseudopotentials [31] for Cu, Au and Pd. This approach is highly successful in calculating observed bulk Cu, Au and Pd properties: the difference between theory and experiment lies within 1% for the lattice constant. Slabs are used for modeling the surface with 8 layer bulk slabs and 5 layer vacuum. Atoms in the top and bottom three layers are allowed to relax while those in the two layers at the center are fixed in bulk positions. The work function is calculated by taking the difference of the Fermi energy and electrostatic potential in vacuum. Details of the calculation method will be published elsewhere [30]. The results of the calculation successfully reproduce the characteristic features of our experimental results: overshooting behavior of Pd overlayers and gradual increase of work function for Au overlayers, although they tend to show larger values than were found experimentally (Fig. 7.18).

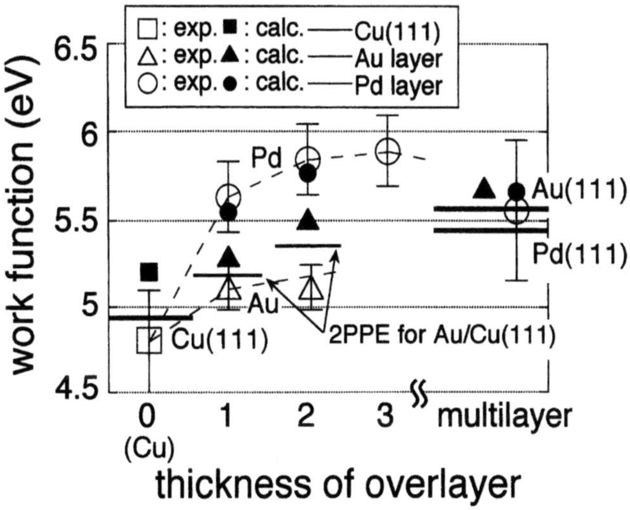

Fig. 7.18. Summary of the experimental and theoretical work function as a function of overlayer thickness. The data obtained by 2PPE [14] are also plotted for comparison, including those of each Au layer on Cu(111) [29]. *Open* and *solid symbols* are used for results of experiments and calculation, respectively

Two possible mechanisms can be considered to account for overshooting behavior in the Pd/Cu(111) system. First, the different atomic density in the overlayer from those of nominal surfaces may account for it. It is known, by the electronic smoothing effect proposed by Smoluchowski [5], that a densely packed surface exhibits a higher work function than a loosely packed surface. If the atomic density in the overlayer is larger than in bulk due to compression, the work function may then become higher and overshooting behavior can be understood in this way. From the STM images, we can determine how much the Pd overlayer is compressed laterally from the nominal surface, and it is concluded that compression is only 2–3%. It is not large enough to explain the observed overshooting behavior, according to the result of the LDA (Local Density Approximation) calculation for bulk materials [3]. Considering that the Au layer has a larger compression factor than the Pd layer and yet does not exhibit overshooting, it is highly unlikely that the observed overshooting behavior could be due to surface compression in the Pd layer.

The other possibility we may consider is a modification of electronic states in thin films, i.e., a quantum scale effect, since the film thickness of the overlayer is smaller than the Fermi wavelength of the metal. Schulte [32] calculated the work function of a thin film suspended in vacuum as a function of thickness, using LDA and the jellium model [34]. He concluded, as shown in Fig. 7.19, that the work function oscillates with periodicity equal to the Fermi wavelength as the film thickness increases, and that the oscillation decays when the work function approaches its bulk value. Under this scheme, the work function of a thin film can be larger or smaller than that of the bulk material, depending on the relationship between thickness, Fermi wavelength

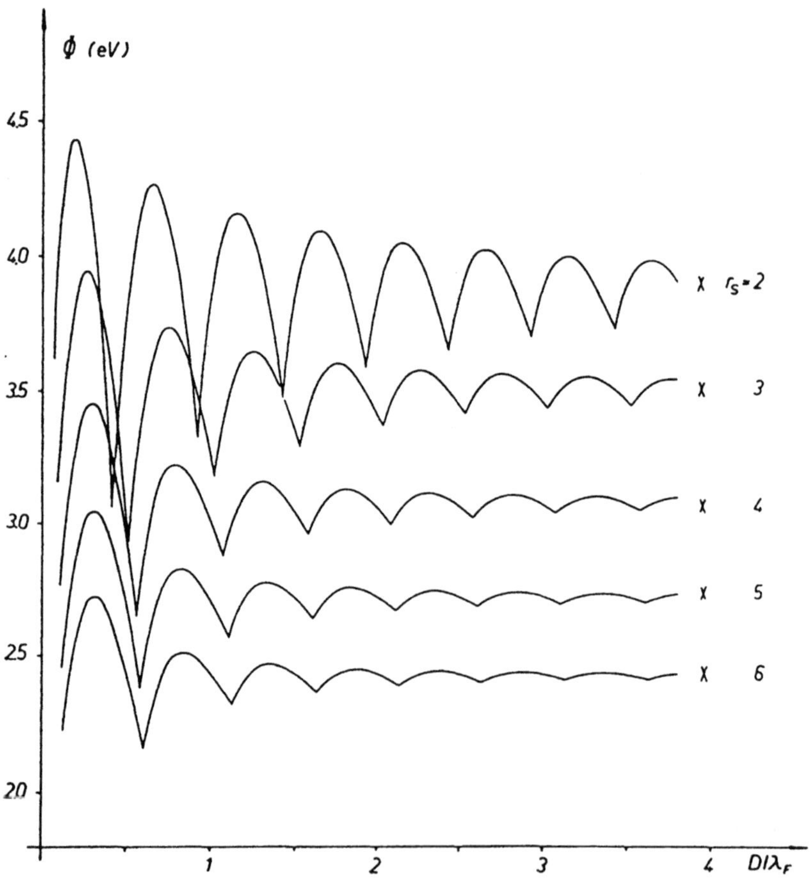

Fig. 7.19. Work function a of self-supported thin film as a function of its thickness, calculated using a jellium model and LDA approximation [32]. It demonstrates that the work function oscillates with thickness

and charge density. Au and Cu are both noble metals and thus have similar electronic structure. On the other hand, Pd is a transition metal and its electronic structure is shifted to higher energy compared with Au or Cu. Hence, a band discontinuity occurs at the interface of Pd and Cu and may result in modulation of the density of states within the overlayer. Indeed, our calculation for the $Pd_2/Cu_6(111)$ system shows that the *d* state level which contributes most to the Fermi level is greatly enhanced in the outermost Pd layer [peak P in curve (a) of Fig. 7.20], while that of the underlying Pd layer is depleted [curve (b) in Fig. 7.20]. This change in the density of states is reflected in the modulation of the work function plotted in Fig. 7.18 (solid circle: theory, open circle: experiment).

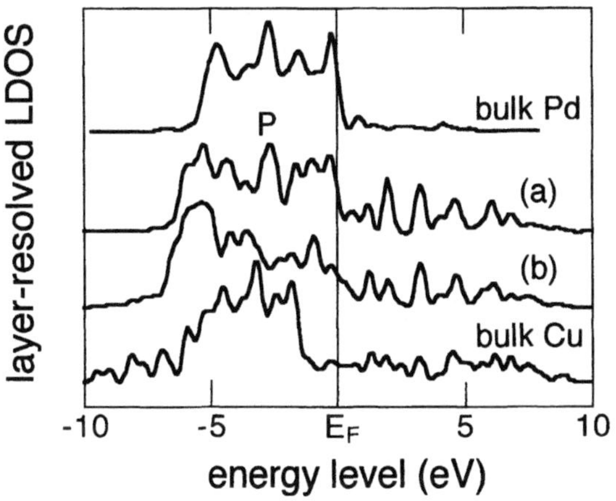

Fig. 7.20. Local density of states of (**a**) the outermost Pd layer and (**b**) the underlying Pd layer for the Pd_2/Cu_6 system, together with that of bulk Cu and Pd. One of the *d* states of Pd, denoted P, is strong in the outermost Pd layer but is diminished in the underlying Pd layer. This modulation results in the work function oscillating or overshooting

7.4.3 Work Function at Step Edges

In the work function images obtained from both the Au/Cu(111) (Fig. 7.14, Fig. 7.15) and Pd/Cu(111) surfaces (Fig. 7.17), we can see dark valleys along steps, indicating that the mesoscopic work function at steps is much lower than it is on terraces. Since the dipole moment is formed at step edges, lowering the potential, the lower mesoscopic work function is very likely due to this. In fact, this had been noticed in some earlier STM work [22,23], although no explanation was given.

In order to see the details, a line scan of the work function crossing a step that separates two Au terraces is shown by a solid line in Fig. 7.21. According to statistics based on more than 100 work function images like Fig. 7.14b, the mean FWHM and depth of the valley for Au–Au monatomic steps are 6.5 ± 1 Å and 0.9 ± 0.3 eV, respectively, while for Cu–Cu monatomic steps they are 10 ± 1 Å and 1.9 ± 0.3 eV, respectively. As mentioned above, dipoles induced by Smoluchowski smoothing [5] at steps are very likely responsible for the formation of such valleys in the mesoscopic work function (Fig. 7.22).

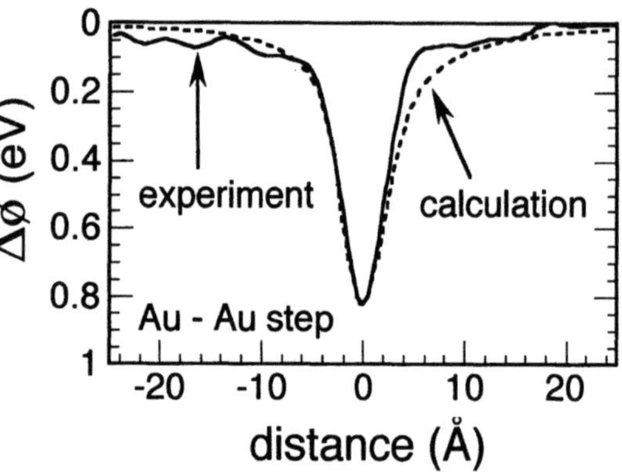

Fig. 7.21. Profile of work function change measured in a direction perpendicular to a Au step. *Solid line*: work function measured experimentally by STM. *Dashed line*: profile calculated by assuming formation of dipole moment array at the step edge and the topographic effect

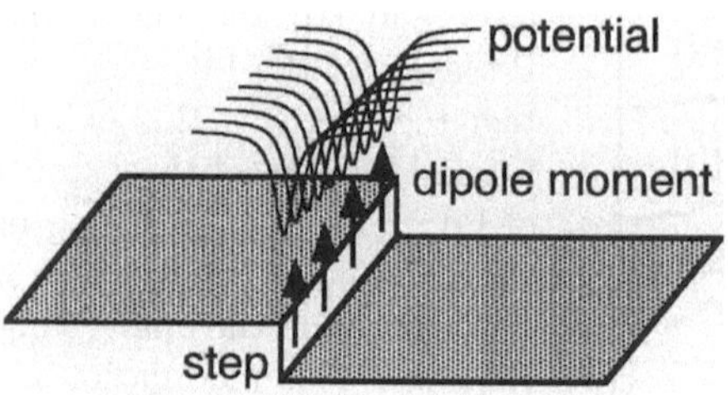

Fig. 7.22. Schematic diagram of potential over the area including a step. The step site produces a dipole moment, forming a potential valley above it

A simple simulation has thus been made, in which an infinite row of equal dipoles is used to simulate a step. If the axis of the dipole row is set along the y-coordinate axis with the positive end of the dipoles pointing to the outside of the surface, i.e., the $+z$ direction, then the local surface potential at a point (x, z) induced by the dipole row is given by

$$D(x,z) = \frac{Q}{4\pi\varepsilon_0} \frac{\ln[(z+l/2)^2 + x^2]}{(z-l/2)^2 + x^2} , \tag{7.1}$$

where Q is the linear charge density and l the distance between positive and negative charges. Let d be the spacing of the step atoms. Then the induced dipole moment can be calculated as $\mu = Qld$ per step atom. We used the constant-current rather than the constant-height mode in our experiment. Therefore, in order to simulate the line scan shown in Fig. 7.21 using (7.1), what we have to calculate is not $\phi_D(x, z_0)$ but $\phi_D(x, z(x))$, where $z(x)$ is the real STM line scan along which the work function was measured. Note that the real line scan is quite different from the schematic step profile, which is expected to be more like the profile of the real step. The reason for this is twofold: Smoluchowski smoothing [5] and, of course, convolution with the tip. However, as pointed out by Binnig and Rohrer [24], the step topography has one more effect on measured values of the work function. This is because what is measured, as mentioned above, is the response $\Delta \ln I$ to the modulation of the gap distance Δs, which ought to be in the direction normal to the surface. If the normal to the surface at a given point is not in the z direction but tilted at an angle θ then the real Δs is reduced by a factor of $\cos^2 \theta$. As a result, the measured local work function at that point is reduced by a factor of $\cos^2 \theta$. Consequently, we must also take this into account in the simulation. The variation of work function around a step is then calculated as

$$\Delta\phi(x) = \phi_{\mathrm{AT}} - [\phi_{\mathrm{AT}} - \phi_D(x, z)] \cos^2 \theta \ , \tag{7.2}$$

where ϕ_{AT} is the measured work function at the terraces separated by the step, and $\phi_D(x, z)$ is the potential of the dipole row along the step and hence is calculated by (7.1). In the calculation, the value of θ at each point was determined from the real STM line scan and the distance l between positive and negative charges was set at the step height (the exact value of l has almost no effect on final results). By optimizing the gap distance and linear dipole moment density, good agreement has been achieved between calculated and experimental curves. The calculated $\phi(x)$ is shown in Fig. 7.2 (dashed line) together with the experimental curve (solid line) for comparison. The values of parameters used in this fitting are 4.9 Å for the tip height and $\mu = 0.12D$ per step atom for the dipole moment. The final result obtained from many similar fittings is $\mu = 0.16 \pm 0.05D$ per step atom, in agreement with the value of 0.2–0.27D per step atom derived by Besocke et al. for a stepped Au(111) surface [7]. It is interesting to see that the influence of a monatomic step on the work function extends to about 10 Å on either side of a step, in agreement with results by Avouris et al. [34], obtained recently from their STM spectroscopic step data. This suggests that step–step interaction may come into play if terraces are narrower than 20 Å in width. A similar simulation has also been carried out for Cu–Cu monatomic steps, and the result is $\mu = 0.5 \pm 0.15D$ per step atom, about twice as large as for Au–Au steps.

7.4.4 Sulfur Adsorption on Pt(111)

Adsorption of atoms on surfaces induces changes in the work function. Adsorbates modify the electronic distribution on the surface, making extra dipole moments there. Once a dipole moment is formed, it can be detected as a change in work function. Dipole moments also affect interactions among the adsorbates. In order to understand these phenomena, it is important to measure the potential or mesoscopic work function above and near the adsorbates. In this study, sulfur atoms were deposited on a Pt(111) surface. Because of their electron affinity, adsorbates form dipole moments whose positive/negative charges are located on the surface/vacuum side, respectively. As this direction coincides with that of the surface dipole, sulfur adsorption induces an increase in the work function. This is in contrast to the case of

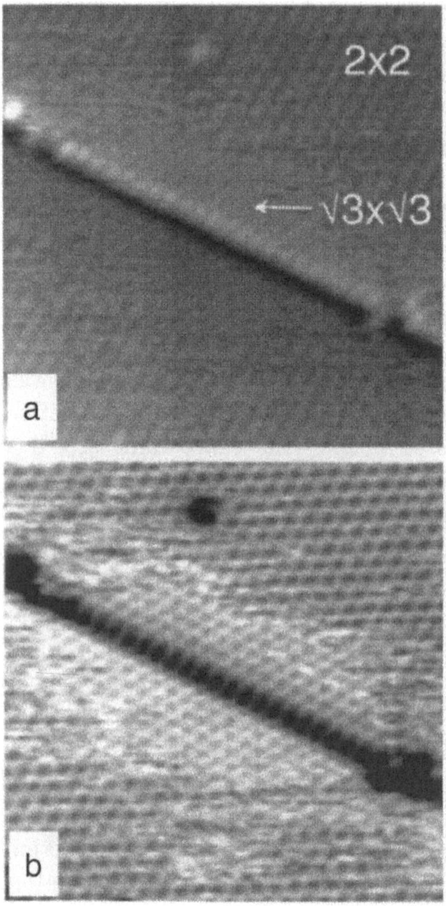

Fig. 7.23. (**a**) STM and (**b**) work function images containing both the 2×2 and $\sqrt{3} \times \sqrt{3}$ structures of S/Pt(111) surface. The size of the area is 120 Å ×120 Å. It is demonstrated that the $\sqrt{3} \times \sqrt{3}$ structure has larger work function than the 2×2 structure

alkali metal adsorption, which lowers the work function. Sulfur atoms were deposited on a cleaned Pt(111) surface by dosing H_2S gas into the UHV chamber. H_2S gas decomposes on the surface, leaving sulfur [35,36]. When the amount of exposure reaches between 1.0 L ($= 1.0 \times 10^{-6}$ torr s) and 4.0 L, a 2×2 structure with an S coverage of 0.25 ML is formed. Further H_2S exposure, beyond the 4.0 L level, produces a $\sqrt{3} \times \sqrt{3}$ structure with a S coverage of 0.33 ML. It is known that sulfur adatoms occupy the threefold fcc sites on the Pt(111) surface [37]. Figure 7.23 shows an STM image and work function image taken at the same place on the S/Pt(111)surface. The level of exposure is about 5 L. The $\sqrt{3} \times \sqrt{3}$ structure is located near the step around the center of the images, surrounded by the 2×2 structure. It is found that the $\sqrt{3} \times \sqrt{3}$ structure nucleates on both sides of the step edge. The corresponding work function image tells us that the work function is higher on the $\sqrt{3} \times \sqrt{3}$ structure than it is on the 2×2 structure. This can be simply explained by the higher dipole moment density in the $\sqrt{3} \times \sqrt{3}$

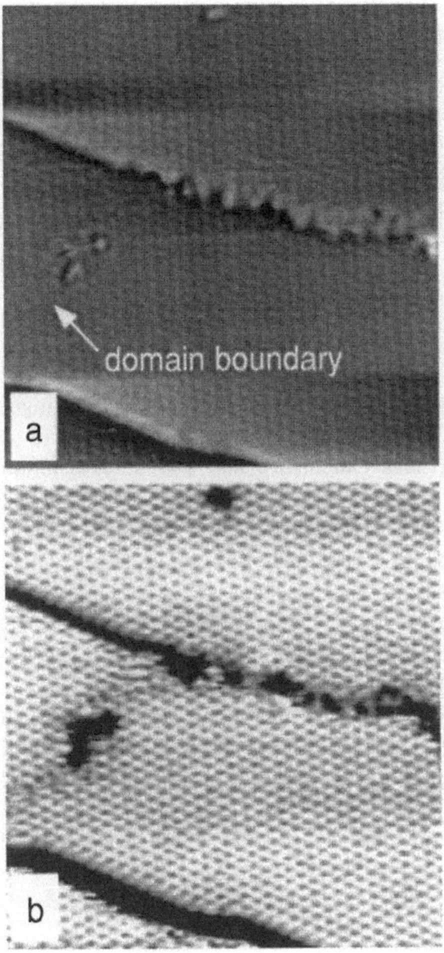

Fig. 7.24. (**a**) STM and (**b**) work function images of the $\sqrt{3} \times \sqrt{3}$ structure of the S/Pt(111) surface. The size of the observed area is 150 Å ×150 Å. Domain boundary and defects are visible in the images. It is shown that the work function around the boundary is slightly lower than the terrace work function

structure. Based on the statistics, we found that the $\sqrt{3} \times \sqrt{3}$ structure has a higher work function by $18 \pm 9\%$ than the 2×2 structure.

One interesting observation as regards this surface is the lower work function around the domain boundary of the $\sqrt{3} \times \sqrt{3}$ structure, as can be seen in Fig. 7.24. At the left side of the middle terrace, the domain boundary is visible in the STM image, and the corresponding work function image demonstrates work function reduction around the boundary. It is curious to note that the region of lower work function has width about 10 Å on both sides of the boundary, while the STM image shows an abrupt rather than a fuzzy boundary. This implies that the electronic effect of the domain boundary spreads around it more than the electronic effect of the structure. The lower work function at the domain boundary is probably due to dipole moment formation, as in the case of step edges. Lower work function values are also observed in the images at defect sites.

7.5 Conclusion

In this review, we have demonstrated that STM enables us to measure the mesoscopic work function and to make its image simultaneously with STM images. Using this technique, we can reveal how atomic and nanometer scale structures on surfaces modify the surface potential and work function. This will be very important for understanding the mechanisms underlying many surface phenomena.

References

1. J. Hölzl and F.K. Schulte, Solid Surface Physics **85**, 1. G. Höhler (Ed.), Springer Tracts in Modern Physics, Springer, Berlin (1979).
2. K. Wandelt, in Thin Metal Film and Gas Chemisorption, 280. P. Wissmann (Ed.) Elsevier, Amsterdam (1987).
3. N.D. Lang and W. Kohn, Phys. Rev. **B1**, 4555 (1970).
4. N.D. Lang, Solid State Physics **28**, 225. H. Ehrenreich, F. Seitz and D.T. Turnbull (Eds.) Academic Press, New York (1973).
5. R. Smoluchowski, Phys. Rev. **60**, 661 (1941).
6. B. Krahl-Urban, E.A. Niekisch and H. Wagner, Surf. Sci. **64**, 52 (1977).
7. K. Besocke, B. Krahl-Urban and H. Wagner, Surf. Sci. **68**, 39 (1977).
8. H. Wagner, in Solid Surface Physics. G. Höler (Ed.) Springer Tracts in Modern Physics, Springer, Berlin (1979).
9. G.A. Haas and R.E. Thomas, J. Appl. Phys. **34**, 3457 (1963); Surf. Sci. **4**, 64 (1966).
10. E. Bauer, M. Mundschau, W. Swiech and W. Telieps, Ultramicroscopy **31**, 49 (1989).
11. M. Nonnenmacher, M.P. O'Boyle and H.K. Wickramasinge, Appl. Phys. Lett. **58**, 2921 (1991).
12. J. Küpers, K. Wandelt and G. Ertl, Phys. Rev. Lett. **43**, 928 (1979).
13. R. Fischer, S. Schuppler, N. Fischer, Th. Fauster and W. Steinmann, Phys. Rev. Lett. **70**, 654 (1993).
14. Th. Fauster and W. Steinmann, Electromagnetic Waves: Recent Developments in Research, 2: Photonic Probes of Surfaces, Chap. 8. Elsevier, Amsterdam (1995).
15. G. Binnig, H. Rohrer, Ch. Gerber and E. Weibel, Phys. Rev. Lett. **49**, 57 (1982).
16. G. Binnig and H. Rohrer, Surf. Sci. **126**, 236 (1983).
17. R. Wiesendanger, Scanning Tunneling Microscopy and Spectroscopy Methods and Applications, 131. Cambridge University Press, Cambridge (1994).
18. G. Binnig, N. Garcia, H. Rohrer, J.M. Soler and F. Flores, Phys. Rev. **B30**, 4816 (1984).
19. J.H. Coombs, M.E. Welland and J.B. Pethica, Surf. Sci. **198**, L353 (1988).
20. M.D. Pashley, J.B. Pethica and J. Coombs, Surf. Sci. **152/153**, 27 (1985).
21. M.C. Payne and J. C. Inkson, Surf. Sci. **159**, 485 (1985).
22. R. Schuster, J.V. Barth, J. Wintterlin, R.J. Behm and G. Ertl, Ultramicroscopy **42–44**, 533 (1992).

23. J.F. Jia, K. Inoue, Y. Hasegawa, W.S. Yang and T. Sakurai, J. Vac. Sci. Technol. **B15**, 1861 (1997).
24. G. Binnig and H. Rohrer, IBM J. Res. Develop. **30**, 355 (1986).
25. T. Sakurai, T. Hashizume, I. Kamiya, Y. Hasegawa, N. Sano, H.W. Pickering and A. Sakai, Prog. Surf. Sci. **33**, 3 (1990).
26. Y. Hasegawa, J.F. Jia, K. Inoue, A. Sakai and T. Sakurai, Surf. Sci. **386**, 328 (1997).
27. I.-W. Lyo and Ph. Avouris, Science **253**, 173 (1991).
28. Y. Hasegawa, I.-W. Lyo and Ph. Avouris, Appl. Surf. Sci. **76/77**, 347 (1994).
29. W. Wallauer and Th. Fauster, Surf. Sci. **331–333**, 731 (1995).
30. Z.Q. Li, K. Ohno and Y. Kawazoe, to be published.
31. D. Vanderbilt, Phys. Rev. **B41**, 7892 (1990); G. Kresse and J. Furthmuller, ibid. **B54**, 11169 (1996).
32. F.K. Schulte, Surf. Sci. **55**, 427 (1976).
33. R.J. Behm, Scann. Microsc. Suppl. **1**, 61 (1987).
34. Ph. Avouris et al., Chem. Phys. Lett. **240**, 423 (1995).
35. J.C. Dunphy, B.J. McIntyre, J. Gomez, D.F. Ogletree, G.A. Somorjai and M.B. Salmeron, J. Chem. Phys. **100**, 6092 (1994).
36. B.J. McIntyre, M. Salmeron and G.A. Somorjai, Surf. Sci. **323**, 189 (1995).
37. K. Hayek, H. Glassl, A. Gutmann, H. Leonhard, M. Ptutton, S.P. Tear and M.R. Welton-Cook, Surf. Sci. **152/153**, 419 (1985).
38. M.D. Thompson and H.B. Huntington, Surf. Sci. **116**, 522 (1982).

8 Scanning Tunneling Microscopy of III–V Compound Semiconductor (001) Surfaces

Qi-Kun Xue, T. Hashizume and T. Sakurai

Summary. This paper reviews advances towards a comprehensive understanding of the geometric and electronic structures and also the chemical properties of the principal reconstructions found on the (001) surface of III–V compound semiconductors, made during the first decade following the invention of Scanning Tunneling Microscopy in the early 1980s. We consider arsenides, such as GaAs, InAs and AlAs, phosphides, such as GaP and InP, antimonides, such as GaSb, AlSb and InSb, and also nitrides (GaN), with special emphasis on GaAs(001).

8.1 Introduction

Scanning Tunneling Microscopy (STM) can, in spite of its conceptually simple operation principle, resolve local electronic structures on an atomic scale on virtually any kind of conducting solid surface, and under various environments, with little damage or interference to the sample [1]. It is now 15 years since the first semiconductor surface – the 7×7 reconstruction of Si(111) – was imaged with STM in real space [2]. Over the years, STM has proved to be an extremely versatile and powerful technique for many disciplines in condensed matter physics, chemistry and biology. At the same time, its application to the area of semiconductors has enjoyed rapid and sustained growth. Early stage investigations concerned geometric and electronic structures of ordered clean surfaces, and also nonperiodic and local surface features such as defects, impurities and atomic steps. It has since contributed to our understanding of stepped surfaces, adsorption-induced structures, surface diffusion, surface reaction, epitaxial growth processes, and so on. All these studies have had a significant impact on our understanding of the fundamental properties of semiconductor surfaces and supplied much of the knowledge required for development of technologically important devices based on these materials. Several comprehensive reviews on applications of STM in this area can be found in literature, for example, 'Scanning tunneling microscopy of semiconductor surfaces' by Neddermeyer [3], and 'Scanning tunneling microscopy of semiconductor surfaces' by Kubby and Boland [4].

In this review, we focus on a more specific area: reconstructions occurring on clean (001) surfaces of various III–V compound semiconductors. These surfaces are interesting and important in that they are the primarily grown surface in the epitaxial growth of these materials. Such a detailed discussion on III–V (001) surfaces was not possible in the reviews mentioned above mainly due to the limited data available at the time they were being prepared. In the seven years following the first Si(111)-7×7 surface image, only two STM studies on the GaAs(001) surface were reported [5,6]. Compared to elemental semiconductor (Ge and Si) surfaces and cleaved III–V (110) surfaces [7], a successful STM study of the compound semiconductor (001) surface is more difficult to carry out because of the sample preparation. Well-ordered Ge and Si clean surfaces can be routinely prepared by Ion Bombardment Annealing (IBA) and/or annealing of commercial wafers that are available in most UHV STM systems. Preparation of a well-ordered and clean III–V (001) surface usually requires additional growth apparatus such as a MBE or MOCVD reactor, but an in situ study with combined growth and characterization systems is a luxury for most STM groups. The As-decapping technique [8] for arsenides or the IBA method for some specific materials may be used to produce a high quality surface, but great care is needed. Group III droplets often form and a non-stoichiometric surface results as the group V component is more volatile. With these methods it is difficult to study As-grown surface and growth phenomena which are more challenging. In addition, compound III–V surfaces involve an additional degree of freedom in chemical species, so that unambiguous STM data interpretation is more difficult and often looks to theoretical analysis that is not widely available.

Continued efforts by scientists in the field have considerably changed this situation. Besides significant progress in the study of GaAs(001) and InAs(001), many previously unexplored (001) surfaces, such as the AlAs(001)-3×2 [9] and GaAs(001)-2×3 surfaces [10] for arsenides, InP(001)-2×4 [11] and GaSb(001)-1 × 5 and 2 × 5 [12] for phosphides and antimonides have recently been investigated by STM with some success. This makes a comprehensive review possible. Interest in this field has been further stimulated by recent rapid development in III-nitrides (GaN, InN and AlN) that have enormous potential for applications in short wavelength photoelectronic devices such as high-brightness light emitting diodes and laser diodes [13,14]. Achievements in preparation of high quality nitride surfaces bring STM into this exciting realm, allowing structural analysis and study of growth processes down to the atomic level.

This paper is organized as follows. After a brief introduction, we give a simple description of the general principles of semiconductor surface reconstruction in Sect. 8.2. In the main part (Sects. 8.3 to 8.6), we will present a comprehensive summary of STM studies of conventional arsenide, phosphide, antimonide and nitride surfaces with emphasis on GaAs. Contrary to the usual review paper, we try to present a detailed analysis of the STM data for each surface and, where possible, to provide details of theoretical calculations. Section 8.7 contains our concluding remarks.

8.2 Semiconductor Surface Reconstruction

8.2.1 Principles of Semiconductor Surface Reconstruction

It is useful to give a brief introduction to two important phenomena, i.e., relaxation and reconstruction, associated with surface formation. When a surface is created, the bonds between atoms are 'cut' and interactions among atoms in the topmost layer and sublayers are modified, due to the absence of neighboring atoms on one side. In order for the system to reach a new equilibrium configuration, rearrangements and displacements of surface layer atoms will occur. The arrangements may result in a pure compression or expansion of interplanar separations of the surface layers in the surface normal direction. This process is called surface relaxation. Surface relaxation does not necessarily induce a change in lateral surface periodicity and is usually inaccessible to STM.

Surface reconstruction, discussed in this paper, refers to a more complicated process in which atomic rearrangements introduce a lateral change in the surface periodicity and often simultaneous lattice relaxation as well. A reconstructed surface has lowered translational symmetry and often different chemical compositions with respect to those of its bulk counterpart. Due to hybridization of atomic orbitals, interatomic bonds in common elemental semiconductors such as Si and Ge, and most III–V compound semiconductors such as GaAs, are tetrahedral. These directional covalent bonds determine the structure of such semiconductors. There are two spin-paired electrons on each bulk bond. With formation of a surface, at least one bond for each atom is cut, producing unpaired 'dangling bonds'. The lack of electron pairing makes these dangling bonds energetically unstable and results in a large free energy. The surface atoms rearrange themselves in such a way that the free energy of the system reaches its new minimum. For a cleaved surface, reconstructions are driven by lowering the energy of bands of surface electronic states associated with rehybridization of the surface dangling bonds. For annealed elemental semiconductor surfaces, complicated surface structures occur which tend to minimize the number of dangling bonds. A variety of nearly degenerate structures is possible, depending on how the system compromises between reduction of dangling bonds and adjustment of the resulting local strain in the surface. In the case of polar compound semiconductor surfaces, the surface composition can vary as well. For a given preparation condition, due to different levels of surface electronic state bands associated with different chemical species, the surface stoichiometry is first selected in order to reach an autocompensated surface with completely filled bands of surface electronic states separated by a gap from completely empty bands. Then the detailed atomic position is determined by minimizing the energy of the filled surface electronic state bands. Despite the manner by which the system searches a new equilibrium, the directional and strong nature of the covalent chemical bonds mean that reconstructions of semiconductor surfaces often manifest themselves in a more complicated way than those on metal surfaces. Rearrangements result in a large surface unit mesh and may extend a few layers deep from the surface. For instance, 49 silicon atoms are involved in the unit cell of the reconstructed Si(111)-7 $\times$ 7 surface. Even for the single clean GaAs(001) surface discussed in this review, nearly a dozen reconstructions have been reported [15].

In search of reasons why semiconductor surfaces reconstruct, during the past decade nearly a hundred tetrahedrally coordinated semiconductor surface structures have been explored both theoretically and experimentally. These extensive efforts have led to the recognition that certain general features of the results can be abstracted and utilized as qualitative 'principles' for semiconductor surface reconstruction [16–18]. These general physical principles are useful in interpreting the major characteristics of known structures, as well as in predicting new surface structures. A detailed discussion of these principles is given by Duke [19]. We briefly describe these principles following his review, and applications to relevant III–V surfaces will be discussed later.

Principle 1. Reconstruction tends either to saturate surface dangling bonds via rehybridization or to convert them into non-bonding electronic states.

Principle 2. In many cases (and all quasi-one-dimensional cases) surfaces can lower their energies by atomic relaxations leading to semiconducting (as opposed to metallic) surface state eigenvalue spectra.

Principle 3. The surface structure observed will be the lowest energy structure kinetically accessible under the preparation conditions.

Principle 4. Surfaces tend to be autocompensated.

Principle 5. For a given surface stoichiometry, the surface atomic geometry is determined primarily by a rehybridization-induced lowering of the surface state bands associated with the filled anion dangling bond orbitals.

8.2.2 General Features of GaAs(001) Reconstructions

We now discuss the typical case of GaAs(001). This is the most commonly used substrate in fabrication of wireless and optoelectronic devices based on III–V compound semiconductors by Molecular Beam Epitaxy (MBE), MetallOrganic Chemical Vapor Deposition (MOCVD) and related techniques. Because of its zinc blende crystal structure with tetrahedral coordination in the bulk, the polar GaAs(001) surface may be terminated with either As or Ga atoms (Fig. 8.1). Depending on the surface coverage and experimental conditions, the surface exhibits a number of reconstructions, starting with the most As-rich phase which has a $c(4 \times 4)$ symmetry, through the 2×4 [including the off-phase $c(2 \times 8)$], 1×6, 4×6, and ending with the 4×2 [including the off-phase $c(8 \times 2)$] Ga-stabilized phase. In addition, several

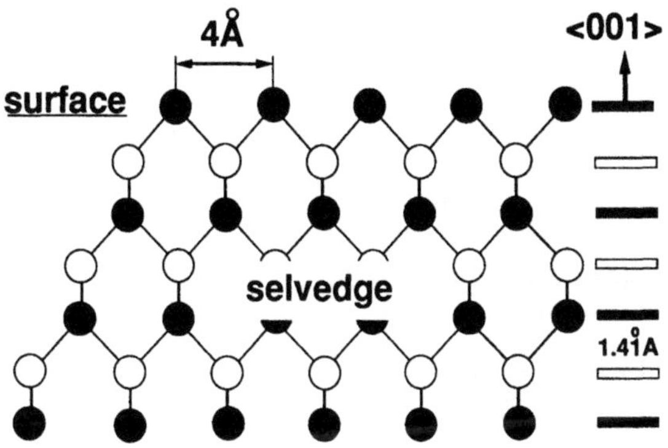

Fig. 8.1. Schematic of GaAs zinc blende structure. *Solid circles* represent As atoms and *open circles* represent Ga atoms. At the bulk terminated (1 × 1) surface each As (Ga) has one unpaired electron with a surface lattice constant of 4 Å along the (001) crystal axis. Alternative As and Ga layers occur with an interplanar distance (from an As layer to a neighboring Ga layer) of 1.41 Å

transient or adsorbate-induced structures such as 2×3, 2×1, $c(6 \times 4)$, 3×1, 3×6 and others have been reported.

Experimental and theoretical studies have resulted in establishment of two basic and general features in the structure of the GaAs(001) surface. As seen below, these two features and the stability of all reconstructions of GaAs(001), including many other low-index compound semiconductor surfaces, can be substantiated by a simple but practical rule – the Electron Counting Model (ECM) [20–22]. This rule is actually an alternative to the guiding principles for semiconductor surface reconstruction listed in the last section. With this rule, we can easily judge whether a structure (model) is a possible candidate for a given reconstruction. Using the GaAs(001) surface as an example, we briefly describe this rule.

The atoms in bulk GaAs are sp^3 hybridized. Two hybridized orbitals, one from each type of atom, combine to form bonding and antibonding orbitals. At the surface, some hybrid orbitals cannot form bonds and therefore, if no reconstruction occurs, the partially filled sp^3 dangling bonds remain. This configuration is energetically unstable. Hence, in order to reduce the number of dangling bonds and decrease the surface energy as required by Principle 2, the adjacent As (Ga) atoms form a dimer along the $[\bar{1}10]$ ($[110]$) direction and the surface reconstructs with the $2\times$ ($\times 2$) periodicity, similar to Si(001). This requirement characterizes the first feature of the surface: dimerization.

The second results from the different dangling bond levels on surface As and Ga atoms. The energy levels of these dangling bonds can be estimated from the energies of the s and p atomic levels from which they are derived [23]. Based on Harrison's estimation, as shown in Fig. 8.2, the dangling-bond energy level $E_{\mathrm{h}}(\mathrm{As})$ of As lies below the valence-band maximum of the bulk

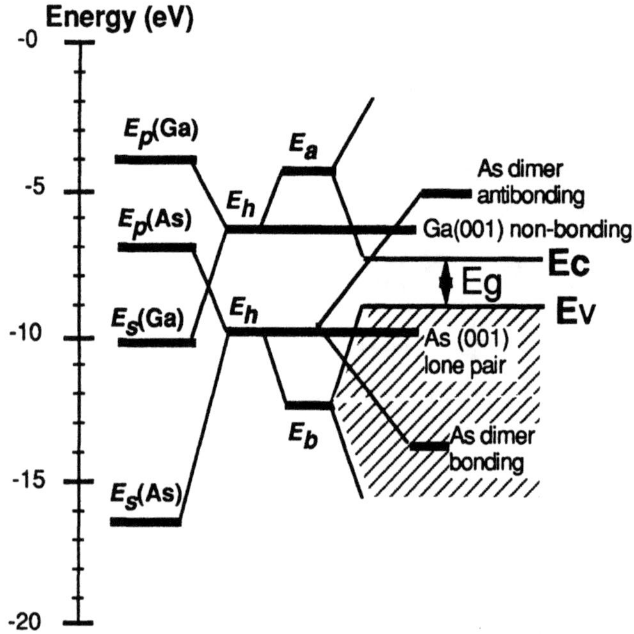

Fig. 8.2. The energy level diagram of GaAs and its application to the (001)-As surface [24]. Beginning with s and p states for the Ga and As, sp^3 hybrids form a bonding (E_b) and antibonding (E_a) combination which broaden into valence and conduction bands, respectively. At the (001) surface, two of the As hybrids are unbonded sp^3 hybrid dangling bonds. Subsequent dimerization produces an As–As dimer bonding and antibonding combination, leaving one As dangling bond per atom at the hybrid energy, E_h. Charge transfer from the underlying Ga into this state produces a filled lone-pair orbital. The data are originally adapted from [23]

semiconductor, and should therefore be filled, whereas the dangling-bond energy level E_h(Ga) of Ga lies above the bulk conduction-band minimum and should therefore be empty. In order to achieve this, electrons transfer from the dangling bonds of the electronegative element (As) to those of the electropositive element (Ga). However, the charge transfer induces additional static electrical capacitance energy due to charge accumulation. The resulting surface becomes energetically unstable unless some of the dimers are removed from the surface to compensate this energy and achieve charge neutrality. Dimer missing is found in almost all studied III–V (001) surfaces as we will see in the review. As required by Principle 4, dimer missing characterizes the second feature of reconstructions on GaAs(001), which is further generalized into the ECM [20]. ECM requires a stable surface structure to be found where the number of available electrons in the surface layer will exactly fill all dangling bond states in the valence band, leaving those in the conduction band completely empty. This rule dictates that there should be no net charge lying on the surface. With further lattice distortion that is characteristic of Principle 5, the surface energy band structure becomes more semiconducting, whereas partially filled dangling bonds may lead to a metallic surface, and the resulting structure is not stable.

To see how ECM works, we consider an example where a surface unit cell consists of D dimers with every Nth dimer missing, i.e., the surface has the $2 \times N$ reconstruction [20]. Assuming that the surface is electronegative-element (anion) terminated, we first count the number of electrons required for this $2 \times N$ reconstruction. Each surface anion dimer requires six electrons, two for a dimer bond and four for two dangling bonds of a dimer, giving rise to a total of $6D$ electrons per surface unit. In addition, $8D$ electrons are required to bond these D dimers to the second layer cation atoms: two electrons for each of the four bonds between an anion dimer in the top layer and the cation atoms in the second layer. The total number of electrons required would be $6D + 8D$. Next, we count the number of electrons available in the reconstructed layers. The number of electrons provided by the top layer anion atoms is $2V_{\mathrm{n}}D$, where V_{n} is the number of valence electrons on the anion atoms. In addition, the number of electrons available from the second layer cation atoms is $2V_{\mathrm{p}}N/2 = V_{\mathrm{p}}N$ (since half of the total $2V_{\mathrm{p}}N$ electrons from the second layer are involved in bonding to third layer atoms), where V_{p} is the number of valence electrons on the cation atoms. For a stable structure the number of electrons required should be equated to the number of electrons that are available, i.e.,

$$6D + 8D = 2V_{\mathrm{n}}D + V_{\mathrm{p}}N \ . \tag{8.1}$$

For GaAs,

$$V_{\mathrm{n}} = 5 \ , \quad V_{\mathrm{p}} = 3 \ , \quad \text{hence} \quad 4D = 3N \ . \tag{8.2}$$

The smallest unit cell that satisfies this condition is the 2×4 unit cell containing three dimers ($D = 3$) with every fourth dimer missing ($N = 4$). If deeper layer atoms are involved in reconstruction, consideration of ECM will give rise to additional two-dimer models [21].

This rule has also been applied successfully to explain the cation atom terminated surface as well as the step structure of GaAs(001) and growth [20]. ECM has been regarded as a general guiding principle for building various III–V (001) surface reconstruction models, with the exception of some specific cases [12].

8.3 GaAs(001) As-Rich Surface

8.3.1 $2 \times 4/c(2 \times 8)$ Phase

Structural Determination

(1) First stage

It is well known that MBE growth of GaAs(001) is usually performed under conditions which lead to an As-terminated 2×4 Reflection High Energy Electron Diffraction (RHEED) pattern. Since these conditions result in the highest quality grown material, the 2×4 reconstruction has been most extensively studied in the past by almost all surface sensitive techniques, after Arthur and LePore demonstrated that GaAs can be grown epitaxially on GaAs by a simultaneous molecular beam method [25].

Historically the first report of the 2×4 reconstruction was on a surface cleaned by argon ion bombardment, observed by Jona using Low Energy Electron Diffraction (LEED). The reproducibility of the result was not very good [26]. Cho performed the first MBE study on 2×4 using RHEED [27]. He suggested that the observed $2\times$ periodicity arises from the fact that neighboring surface atoms are pulled together to form a pair in the direction of their dangling bonds. A similar phenomenon occurred on Ge and Si (001) surfaces [28], and the $c(2 \times 8)$ is probably due to the out-of-phase arrangement of the rows of these paired atoms. Later Cho reexamined this 2×4 as well as the Ga-rich 4×2 phase and found that the surface has the $2\times$ periodicity in the direction parallel to planes containing dangling bonds in the surface, further reinforcing the bond-pairing mechanism [29]. He speculated on an additional mechanism such as Fermi surface instabilities or surface vacancies for the fourfold periodicity, different from sterically forced buckling for the Ga-terminated 4×2 and $c(8 \times 2)$.

The pairing scheme for the $2\times$ periodicity was supported by the first LEED study of the MBE-grown surface, made by Bommel and Crombeen [30]. Over the ten years from 1978 through 1987, studies using diffraction techniques such as RHEED and LEED, photoemission, AES (Auger Electron Spectroscopy) and EELS (Electron Energy Loss Spectroscopy) etc., were conducted to characterize surface composition and structure [31–41]. Most surface reconstructions of the (001) surface of GaAs were discovered and the surface phase transition between these phases was well understood due to significant development of the MBE technique over this period. Further RHEED analysis by Joyce et al. [40] explicitly demonstrated for the first time the relationship between the 2×4 and $c(2 \times 8)$ phases. Theoretical calculations on As(001) surfaces showed that the formation of As dimers is energetically favorable by releasing about 2 eV per bond [42–44]. The photoemission intensity as a function of polar angle from the 2×4 surface state at 1 eV below the valence band maximum was interpreted in terms of surface As dimers [36,38], yielding additional evidence of the bond-pairing scheme.

Thus, the dimerization mechanism of the 2 × 4 surface seems to have been well established. However, the relatively large size of the 2 × 4 unit cell has greatly hindered further progress in determining the atomic geometry of the surface. In particular, there is no satisfactory explanation for the fourfold periodicity [36,39]. This period (before 1987) may be regarded as the first stage in study of the 2 × 4 surface. During this period, experimental structural determination of the surface was mainly performed by diffraction techniques. The 2× periodicity was attributed to bond pairing.

(2) Second stage

As seen in the following, this stage (from 1987–1993) is characterized by the indisputable establishment of the most fundamental feature of the 2×4 reconstruction – the As-dimer-missing mechanism. STM was critical in establishing the role of such a mechanism.

In 1987, Chadi examined the structure of the 2 × 4 surface by tight-binding based total energy calculations. Instead of assuming asymmetric As dimers for explaining the 4× periodicity along the [110] direction [36,39], he introduced an idea of 'As-dimer vacancy' to explain the 2 × 4 symmetry of the surface [21], as speculated by Cho [29]. Two different structures each containing three As dimers per unit cell, as shown schematically in Fig. 8.3, were found to be stable. He also calculated the surface state energy for these structures, obtaining good agreement with data obtained from photoemission

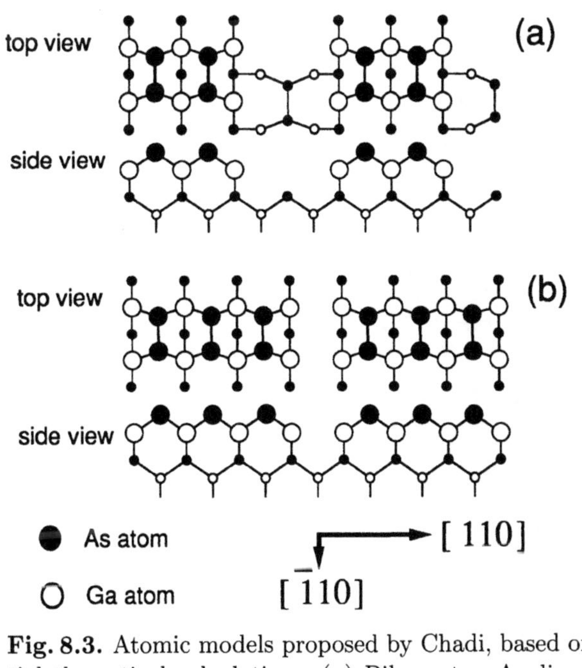

Fig. 8.3. Atomic models proposed by Chadi, based on tight-binding pseudopotential theoretical calculations. (**a**) Bilayer two-As-dimer model, (**b**) three-As-dimer model [21]

measurements [38]. From his structural models, the $c(2 \times 8)$ can easily be constructed with only a lateral shift of the dimer by a 1×1 lattice constant every four rows. The resulting $c(2\times8)$ is even found to be slightly more stable than the 2×4 phase. Systematic theoretical analysis by Qian, Martin and Chadi [45] further indicated that the dimerization of As (Ga) always gives rise to the most stable configuration and should be included in any plausible structural model to be considered. In addition, the asymmetric As dimer is unstable against the symmetric one, in contrast to the case of the Si (001) surface.

Based on a HREELS study of hydrogen chemical adsorption on different GaAs(001) surfaces, Frankel et al. proposed three models to explain their HREELS results [46]. The first one is the same three-As-dimer model proposed by Chadi (see Fig. 8.3b), and the other two are shown in Fig. 8.4. Within the limitations of the HREELS results, they were not able to distinguish between the simple removal of As_2 species from the surface and removal followed by surface diffusion. Both structures shown in Fig. 8.4 are therefore possible. Based on these models, Farrell and her colleagues [22], proposed a growth mechanism to explain atomic processes whereby MBE takes place. This mechanism is attractive in terms of improving growth techniques, e.g.,

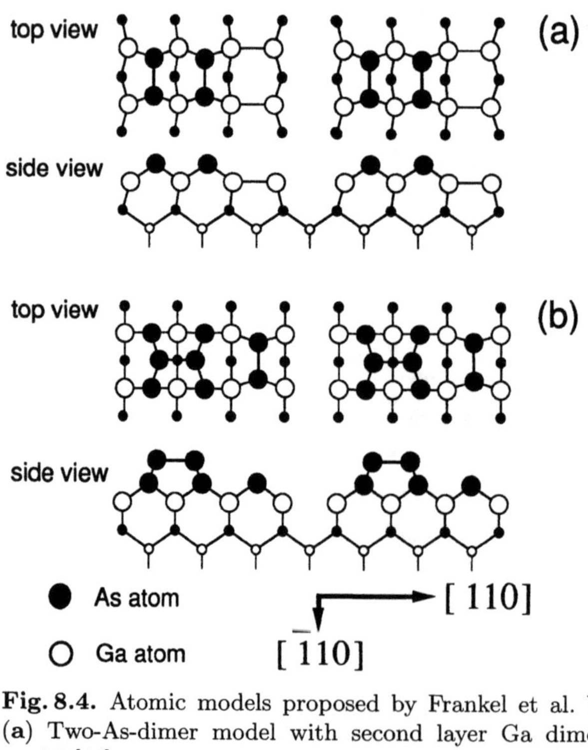

Fig. 8.4. Atomic models proposed by Frankel et al. based on HREELS studies. (**a**) Two-As-dimer model with second layer Ga dimerized. (**b**) Extra-As-dimer model [46]

substantially reducing substrate temperature and optimizing the resulting materials.

The first successful STM observation of the GaAs(001) 2×4 As-rich reconstruction was conducted by Pashley et al. in 1988 [5]. They prepared the sample in an ex situ MBE system, capping the sample surface with As for protection prior to transferring the sample through air to the STM apparatus. The As overlayers were removed by brief heating before the STM study. STM images directly confirmed that the fourfold periodicity arises from a regular array of missing arsenic dimers, and that the $c(2 \times 8)$ domain is due to the antiphase arrangement of the 2×4 subunit [21,40]. Their images agree with the three-As-dimer model containing three As dimers with every fourth missing, as proposed by Chadi (Fig. 8.3b) [21]. Later on, by combining the STM chamber and MBE growth apparatus together, Biegelsen et al. performed the first in situ STM investigation of various reconstructed GaAs(001) surfaces [6]. Their observation on the 2×4 surface basically agrees with the three-As-dimer model. But they also found that when the surfaces are annealed longer or at higher temperatures, or grown with lower As_4/Ga flux ratio, the 2×4 cells possess only two neighboring As dimers with two missing dimers [47]. This is shown in Fig. 8.3a. It implies varied stoichiometry of the 2×4 surface. Despite such uncertainties, these two pioneering STM studies, together with previous experimental and theoretical work, indisputably established the most important features of the 2×4 As-rich reconstruction: the $2\times$ periodicity along the $[\bar{1}10]$ direction and $4\times$ periodicity along [110] arise from dimerization of the As atoms and regularly arranged dimer missing, respectively.

Following the success in STM study of GaAs(001) surfaces by these two groups, many other groups conducted STM investigations on the MBE grown GaAs(001) surface from 1990 to 1994, in order to understand the atomic structure [24,48–57] and other properties of the surface, such as defects and Fermi level pinning [48,58–61], morphology [49,62,63] and step structure [49,62,64,65]. These studies have further substantiated the dimer-missing hypothesis, the most basic feature of the surface. Although most STM observations show that the 2×4 surface unit cell contains two As dimers, there is general acceptance that surfaces prepared under different MBE growth conditions or different sample preparation methods may have a different structure, e.g., containing three As dimers per unit cell. Recently, Falta et al. performed an MEIS (Medium Energy Ion Scattering) experiment on ex situ (decapped) MBE-grown GaAs(001) surfaces and proposed a radically new model in which the first layer of the (2×4) phase contains both Ga and As atoms on the basis of the three-As-dimer model (Fig. 8.3b). This is in contrast to the commonly accepted purely As-terminated models [66].

According to the proposed atomic models, surface As coverage ranges from 0.5 ML to 0.75 ML. Since the 1970s, there have been many attempts to measure the surface composition of various GaAs(001) surfaces, as witnessed by the extensive literature [31,32,37,39,67–70]. The results from these studies reveal a complicated situation. The first AES measurement on the 2×4 surface gave an As coverage of 0.61 ML [32]. A value of 0.89 ML is deduced from angle integrated photoemission measurements [37]. The As coverages as determined by total energy calculations can be 0.5 ML and 0.75 ML [68]. The latter is consistent with a report of 0.75 ML by Secondary Electron Intensity (SEI) [69]. Careful RHEED spectrum beam intensity analysis by Deparis and Massies [70] indicates that there may exist a varied surface composition between 0.55 ML and 0.89 ML. It is believed that the different results are due to the existence of the 2×4 surface over a wide range of the stoichiometry associated with different structures, as indicated by the models [21,46]. Understanding of these varied compositions requires unambiguous determination of surface structure(s).

One important study during this period was the RHEED investigation carried out by Farrell and Palmstrom [71], in which they deduced quantitative information about surface stoichiometry by diffraction techniques, and were able to make a direct connection between the RHEED pattern and surface stoichiometry. By carefully monitoring the $[\bar{1}10]$ direction diffraction pattern during annealing, they found that there should exist three different phases (the α, β and γ phases), corresponding to characteristic fractional order (1/4th, 2/4ths and 3/4ths) spot intensities. The α phase occurs at the highest substrate temperatures with relatively weak 2/4ths spot intensity compared with the 1/4th and 3/4ths intensities, while the γ phase is formed in the lowest substrate temperature range and its 2/4ths spot is almost absent. The β phase exists between these two and the 2/4ths spot is strong, almost equal to the 1/4th and 3/4ths spots. They performed kinematical calculations to analyze these data and proposed the outermost unit structure of the α, β, γ phases as shown in Fig. 8.5. The α phase is made of two As dimers with second layer Ga dimerization (Fig. 8.5a) [46]. The unit structure of the β phase is Chadi's three-As-dimers (Fig. 8.5b), and the γ phase is assigned to the model which has an extra As dimer sitting on the β phase unit (Fig. 8.5c). Based on this scheme, the As coverage is 0.5 ML for the α phase, 0.75 ML for the β phase and 1.0 ML for the γ phase, consistent with that expected from the preparation conditions for each phase. Although this scheme is more complicated than previously expected, for the first time it successfully explained the various reported surface compositions and the different models proposed. However, a precise appraisal of the scheme requires additional experimental and/or theoretical evidence.

(3) Third stage (1993–present)

During this period, our understanding of the structure of the 2×4 surface has been considerably improved by advanced theoretical studies [72–79]. The most important and common conclusion from these studies is that the bilayer two-As-dimer model (see Fig. 8.3a) proposed by Chadi [21] is the most

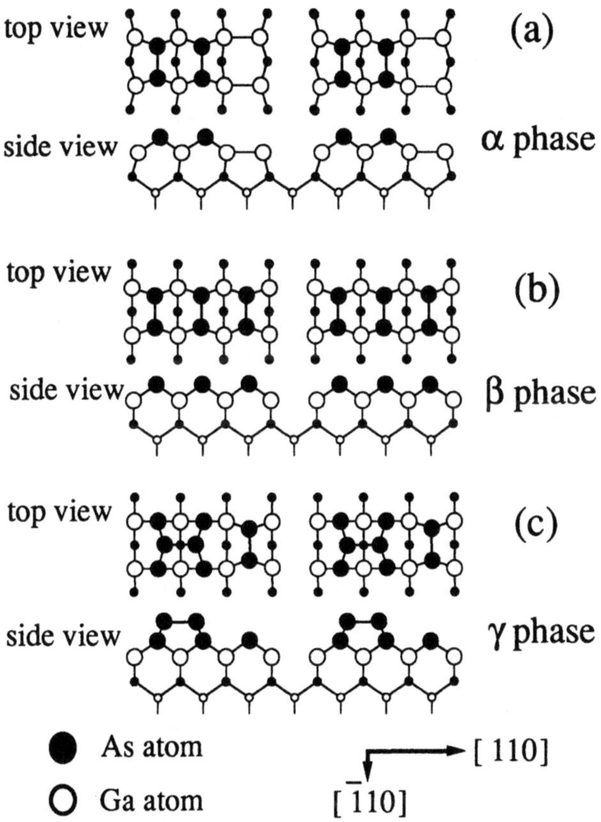

Fig. 8.5a–c. Farrell and Palmstrom's atomic models for the α, β, γ phases of the 2×4 surface based on kinematic RHEED analysis [71]

stable structure energetically, and the two-As-dimer structure (see Fig. 8.4a) proposed by Frankel et al. [46] stabilizes in a limited range of the As chemical potential. The three-As-dimer structure is ruled out as a possible equilibrium phase, and is only accessible kinetically.

In order to clarify the experimental uncertainties mentioned in the discussion of the second stage and to establish reliable structural models, Northrup and Froyen performed a series of first-principles total energy calculations [72]. The formation energy for a variety of possible structural models proposed (see Figs. 8.3, 8.4 and 8.5) for the GaAs(001) surfaces was calculated as a function of the Ga chemical potential (μ_{Ga}) using the first-principles pseudopotential method and the local density functional approximation. For each value of μ_{Ga} within its allowed range, they determined the structure having the lowest formation energy. The main results for As-rich 2×4 are summarized as follows:

- the three-As-dimer structure with one As dimer on the top is energetically unfavorable against dissociation into the three-As-dimer (Fig. 8.3b) and $c(4 \times 4)$ structures, and thus should be ruled out;
- the three-As-dimer structure proposed by Chadi (Fig. 8.3b) is only stable in a limited range of the allowed chemical potential and becomes unstable in As-rich conditions;
- the two-As-dimer structure with the second layer Ga-dimerized in the As-missing scheme proposed by Frankel et al., based on HREELS studies, is stable in the less As-rich limit, but for formation of second layer Ga bonds, a significant lattice relaxation is required, which pushes the top layer As further toward the vacuum direction and reduces lateral As dimer separation;
- the replacement of first layer As by additional Ga, as proposed by Falta et al. [66], is unlikely since it changes the stoichiometry and nature of backbonds. The resulting energy is much higher than that of the three-As-dimer structure, and hence the mixed model can be safely removed.

All these were basically reproduced by Zhang and Zunger using the LCSM (Linear Combination of Structural Motif) method [79].

Northrup and Froyen further examined the stability of another two-As-dimer structure (referred to as the bilayer two-As-dimer model) as shown in Fig. 8.3a [73]. In order to determine the relative stability of the bilayer two-As-dimer and three-As-dimer models, which was impossible using Chadi's calculations [21], they found that electrostatic interactions between the charged building blocks play an important role in determining equilibrium structure, because there is charge transfer from Ga to As. Although the local bonding environments of each atom in the bilayer two-As-dimer and three-As-dimer models are essentially identical, the arrangement of charge within the unit cell is significantly different. Their calculations favor the bilayer two-As-dimer model and indicate an energy difference as large as 0.05 eV per 1×1 unit cell for the entire allowed range of the chemical potential, ruling out the three-As-dimer model as a possible equilibrium phase. In addition, the stability of the bilayer As-dimer structure greatly reduces the chemical potential window for the more Ga-rich two-As-dimer structure (see Fig. 8.4a). These results are in good agreement with most STM observations available at that time; the top layer surface prevalently contains two As dimers in a unit cell.

Another first-principles total energy investigation of the 2×4 surface was performed by Ohno [74,76,80]. This appears in Chap. 1 and will not be discussed in detail here. He found that the bilayer structure does indeed become lower in energy by 0.01 eV per (1×1) unit than the three-dimer structure. However, as redimerization of the third As atoms leaves no As missing in the bilayer two-As-dimer model, the surface stress induced by formation of As dimers is stronger and difficult to relax. As a result, the energy decrease due to electrostatic interaction, as suggested by Northrup and Froyen [73], may be partially canceled by the energy increase from surface stress, so that the difference in surface formation energies between these two structures is smaller than that obtained by Northrup and Froyen [73].

Since these two total energy calculations were intentionally performed to determine the relative stability of a variety of structures as a function of As (or Ga) chemical potential and thus to establish reliable atomic models for the 2×4 surface, the detailed geometry and electronic properties were not discussed as a main part of the work. Recently, Schmidt and Bechstedt have conducted converged first-principles total energy calculations and presented comprehensive sets of geometric parameters and resulting bound surface states for several possible structures [78,81]. The formation energies they obtain for the various structures are in good agreement with those obtained by Northrup and Froyen [73] and by Ohno [76]. The two-As-dimer model with second layer dimerized Ga (Fig. 8.4a) represents the ground state geometry for a small range of the As chemical potential. Under more As-rich conditions, the bilayer two-As-dimer structure is the most energetically stable and has the lower energy by 0.045 eV per (1×1) unit cell compared with the three-As-dimer structure.

The fully relaxed atomic structures for three different models and corresponding geometric parameters obtained by Schmidt and Bechstedt are shown in Fig. 8.6 and Table 8.1, respectively. The main conclusions can be summarized as follows. The minimum energy configuration has asymmetric As dimers only for the bilayer two-As-dimer structure with a buckling of

Table 8.1. Geometric parameters (in [Å]) for the relaxed GaAs(100)-(2×4) β, $\beta 2$, α and α' structures, according to Fig. 8.6

GaAs(100)-(2 × 4)	β	$\beta 2$	α	α'
$\Delta_{a1,x}$	2.50	2.50	2.50	2.50
$\Delta_{a2,x}$	2.50	–	2.50	2.33
$\Delta_{b1,x}$	3.65	3.65	3.85	3.66
$\Delta_{b2,x}$	3.49	3.52	3.64	3.58
$\Delta_{b3,x}$	–	–	3.50	–
$\Delta_{b4,x}$	–	–	3.66	–
$\Delta_{c,x}$	–	2.52	–	–
$\Delta_{d,x}$	–	3.70	–	–
$\Delta_{a,y}$	3.80	3.82	3.92	3.87
$\Delta_{b,y}$	–	–	2.50	–
$\Delta_{ab,y}$	1.42	1.41	1.50	1.39
$\Delta_{ab',y}$	–	1.44	1.73	–
$\Delta_{ac,y}$	–	5.92	–	–
$\Delta_{a,\perp}$	0.02	–	0.06	–
$\Delta_{b12,\perp}$	0.24	0.28	0.20	0.27
$\Delta_{b23,\perp}$	–	–	0.19	–
$\Delta_{b34,\perp}$	–	–	0.22	–
$d_{ab,\perp}$	1.47	1.49	1.42	0.88

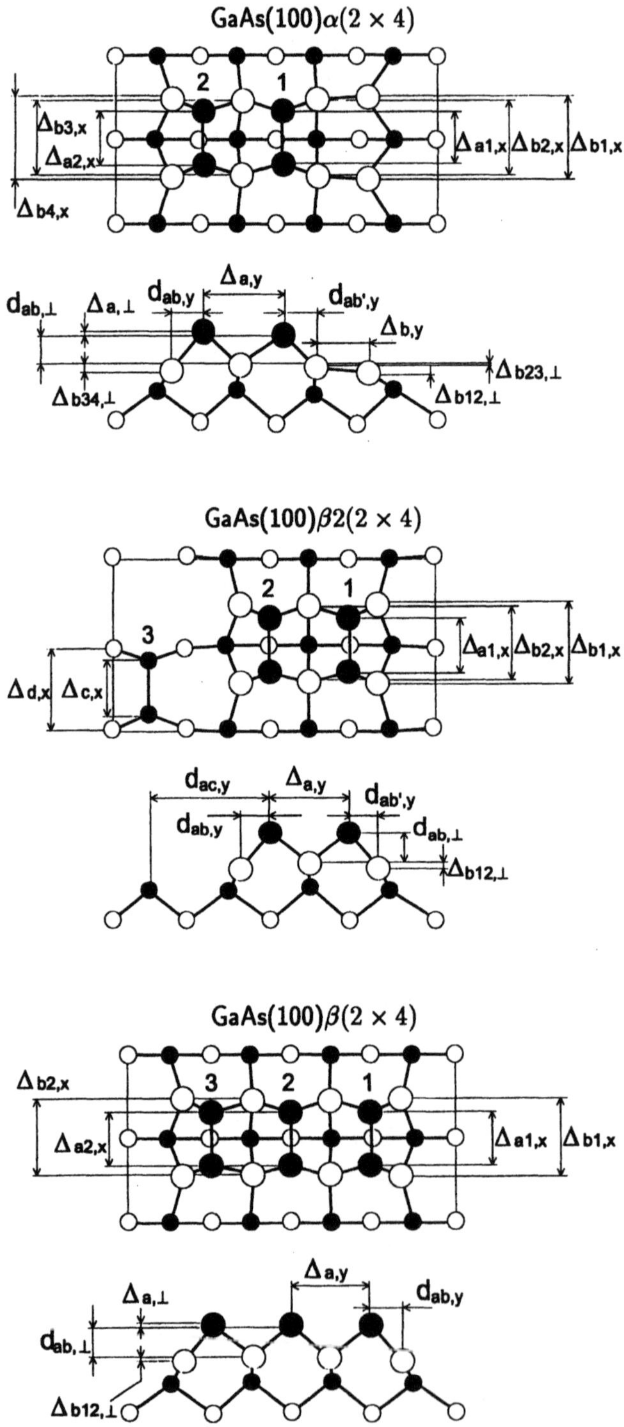

Fig. 8.6. Relaxed structures of the GaAs(001)-2 × 4 surface calculated by Schmidt and Bechstedt [78]

0.02 Å. The As–As dimer length for all three structures is basically the same and is 2.5 Å. This result is very close to the value of 2.51 Å for the threefold coordinated bulk As [82] and in good agreement with a value of 2.52 Å by Qian, Martin and Chadi [45]. It is also comparable to the experimental results: 2.2 Å [83], 2.4 Å [84], 2.73 Å [85], 2.89 Å [86], and other theoretically predicted values: 2.39 Å by Srivastava and Jenkins [87] and 2.60 Å by Shiraishi and Ito [88]. The threefold coordinated second layer Ga bonded to the As dimer are displaced from their ideal lateral position by approximately 0.65 Å toward the As dimers. This leads to the upward movement of the As dimers and to steepening of the dimer block. As a result, the minimum interplanar distance d_{ab} between the first and second layers changes to 1.42–1.49 Å from the bulk value of 1.39 Å, and the As dimer–dimer distance along the 4× direction becomes 3.8–3.9 Å, slightly smaller than the surface lattice constant of 4 Å.

In addition to these features common to all three structures, the calculated second layer Ga–Ga bond for the two-As-dimer structure has a length of 2.51 Å, close to the sum of the Ga covalent radii (2.52 Å). Schmidt and Bechstedt also discuss the details of the electronic structures [78]. In all three cases, they found that the bound surface states are characterized by dimer bonding π and antibonding π^* combinations of the P_z orbitals localized at the As dimers and near the bulk VBM in energy, and empty Ga dangling bonds near the bulk CBM. This implies that the fundamental gap of GaAs is nearly free of states. Experimentally observed Fermi level pinning [35] may thus be due to intrinsic defects on the surface, such as kinks [58]. Details of the surface band structures are presented in Figs. 3, 5 and 7 of [78].

The charge density distributions for three different models are shown in Fig. 8.7a, b and c, respectively [78]. For the three-As-dimer structure, the charge density above the middle dimer is more prevalent (see Fig. 8.7c), in agreement with Ohno's calculations [74]. Thus, if the surface contains three As dimers in a unit, the features for the middle dimer and two other dimers in STM images are expected to be non-equivalent. In the case of the two-As-dimer structure with dimerized Ga in the second layer (Fig. 8.7a), the HOMO state (V1) is localized at the As dimer 1 that corresponds to the antibonding π^* combination of the P_z orbitals of the dimer. A state (V2) below HOMO corresponds to the bonding π combination of the P_z orbitals also localized at As dimer 1, while V3 is related to an antibonding π^* combination of the P_z orbitals at As dimer 2.

All these are accessible by STM. It is expected that tunneling from these states makes the main contribution to the bean-like features observed in STM images. The As dimer close laterally to the rebonded Ga dimer in the second layer should be imaged slightly brighter. Figure 8.7b shows the charge density distribution for the bilayer two-As-dimer structure. Here the difference with the Ga-rich two-As-dimer model is that the states localized at the two top-layer As dimers have nearly identical charge densities. We therefore

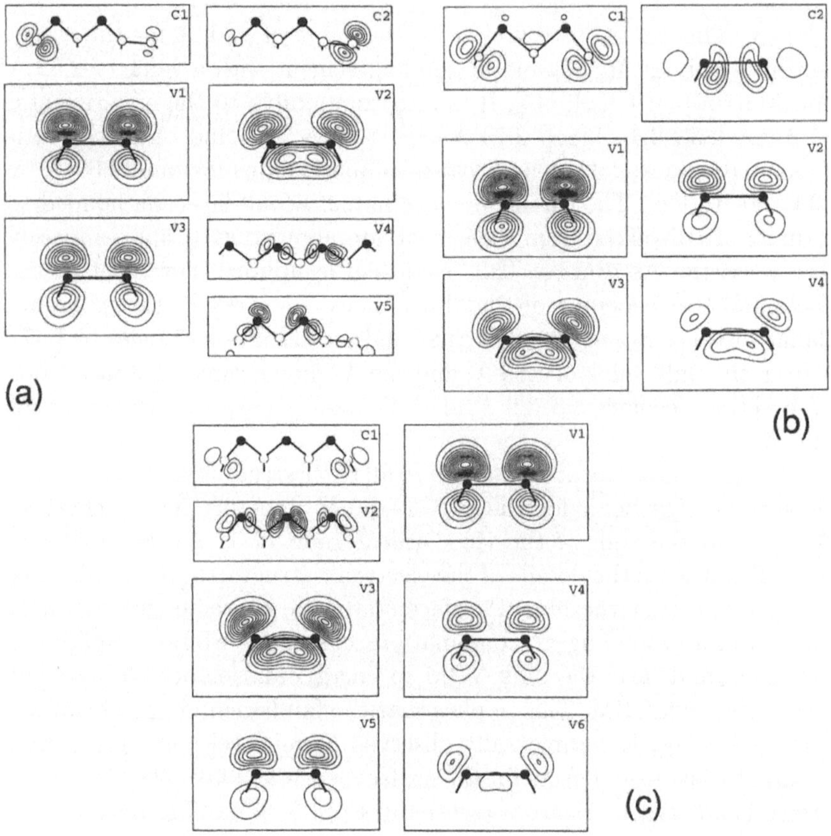

Fig. 8.7. Charge density distributions for the GaAs(001)-2×4 structures calculated by Schmidt and Bechstedt (**a**) for the two-As-dimer structure with the second-layer Ga dimerized in the missing row, (**b**) for the bilayer two-As-dimer structure with third-layer As dimerized and (**c**) for the three-As-dimer structure [78]

expect symmetric features for two surface dimers in STM images along the [110] direction. Considering the fact that the characteristic RHEED pattern of the 2 × 4 can be used as a fingerprint, Hashizume et al. [89,90] performed the first systematic MBE-STM study of the α, β and γ phases of the surface, and attempted to establish a clear connection between the diffraction pattern and corresponding atomic structure for each phase in order to confirm the well accepted three phase scheme suggested by Farrell and Palmstrom [71]. The characteristic RHEED patterns and corresponding debroadened spot intensity profiles, measured using a photomultiplier with pinpoint fiber for the α, β and γ phases, are shown in Fig. 8.8. Figures 8.9a, b and c show typical color-code STM images of the (2 × 4)-α, β and γ phases, respectively, with insets showing enlarged small area images and typical depth profiles measured along the [110] direction. When the resolution of some STM im-

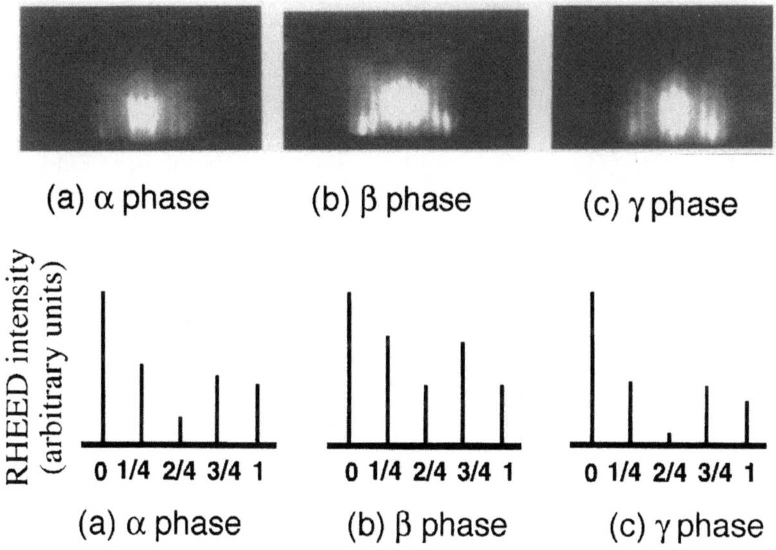

Fig. 8.8. Experimental RHEED patterns for the (**a**) α, (**b**) β and (**c**) γ phases and their debroadened spot intensity profiles for (**a**) α, (**b**) β and (**c**) γ phases [89]. The results are in good agreement with the description by Farrell and Palmstrom [71]

ages is high enough to resolve individual As dimers [5,6], the enlarged images and contour profiles (Fig. 8.9) clearly show that the 2×4 units at the outermost surface layer of the α, β and γ phases all consist of two As dimers and two dimer vacancies. Regardless of the preparation method, almost all STM images taken after 1990 show the dominant two-dimer structure at the top layer of the surface [8,47–53,56,91–96], with the exception of the study by Broekman et al. which shows a clear three-As-dimer structure [97]. (This three-dimer structure will be discussed below.) Wassermerier et al. reported the first empty states STM image of the 2×4 surface of p-type doped samples [93]. They found that the empty states images are tunneling from the same As position as the filled states. This has been attributed to tunneling through empty As orbitals that lie in the conduction band with the energy degeneracy of the empty Ga atomic orbitals and As molecular orbitals (the high energy antibonding orbitals) [98].

Besides the results for the top layer structure, Hashizume et al. also analyzed the surface ordering of each phase [89,99], and found that the β phase (Fig. 8.9b) is the most ordered with a 2×4 domain size of about 300 Å along the $[\bar{1}10]$ direction and 3 000 Å along the [110] direction [99]. The domain size of the α phase is typically 60 Å and 500 Å in the and $[\bar{1}10]$ and [110] directions, respectively. In the γ phase, the kink density is similar to that of the α phase but kinks distribute randomly, and the domain size in Fig. 8.9c is approximately 60 Å by 100 Å. Figure 8.10 summarizes the depth profile in the As dimer vacancy measured from the position of As dimers at the outermost

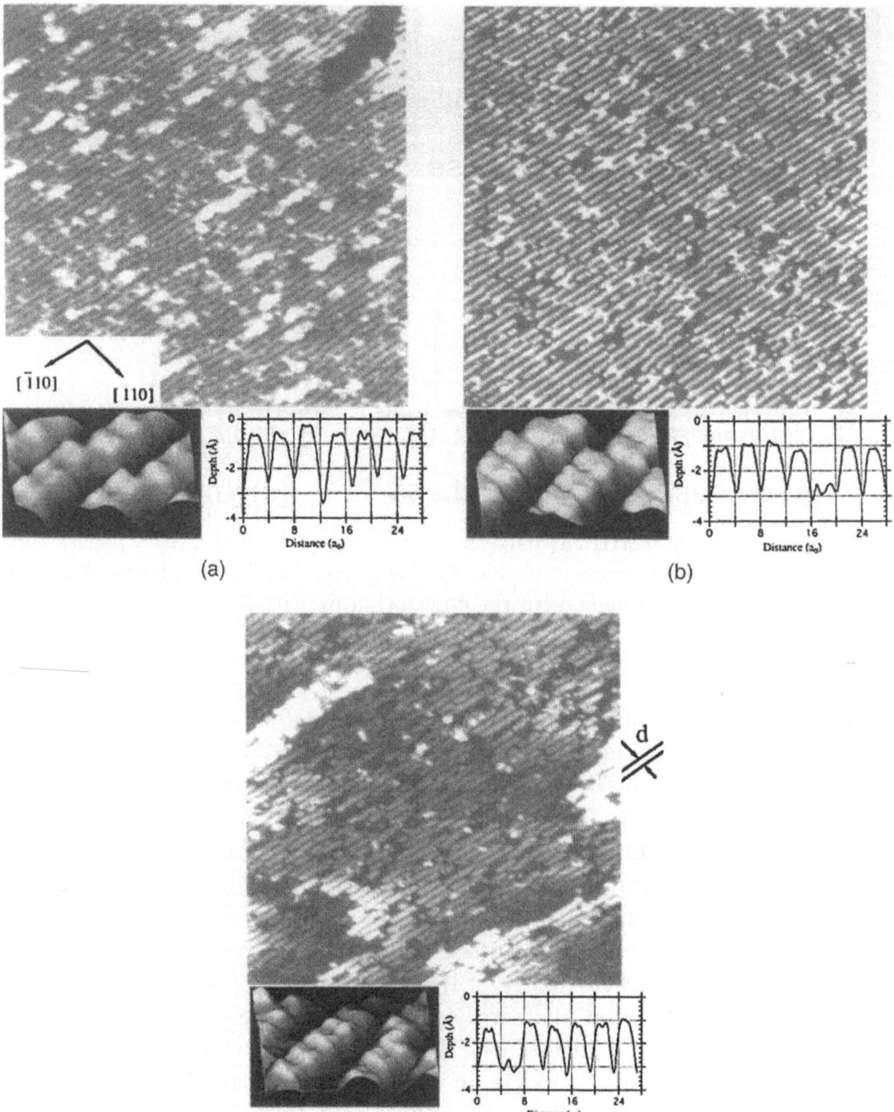

Fig. 8.9. Typical 3-dimensional display STM images of the (**a**) α, (**b**) β and (**c**) γ phases of the GaAs(001)-2×4 surface (800 Å × 800 Å) together with *insets* showing the magnified images to display the details and their depth profiles measured along the [110] direction. In the STM images, it is evident that the bright lines are running along the [$\bar{1}$10] direction, separated typically by $4a_0$ along the [110] direction. All (2 × 4) unit cells in these three phases consist of two As dimers and two dimer vacancies [89]

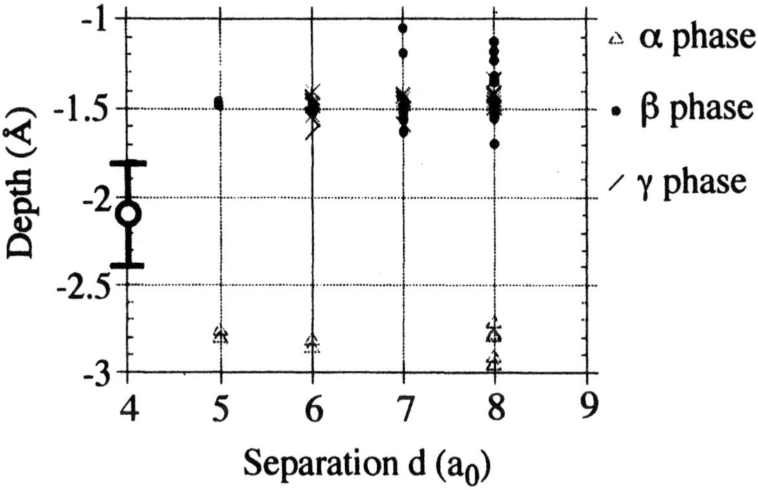

Fig. 8.10. Plot of the depth of the dimer vacancy region for three different phases of the GaAs(001)-2 × 4 surface, measured from the position of the top layer As dimers, as a function of the separation d (in units of a_0) [89]

surface layer as a function of separation d between neighboring (2 × 4) units (in units of a_0). The most important result is that for separation $d > 5a_0$. Here, they observed the reasonable 2.8 Å depth in the α phase, while the depth in the β (γ) phase was about 1.4 Å, one half of the bilayer step height. Observation of the $H/2$ depth for the β and γ phases implies that there must be some unknown subsurface structure in the vacancy region in the presence of these phases.

In order to fully understand the second and third layer structures of the α, β and γ phases, Hashizume et al. calculated the RHEED spot intensities for the possible (2 × 4) models using the dynamical theory developed by Ichimiya [100–102]. The results of calculated spot intensities for the various models proposed are summarized in Fig. 8.11. The result of the kinematical calculation for this model, taking the second layer into account, is shown in Fig. 8.11e–g [71]. By comparing the calculations (Fig. 8.11) with experimental intensity profiles (Fig. 8.8), they concluded that:

- the α phase corresponds to the two-As-dimer model with second layer Ga dimerized, which was proposed by Frankel et al. [46], with relaxation introduced by Northrup and Froyen [73];
- the β phase corresponds to the bilayer two-As-dimer model proposed by Chadi [21];
- the γ phase is the locally ordered β phase with open areas carrying a disordered As double-layer structure similar to that of the $c(4 \times 4)$ phase [103,104].

Figure 8.12 summarizes the structural models for the α, β and typical γ phases proposed by Hashizume et al. [90]. The two-As-dimer structures for

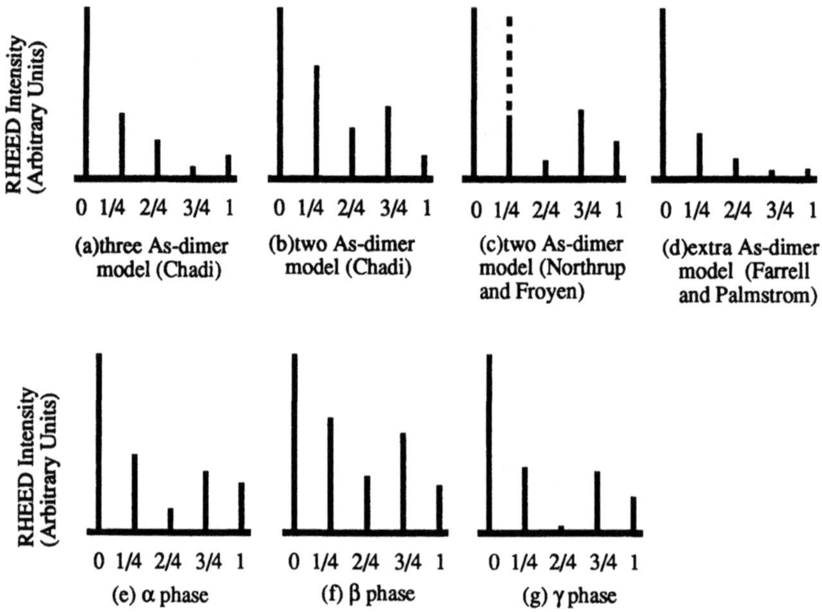

Fig. 8.11. Theoretical RHEED spot intensity profiles for various models of the GaAs(001)-2 × 4 surface. (**a**) to (**d**) were calculated using the dynamical theory discussed in the text. (**e**) to (**g**) are based on the kinematical calculation [89].

the α and β phases were later confirmed by a grazing incidence XRD study of the GaAs(001) As-rich 2 × 4 reconstruction [105]. The stability of the mixed structure for the γ phase was theoretically justified [79]. These results agree well with the main conclusion of the theoretical studies. They unambiguously demonstrate that the bilayer two-dimer structure is the most stable and can indeed be used as a reliable model for this extensively studied 2 × 4 surface (the β phase).

As for the three-dimer structure, Broekman et al. [97] did observe an STM image containing three clearly resolved and separated bean-like features in a unit cell. This indicates that the three-dimer structure does exist. It occurs as a metastable structure in certain regions, such as near step edges or defects, and only occupies a small portion of the surface area under unique preparation conditions, as suggested by Ohno and also by Schmidt and Bechstedt in their theoretical calculations [76,78]. However, a mere observation of such a structure may not imply that it represents an equilibrium phase, i.e., the most stable β phase. For identification of the most stable β phase, the strong 2/4th fractional diffraction feature should be used as a fingerprint [71], and its preparation should be carried out under near-equilibrium conditions. (For example, the β phase is usually prepared by annealing at growth temperature under the As flux for several minutes.)

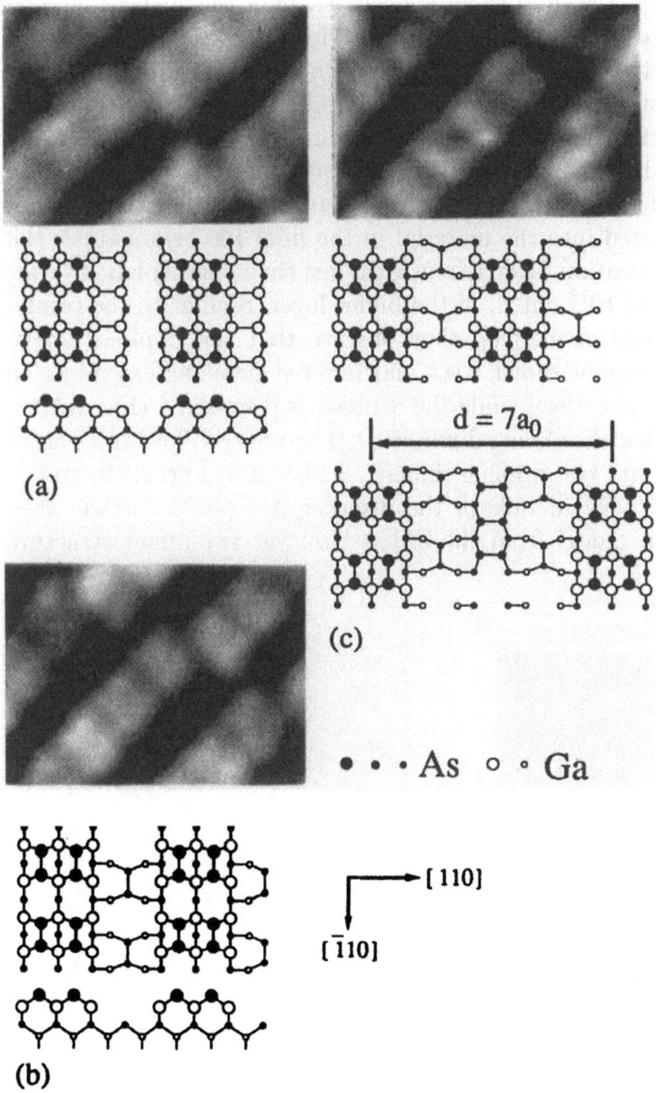

Fig. 8.12. Typical STM images and proposed structure models for the (**a**) α, (**b**) β and (**c**) typical γ phases of the GaAs(001)-2 × 4 surface [89]

Disordering Mechanism. According to the study by Hashizume et al., the structural disordering in the γ phase is responsible for absence of the 2/4th intensity. This can also be an electronic effect induced by heavy n-type doping of the sample and has been used to study the Fermi level pinning effect by Pashley and Haberern [48,58]. By studying the dependence of kink density on doping level and the doped-layer thickness, they found that the kink density is exactly equal to the number of doped atoms in depletion layers, which are donors, until the Si concentration reaches 1×10^{19} cm^{-3}. Kinks therefore act as surface acceptors and each of them is occupied by a single electron. The formation of these acceptors allows electrons to drop from the Fermi level to the low energy acceptor states, eventually resulting in Fermi level pinning.

A careful STM study of the disordering mechanism in the γ phase of the 2×4 As-rich surface has recently been performed by Avery et al. using STM observation and tight-binding calculations [94]. In order to prevent the formation of kink defects due to charge compensation discussed above, no Si dopant was incorporated into the material in the final 100 layers above the buffer layer. This prevention layer is much thicker than the depletion layer, of low Si concentration 10^{18} cm^{-3}, of the buffer layer. Similar to the results obtained by Hashizume et al. [90], they observe that the β phase image is characterized by straight dimer rows and ordered domains extending to several hundred Å (Fig. 8.13a), while the γ phase is disordered (Fig. 8.13c), leading to the 1/4th features being diminished. The unit cell for both phases contains two dimers and two missing dimers, as shown in Fig. 8.13b and d. To establish the structural details of the disordered γ phase surface, they proposed a transition model from the ordered bilayer two-dimer structure

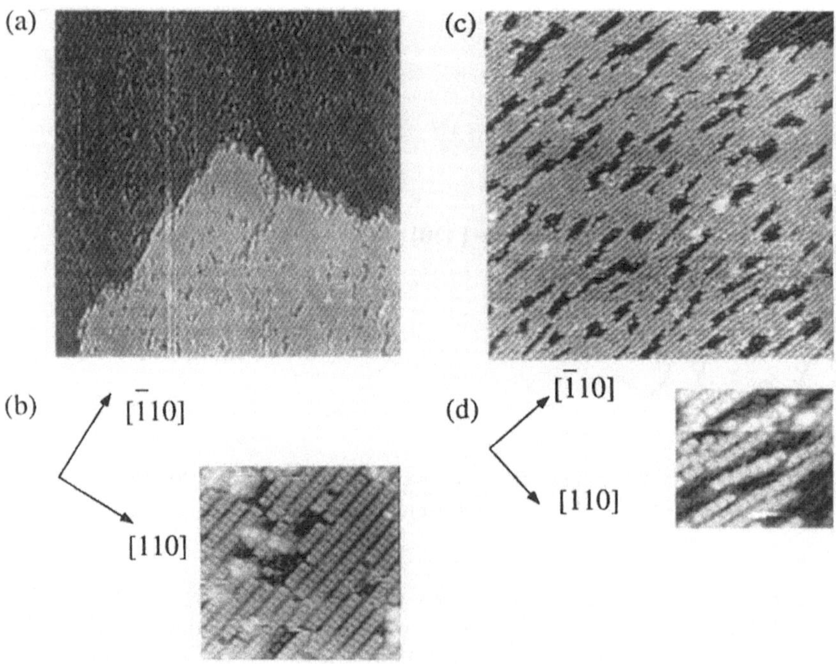

Fig. 8.13. Filled state STM images for (**a**), (**b**) the β phase and (**c**), (**d**) the γ phase of the GaAs(001)-2×4 surface. Image dimensions are (**a**) 1200 Å×1200 Å, (**b**) 200 Å×200 Å (**c**) 1000 Å×1000 Å , and (**d**) 190 Å×190 Å. The depth of the missing dimer trenches is 1.4 Å. The depth of the holes in (**c**) is 2.8 Å, i.e., one bilayer along the GaAs(001) crystal axis [94]

to its disordered one. To do this, they first employ an indirect method to evaluate the kink energy for different structural models, defined by

$$E_{\text{kink}} = \frac{E_{\text{bind}}^{\text{kinked}} - E_{\text{bind}}^{\text{perfect}}}{N_{\text{kinks}}} \,. \tag{8.3}$$

The calculation results show that the kink energy (1.02 eV) of the bilayer two-dimer structure is almost ten times higher than that for the two-dimer structure with second layer Ga-dimer or the three-As-dimer structure, due to breaking of the third layer As dimer in order to create a kink (Fig. 8.14a). This explains the high quality ordering in the β phase. They further found that when additional As atoms are allowed to adsorb on the β phase surface, the lowest energy adsorption site for an As atom is the vacant second layer Ga site, which results in a hetero-bond to one of the three coordinate Ga atoms in the second layer (see Fig. 8.14b). The kink energy for the configuration shown in Fig. 8.14b is -0.20 eV, indicating that the presence of extra As kinks spontaneously formed in the missing dimer row leads to the formation of the disordered γ-2×4 phase (Fig. 8.14c). This process is shown to occur

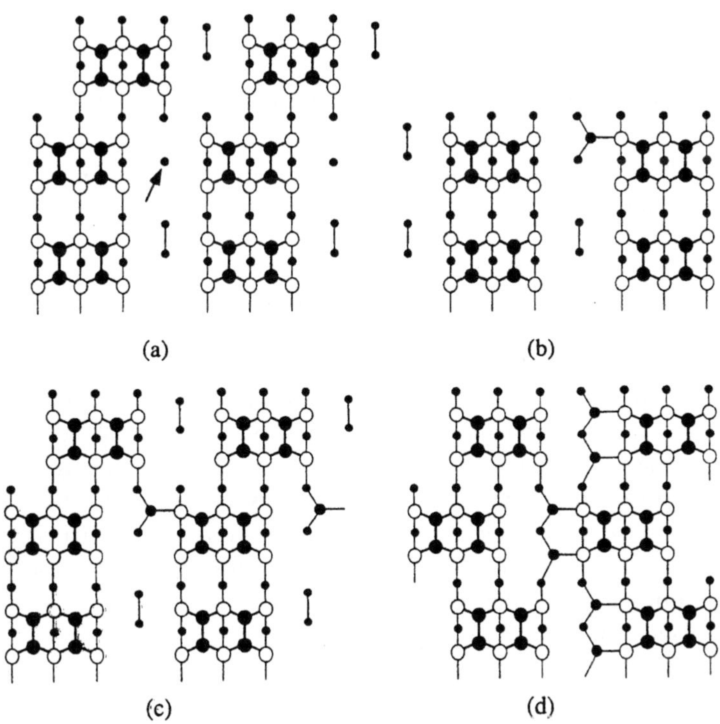

Fig. 8.14. A model based on tight-binding calculations showing the transition with additional As from the β phase to the γ phase GaAs(001)-2×4 surface. The As occupies vacant Ga sites in the missing dimer trenches of the initial bilayer two-As-dimer structure leading to disorder via kink formation. *Filled circles* represent As atoms and *open circles* represent Ga atoms [94]

for a range of extra As coverage from 0.016 to 0.25 ML, which gives rise to an upper limit for As coverage of about 1 ML in the γ phase [90]. In addition, they emphasize that the additional As prefers to be isolated, i.e., randomly distributed. The structure with a single kink per As atom is more stable than one with two As atoms per kink (Fig. 8.14c). This implies that As tends to form a kink for coverages up to 0.125 ML, whereas from 0.125 to 0.25 ML it simply fills the trenches without forming more kinks. The kink structure involves a single additional As atom, and may be a good acceptor since there is a deficiency of one electron per kink. Therefore, Avery et al. have speculated that the present kink forming mechanism and the one induced by Si doping may be closely related, and that the Si-doped surface should contain more As than the unkinked surface [58].

Step Structure. Using RHEED, Pukite et al. [106] have differentiated two types of step on the vicinal surface miscut towards (111)-A and (111)-B, respectively. In this study, they found that steps (called A steps) running parallel to the 2×, i.e., $[\bar{1}10]$, direction are straight over long distances, whereas steps (called B steps) along the 4× direction are highly kinked. This conclusion was confirmed experimentally using STM by Pashley et al. [64]. The STM images also revealed that the kink always occurs in units of the complete 2×4 unit cell. The origin has been analyzed in terms of ECM by Pashley [20]. The 2×4 unit cell is energetically stable because there are exactly the ideal number of electrons to allow all As dangling bonds to be filled with two electrons, and all Ga dangling bonds to be completely empty. Island structures made up of defect-free 2×4 unit cells extending in the 2× direction satisfy ECM, and should thus be stable. These correspond to structures that grow out from the B steps, resulting in ragged B-type step edges [64].

Heller and Lagally studied the surface morphology of the nominally flat GaAs(001) surface using STM [49]. Their samples were prepared at low growth rate (0.05 μm/h) and low growth temperature of 610°C, followed by 15 min annealing at the growth temperature. The STM images show that the grown front consists of a multi-level system of terraces that are elongated parallel to the 2× direction, occupying six or seven levels over a distance of 8 000 Å along the 4× direction, but only one level over a distance of 6 000 Å in the 2× direction. Assuming that the observed structures nearly represent the equilibrium configuration, the results immediately suggest a B-step to A-step energy anisotropy of $\sim 10:1$.

With this step energy anisotropy, one can easily understand the different step structures on the vicinal surfaces [106]. Kinks in B-step edges are A-step and more easily made since the A-step site energy is low, while kinks in A-step edges are B-step and more difficult to form due to high formation energy of the B step. However, since the morphology of the surface they studied corresponds more or less to a growth shape, the observed terrace anisotropy also indicates a possible sticking coefficient anisotropy with preference for

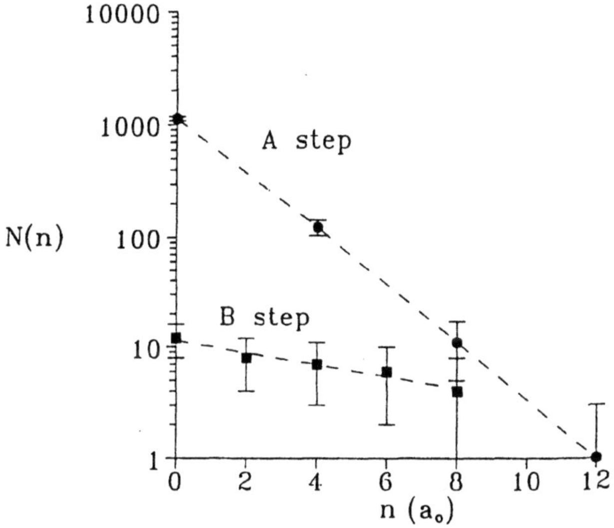

Fig. 8.15. Kink length populations for A and B steps on the GaAs(001)-2 × 4 surface measured using STM by Heller, Zhang and Lagally. *Dashed lines* represent a best fit exponential to the data. Kinks in A steps appear only in multiples of $4a_0$, while kinks in B steps appear only in multiples of $2a_0$ [65]

sticking at B steps, and/or diffusion barrier anisotropy, and only represents the upper limit of the equilibrium shape anisotropy [49]. The same group also used STM images of the equilibrium structure of A- and B-type steps on 0.5°-off (toward $[\bar{1}10]$ or [110]) vicinal GaAs(001) substrates to make quantitative estimates of step and kink energies [65]. They counted the number of kinks $N(n)$ of varying lengths n as well as the distribution of separations s between kinks in STM images of A and B steps. Here the length n of a kink is defined in terms of the number of units between an inward corner and an outward corner. For the A step, n occurs in units of $4a_0$ with s in units of $2a_0$, and conversely for the B step. Figure 8.15 shows the distribution of kink length $N(n)$ in A and B steps. This distribution allowed them to estimate the energy ε of a kink of given length in the A- and B-type steps. The results are shown in Fig. 8.16, in which the slope of the lines for the A step is the energy ε_B required to form B steps and vice versa [65]. The intercept, ε_C, is an effective corner energy that arises because corners are always present regardless of the length of the kink.

Taking 573 K as a low limit of the temperature for mobile kinks, they deduced a value of 5 ± 3 meV$/a_0$ for ε_A, 30 ± 5 meV$/a_0$ for ε_B and 35 ± 15 meV for ε_C, respectively. Thus the energy anisotropy ratio of the A step over the B step is 1 : 6. In addition, this statistical analysis demonstrates a short-range nearest-neighbor kink repulsive interaction due to charge transfer from Ga to As. The strength of the interaction is estimated to be 20 meV for opposite-

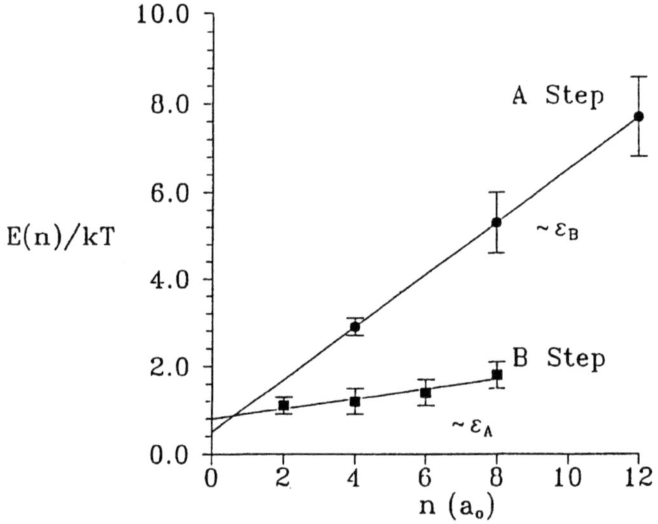

Fig. 8.16. Measured energy $E(n)/kT$ of a kink of length n atoms in the A- and B-type steps of the GaAs(001)-2 × 4 surface. *Solid lines* represent least-squares fits of form $E(n)/KT = n\varepsilon_{\text{step}}/kT + \varepsilon_C/kT$, where $\varepsilon_{\text{step}}$ is the energy cost per unit length to create an A or a B step from a flat terrace. By creating kinks in the A-type steps, segments of the B-type step are formed, and vice versa. The *slope* of the line for the B step gives the energy per unit length of the A step and vice versa [65]

sense kinks separated by $[\bar{1}10]$, which effectively eliminates all opposite kinks at $[\bar{1}10]$, as observed in their STM experiment.

8.3.2 $c(4 \times 4)$ Phase

The $c(4\times4)$ reconstruction can be observed during MBE growth of GaAs under rather extreme conditions of low temperature and high As flux. Otherwise it can be produced by cooling down the 2×4 surface in an As flux. The chemical composition and structure of the $c(4\times4)$ phase have been previously studied by LEED and AES [32], photoemission [37,39,67], RHEED [69,70,107], SEI [69], and work function measurement [108]. A wide range of surface As coverage from 0.86 ML to 1.75 ML has been reported. However, most previous studies have not identified the structure. This is partly because of the appearance of the fourfold symmetry that the reconstruction exhibits, which seems to contradict the 'native' twofold symmetry supposedly induced by the dangling bond orientation on the surface. The first theoretical study of the atomic structure of the $c(4 \times 4)$ reconstructed GaAs(001) surface was performed by Chadi et al. using an energy-minimization approach [42]. On the basis of the dimerization scheme, and using the previously established fact that the surface is stoichiometrically As-terminated, they suggested that two distinct types of As–As dimer are involved. The first is the asymmetric or

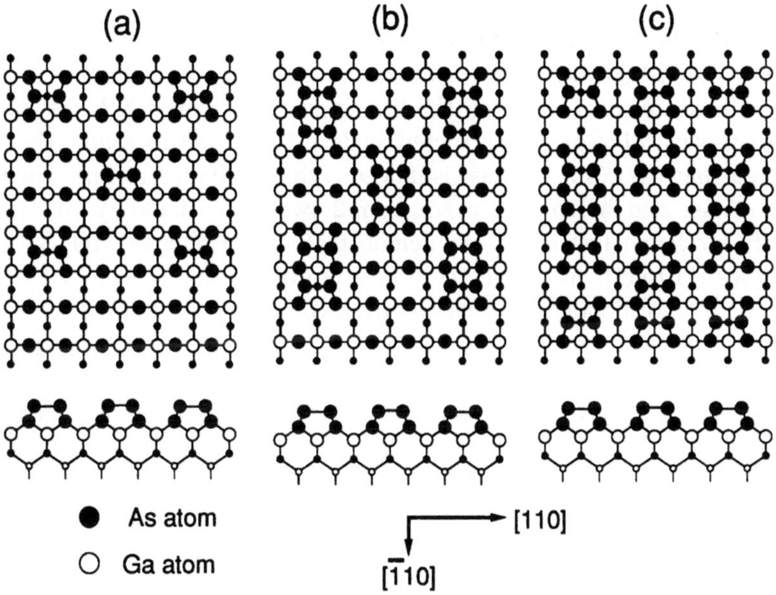

Fig. 8.17. Ball-and-stick models proposed for the GaAs(001) $c(4 \times 4)$ reconstruction: (**a**) and (**b**) by Larson et al. [67], (**c**) by Sauvage-Simkin et al. [109]. All three structures contain a complete layer of As atoms adsorbed with extra amounts of As. Additional As coverages for the structures in (**a**), (**b**) and (**c**) are 25%, 50% and 75%, respectively

noncoplanar dimer and the second is symmetric (coplanar). This theoretical model of the $c(4 \times 4)$ reconstruction contains equal amounts of these two types of As dimer on the bulk-like Ga-terminated surface. The surface constructed from these dimers has been shown to be semiconducting and stable. Nevertheless, the surface As coverage is limited to one monolayer of As. This model was questioned by a combined study using RHEED with angle-resolved photoemission and core-level photoemission by Larson et al. [67]. They found that the measured high surface As coverage (more than one monolayer) together with amorphous As-3d features and more importantly an absence of asymmetric dimer bond band, cannot be properly described in the model by Chadi et al. [42]. This analysis led them to identify the $c(4 \times 4)$ as an overlayer derived phase, based on chemisorbed As trigonally bonded to one complete layer of As atoms. The ball-and-stick models proposed by Larson et al. are shown schematically in Fig. 8.17 [67]. Both structures have the twofold symmetry, but the unit cell is square.

The models satisfy ECM, assuring the experimentally observed semiconducting property. The resulting extra As coverages for the two structures are 1/4 and 2/4 ML, respectively. They further emphasize that these two structures should not be regarded as distinctly different models, but that both can coexist as the amount of adsorbed As changes. This explains the existing wide range of surface As coverage for the given $c(4\times4)$ diffraction symmetry. According to the model, the maximum As coverage is 1.5 ML, corresponding to a surface covered by complete layer of the structure shown in Fig. 8.17b.

Sauvage-Simkin et al. performed the first grazing incidence XRD study of the MBE-grown $c(4 \times 4)$ reconstructed surface [109]. In this study, they demonstrated that the $c(4\times4)$ phase is indeed a chemisorbed structure with two layers of As, and belongs to a true $mm2$ two-dimensional symmetry group. Such a scheme has also been demonstrated to occur in MOCVD-grown samples [110,111]. Although the four-atom-block model proposed by Larson et al. [67] is still valid, in order to fully explain their diffraction data, they pointed out that varying As-dimer ordering is possible. An alternative model was proposed, which is shown in Fig. 8.17c. In reality, the $c(4\times4)$ surface adjusts its structure between these two extremes (Fig. 8.17b and c) so as to accommodate different amounts of As. This six-atom cluster structural unit arranged in three pairs was further examined by two first-principles total energy studies [72,76]. Both studies concluded that under As-rich conditions the $c(4\times4)$ structure shown in Fig. 8.17c is always energetically favored. This has now been accepted as the standard model for the $c(4 \times 4)$ reconstruction. Kamiya et al. suggested that the two-extra-As-dimer structure shown in Fig. 8.17b can be better regarded as a transition phase [111] and may be locally ordered. The fluctuation in surface coverage (less than the ideal 1.75 ML expected from the model shown in Fig. 8.17c) may be due in part to a locally existing two-dimer structure for the 2×4 ordering, which is not accessible to all diffraction techniques.

Unambiguous confirmation of the square arrangement of the As clusters forming the $c(4 \times 4)$ symmetry was achieved by the first in situ MBE-STM study of GaAs(001) [6]. The high resolution STM images obtained by Biegelsen et al. clearly show that each cluster block consists of three dimers. These dimers are found to be aligned perpendicular to the As dimers on the 2×4 surface, indicating that they correspond to the chemisorbed structure. Since only three-dimer blocks were observed, Biegelsen et al. explained that a wide range variation in the composition of the $c(4\times4)$ phase is due to occurrence of the local 2×2 structure, and possibly due to the mixture of As and Ga atoms in the underlying layer. These basic features of the $c(4\times4)$ surface have also been observed by Hashizume et al. [90], Tanaka et al. [112] and Avery et al. [113] with slightly different preparation methods. In Fig. 8.18, we show the high resolution images of the $c(4 \times 4)$ reconstruction obtained by Avery et al. [113].

Here, the six-atom unit cell suggested by Sauvage-Simkin et al. [109] is imaged as a bright rectangular block and, within a block, six individual features are visible (see Fig. 8.18a). Another important observation from this study is that no image shows either a single or double pair of As atoms aligned along the [110] direction. This excludes the possibility of two-atom and four-atom units occurring as suggested by Larson et al. [67] (see Fig. 8.17a and b). The

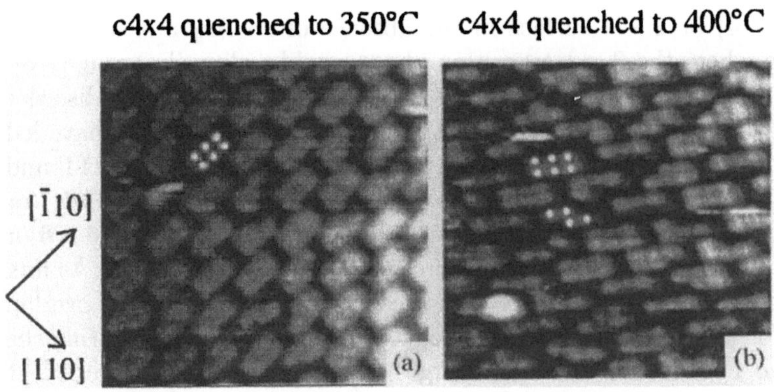

Fig. 8.18. Filled state (−3 V, 0.5 nA) STM images of the GaAs(001)-$c(4 \times 4)$ surface prepared by quenching an initial As-terminated 2×4 surface to (**a**) 350°C (70 Å×70 Å) and (**b**) 400°C (80 Å×80 Å) in the presence of an As_4 flux. Identical RHEED patterns were obtained in both cases [113]

wide range in surface composition reported before is not due to fundamentally different structural units, but rather to an increasing number of missing atoms from the basic six-atom unit, as shown in Fig. 8.18b. This is probably the most reasonable explanation of the varying surface composition reported by different groups. It is noteworthy that the central dimer of each three-dimer block appears darker than the other two. We will discuss this case in the section on InSb(001)-$c(4 \times 4)$.

In summary, the structure of the GaAs(001)-$c(4 \times 4)$ As-rich phase can be best described by the model proposed by Sauvage-Simkin et al. [109]. The surface coverage of the ideal $c(4 \times 4)$ structure is 1.75 ML As. The varied surface coverages measured experimentally are due to desorption of atoms from this ideal structure. This desorption takes place by removal of atoms from the dimers, not by removal of dimers from the unit, implying that the one- or two-dimer models shown in Fig. 8.17a and b may not be stable.

8.3.3 1 × 6 Phase

The GaAs(001)- 1×6 phase was first reported on a surface cleaned by argon-ion bombardment using LEED by Jona [26]. At that time, this structure was called GaAs(001)-6 because apparent diffraction spots were observed only for diffraction along the [110] direction. Using RHEED, Cho investigated the first MBE-grown GaAs(001)-6 structure where the surface exhibits 6×6 diffraction patterns in RHEED, although a 2×6 structure was also observed [27,114]. Generally, the clean (001) surface sputtered with argon ions and annealed at around 500°C displays a sharp 1×6 LEED diffraction pattern [32,33,35]. This structure can also be obtained by simply cooling down the Ga-rich 4×2 surface under a low As background pressure (5×10^{-9} torr) [33], or by heating the $c(4 \times 4)$ or 2×4 surface in the absence of an As_4 flux above the temperature where the 3×1 transition phase would ordinarily occur [108]. In any case the LEED pattern is always accompanied by $(1/6)nY^*$ disorder lines or $2 \times /3\times$ intermediate streaks. The $(1/6)nY^*$ disorder lines have led several workers to attribute this surface to a 6×6 phase, e.g., Cho [114] and Kuball et al. very recently [115]. However, the occurrence of the $(1/6)nY^*$ or $2 \times /3\times$ streaky lines implies that the surface does not have a true 6×6 or $2 \times 6/3 \times 6$ translational symmetry. Moreover, the temperature and As flux windows for the 1×6 ordering are found to be very small and often overlap with those for the 3×1 and 4×6 phases [108]. Furthermore, considering the fact that the 1×6 structure only emerges when the As flux is insufficient to form the 2×4 reconstruction [111], it should not represent a thermodynamically stable phase. It is therefore more appropriate to describe it as a non-equilibrium or 'transient' structure. This may be the reason why it was not considered in earlier theoretical work [72,74]. Any successful structural model should include a structure disorder from which the intermediate streaks result.

Real-space STM images have played a key role in understanding this surface [6,8,115,116]. Basically identical images were also observed for the MOVPE-grown samples by Li et al. [96]. These STM images show an apparent $\times 6$ periodicity characterized with alternative zigzagging bright rows and missing vacancy rows running along the $[\bar{1}10]$ direction, similar to what is observed for the 2×4 surface. These bright rows have been identified as tunneling from the first layer As dimers [6,8,115–117], suggesting that the surface is As-terminated. Typical filled and empty state images are shown in Fig. 8.19a and b. However, the distribution of As dimers in the row is found to be intrinsically irregular (see the zoom-in image in Fig. 8.19c). The stacking disorder in the $[\bar{1}10]$ direction must be such that the LEED observation always shows streaks corresponding to 1×6 [6], rather than spots.

The ball-and-stick model for the ideal 2×6 ordering is shown in Fig. 8.20a. The appearance of both As and Ga dimers in this structure has been independently confirmed by RDS [111] and infrared study [118]. For the surface that exhibits 1×6 LEED spots with $(1/6)nY^*$ streaks, Kuball et al. reported STM images showing the $6\times$ periodicity for the features in the missing dimer rows [115]. They proposed a structure model based on the findings that the top-layer surface consists of regular two-As-dimer rows, but that there is $6\times$ ordering in the missing row region (Fig. 8.20b). Xue et al. also examined the second layer structure by STM and found that the features in the missing dimer row region display a well defined $2\times$ ordering (see Fig. 8.21). These studies indicate that the surface structure is actually rather complicated. Since both structural models shown in Fig. 8.20 eventually result in sharp 2×6 or 6×6 LEED spots rather than disordered streaks, the actual atomic arrangement/ordering and surface coverage in the surface layers may

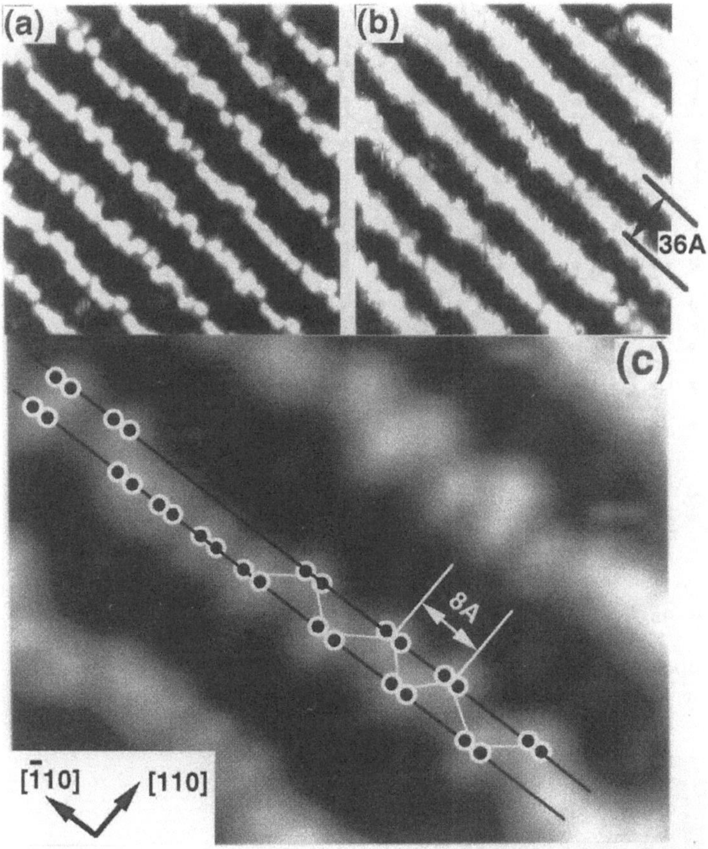

Fig. 8.19. Dual bias STM images of the GaAs(001)-1 × 6 surface: (**a**) filled states at −2 V, (**b**) empty states at +2.5 V. They clearly show that the bright lines are tunneling from the same As regions. (**c**) The zoom-in filled state (−2.5 V) image (60 Å×85 Å) of the surface, displaying the irregular distribution of As-dimers in a row [116]

depend strongly on growth kinetics that the preparation conditions facilitate. This further suggests the non-equilibrium nature of the 1 × 6 surface and explains the wide range of surface compositions reported by different groups [32,33,35,37,108,119]. Zhang and Zunger calculated the formation energy of the model proposed by Biegelsen et al. [6] and pointed out that this 2 × 6 structure is only 0.03 eV higher in energy than the stable two-Ga-dimer structure for the α phase, implying that it is metastable. Consequently, a unique model for this transient 1 × 6 phase may not exist.

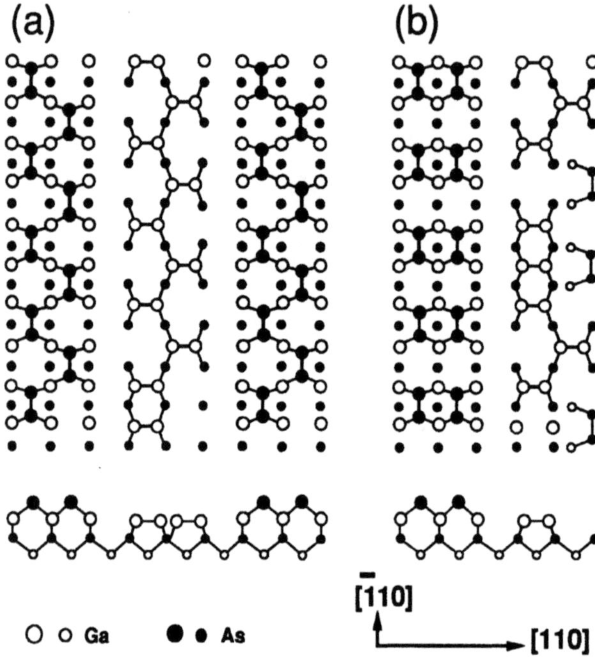

Fig. 8.20. The proposed models for the GaAs(001)-1 × 6 surface (**a**) by Biegelsen et al. [6] and (**b**) by Kuball et al. [115], respectively, based on STM observations

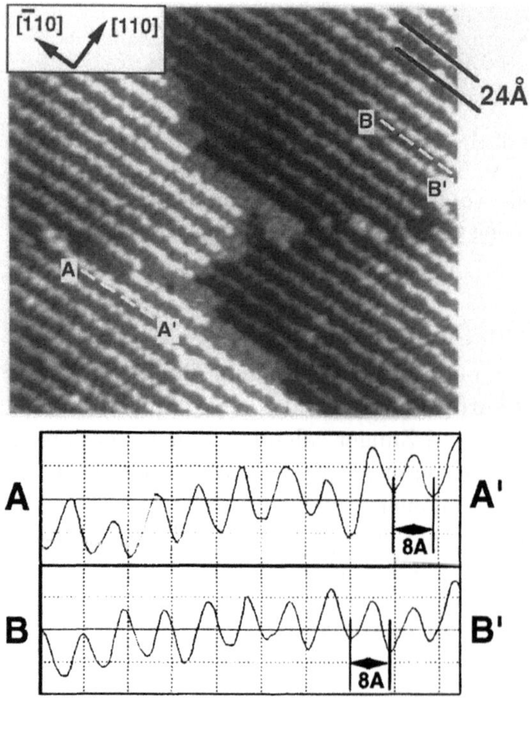

Fig. 8.21. Typical occupied states image of the GaAs(001)-2×6 surface (420 Å×540 Å) prepared by UHV annealing of the 2×4 surface at 600°C for 10min. The *lower panel* shows two cross-section profiles (AA′) and (BB′) parallel to the As row direction as depicted in the figure. The cross-section clearly reveals that the arrangement of Ga dimers in vacancy regions has a 2× periodicity along the [$\bar{1}$10] direction [116]

8.4 GaAs(001) Ga-Rich Surface

8.4.1 $4 \times 2/c(8 \times 2)$ Phase

The first report of the $4\times2/c(8\times2)$ phase also appeared in the mid-1960s in a LEED study of the surface cleaned by argon ion bombardment [26]. Cho [27] studied the first MBE prepared $4 \times 2/c(8 \times 2)$ using the RHEED technique. This surface was further examined by Arthur [31]. The $4 \times 2/c(8 \times 2)$ phase was prepared using low As_2/Ga ratio or heating the 2×4 surface in vacuum, suggesting that the surface is Ga-stabilized. Since the RHEED patterns for the 2×4 and 4×2 are perpendicular to one another, Cho concluded that the twofold periodicity of the $4 \times 2/c(8 \times 2)$ phase, parallel to the direction $[\bar{1}10]$ of Ga dangling bonds, is also due to the pairing mechanism [114], supported by Bommel and Crombeen's LEED experiment [30]. In the intervening years the Ga-rich $4 \times 2/c(8 \times 2)$ surface has been extensively studied experimentally [32–35,37,120,121]. However, no further model concerning the surface structure has been suggested. In particular, there exists no satisfactory explanation on the fourfold periodicity, besides Fermi-surface instabilities or surface vacancies suggested earlier [21,114].

In 1987, based on a HREELS data for an H adsorption experiment, Frankel et al. [46] proposed the first structural model (see Fig. 8.22a), concluding that the fourfold ($4\times$) periodicity is due to regular Ga-dimer missing, the same as the situation on the As-rich 2×4 surface, except that now the surface dimers are rotated by 90° along the (001) crystal axis. The unit cell consists of three Ga dimers and one missing Ga dimer. This three-dimer model contains 3/4 of a monolayer (ML) of the Ga dimers on top of a full monolayer As-terminated surface, in good agreement with existing LEED, AES, RHEED and photoemission results [32–35,45]. In particular, the dimer- and dimer-vacancy related properties are corroborated by all later studies using various techniques, such as RDS (the measurements were performed during MBE and MOCVD growth) [110,111], Temperature-Programmed Desorption (TPD) [119], RHEED intensity measurements [70], LEED/AES [119,122,123], X-ray photoemission [122], work function measurements [108], and SEI analysis [69]. It is noteworthy that a completely identical structural model was proposed by Hou et al. in an independent study of the clean InP(001) In-rich 4×2 surface [124]. These two studies are significant since the model includes almost all structural features associated with the experimental data, establishing the basis from which further models are constructed.

Direct evidence for the dimer vacancy model of the 4×2 phase was first established by Biegelsen et al. in their in situ STM observation in 1990 [6]. Their samples were prepared by annealing in UHV at 690°C. The filled states images are not consistent with the model proposed by Frankel et al. [46], showing a two-dimer top layer structure. The transition from the 4×2 unit cell to the $c(8 \times 2)$ arrangement is also evident in their image [6]. They

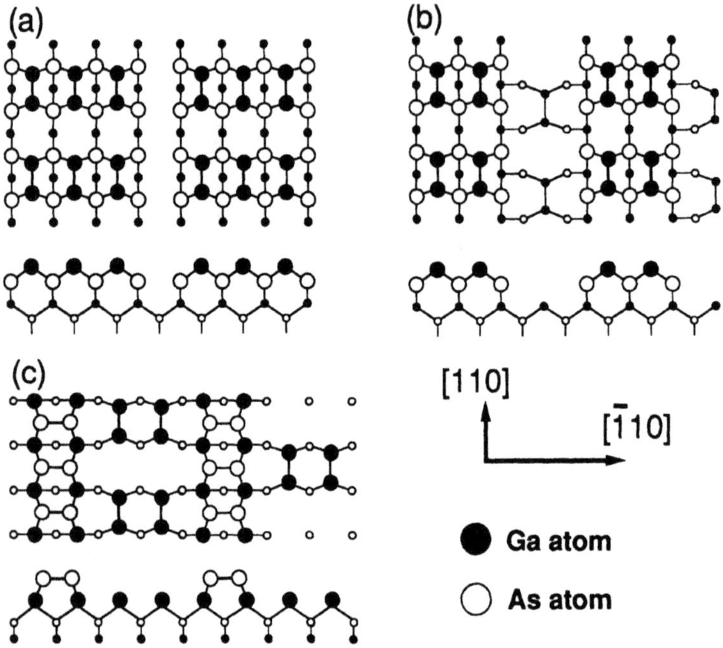

Fig. 8.22. Ball-and-stick models for the GaAs(001)-4×2 reconstructions. (**a**) Three-Ga-dimer model by Frankel et al. [46]. (**b**) Ga-bilayer model by Biegelsen et al. [6] which has two Ga dimers at the top layer and one Ga dimer at the third layer. (**c**) As model proposed by Skala et al. [125]

proposed a bilayer model, shown in Fig. 8.22b, with unit cell consisting of two Ga dimers at the top layer and one Ga dimer at the third layer, which is exposed in the missing row. The models shown in Fig. 8.22a and b both satisfy ECM, giving rise to surface Ga coverages of 0.75 ML and 0.5 ML, respectively.

Later, atomically resolved images were successfully obtained by Skala et al. [125] and Xue et al. [126]. The model that Skala et al. proposed (referred to hereafter as the As model) consists of two As dimers in the first layer and two Ga dimers in the second layer in a unit cell, which is radically different from all other Ga-rich models (Fig. 8.22a and b). Another alternative model was suggested by Moriaty et al. [91]. A problem with this model is that the features for the top-layer Ga dimers should be absent, while the third-layer Ga dimers that are lower than the top-layer Ga by 2.8 Å should be resolved in filled states images [127]. To summarize these STM studies, the images consistently reveal the two-Ga-dimer related structures. However, a generally accepted model or satisfactory explanation of the images is lacking.

If one adopts the general belief that the 4 × 2 surface is terminated with the Ga dimer [6,46] and that the necessary charge transfer takes place according to ECM [20], this surface is very puzzling as regards interpreting the STM images. The filled dangling bond orbital of the Ga dimer located in the first layer has a relatively low Local Density of States (LDOS) which possesses a large energy dispersion from −1.9 to −0.5 eV [128]. The occupied As dangling bond orbitals have predominantly high LDOS, but the As atoms are geometrically placed in the second layer, lower than the first layer Ga by as much as 1.4 Å. At the negative bias voltage of −2 eV used in the STM studies mentioned above, both of these Ga and As orbitals overlap to some extent and are accessible to the tunneling process, so that they are imaged by STM. An appropriate interpretation of the STM image requires in-depth knowledge of the surface charge density distribution unique to the models.

Another important fact is that, unlike the As-rich 2×4 surface where the 2/4th fractional diffraction feature can be used to discuss the detailed structural differences, there is no characteristic RHEED spectrum beam intensity behavior which can be used as a fingerprint [22,71]. The intrinsic difficulty in preparation of the single-phase 4×2 infers that this phase probably has a very narrow composition range with which the structure is stabilized. This was also suggested by the soft X-ray photoemission study by Vitomirov et al. [122], where the measured As/Ga 3d core-level intensity ratio remained constant with the 4×2 surface. Therefore, the structure of the 4×2 phase may not be too complicated, and a well defined single-domain atomic arrangement should be able to account for the surface layer. Theoretical work played a crucial role in structure determination. Northrup and Froyen [72] calculated the surface energy as a function of the Ga chemical potential for a variety of possible structural models. The calculations indicate that the three-Ga-dimer model shown in Fig. 8.22a is unstable with respect to the two-As-dimer structure for the 2×4 (Fig. 8.6a) even in the most Ga-rich environments, and it is thus removed from consideration as a thermodynamically stable structure. They found that the substitution model by Falta et al. [66] is not likely in the entire As/Ga ratio composition. Calculations show that the minimum energy cost is 0.67 eV for each As–Ga replacement [72]. Northrup and Froyen also calculated the formation energy of the As-dimer structure proposed by Skala et al. [125]. As the structure by Skala et al. requires rather long range (approximately 5 Å) lateral and vertical charge transfer in order to achieve charge neutrality, this model does not represent an energetically stable phase in the Ga-rich limit [73]. Among all available models, the bilayer Ga-dimer structure is the lowest in energy for the entire range of the Ga chemical potential [72,73,76]. So far, the bilayer two-Ga-dimer model is the only possible candidate among the available models. Otherwise some completely new model is required.

Typical large scale filled states images obtained by Xue et al. are shown in Fig. 8.23a and b [129]. The image of the 4×2 surface contains bright lines running along the [110] direction with a uniform spacing of 16 Å in the $[\bar{1}10]$ direction, corresponding to the 4× periodicity of the surface [6,8,91,125,126]. In the high resolution filled states image of the 4×2 surface shown in Fig. 8.24 [125,126], each bright line in Fig. 8.23a and b is resolved as a dou-

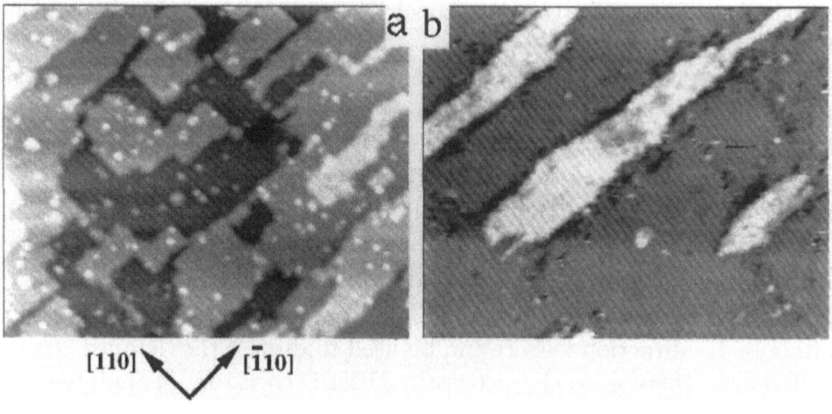

Fig. 8.23. Large scale filled states STM images (both at −2 V) of the GaAs(001) Ga-rich 4 × 2 reconstruction prepared (**a**) (875 Å×720 Å) by UHV annealing of the 1 × 6 surface at 590°C for 15 min, (**b**) (1040 Å×580 Å) MEE growth at 500°C with As_4/Ga flux ratio of 8 and shutter opening times of 2 s for both Ga and As [129]

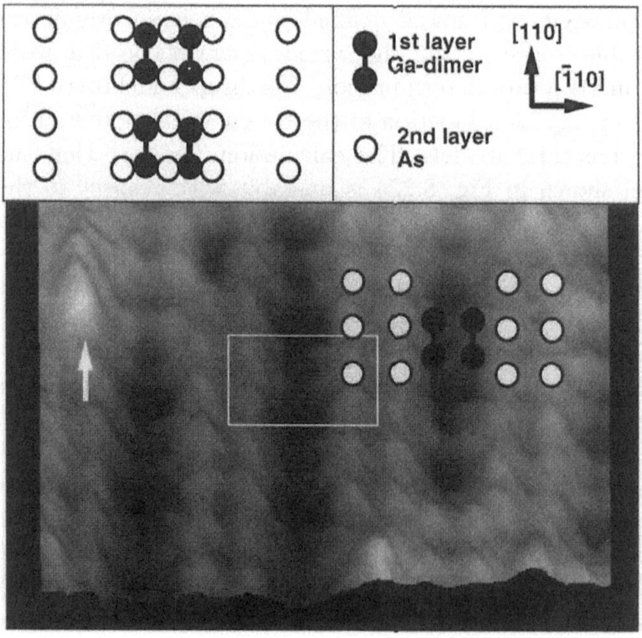

Fig. 8.24. Zoom-in filled states image with details of the GaAs(001) $4 \times 2/c(8 \times 2)$ phase. The *arrow* indicates a Ga cluster around which the Ga dimers are imaged with enhanced brightness. The 4 × 2 unit cell is highlighted with a *white rectangle* and the second layer As atoms and first layer Ga dimers imaged are highlighted by *white* and *black dots*, respectively. The schematic, to the same scale as the image, in the *upper panel* of the figure, is the two-Ga-dimer model proposed by Biegelsen et al. [6]. By comparing this schematic with the highlighted image, the discrepancy between model and image is immediately evident (see text) [126]

blet, consisting of a pair of rows separated by 5.1 Å along the [$\bar{1}$10] direction, whereas the row itself is a chain of bright protrusions separated by 4 Å along the [110] direction. The weak features (blue) always couple together to form a pair-like structure parallel to the bright (yellow) rows. The separation between the neighboring pair-like features along the [110] direction is 8 Å resulting in $\times 2$ symmetry.

There are some difficulties in interpreting the image with respect to the bilayer two-Ga-dimer model. An obvious problem is that the faint pair-like features are located in between the protrusions forming the bright rows along the [$\bar{1}$10] direction, while the Ga and As atoms line up geometrically in the model (see Fig. 8.24), if the faint paired features are assumed to be related to the first-layer Ga dimers and the individual humps to the second-layer As.

On the other hand, the As-dimer model [125] also reveals some weaknesses in interpreting the image, since

- the sample preparation process favors a more Ga-rich surface than the 1×6, whereas the surface coverage expected from the As model is more As-rich than the 1×6;
- the As model cannot explain the measured height difference (only 0.6 Å) between the bright row and the faint pair-like feature.

Xue et al. also reported the empty states image of the 4×2 (Fig. 8.25) [130]. Based on the out-of-phase arrangement of the humps and their separation of 8 Å along the [110] direction, each bright hump in Fig. 8.25 can be assigned to tunneling from unoccupied states of the first-layer Ga dimer, supporting the bilayer Ga-dimer model. If the surface consists of the top-layer As dimers, it should be observed, as in the case of the 2×4-As surface [50].

In order to resolve these discrepancies, Ohno performed a first-principles total energy calculation of the surface charge density distribution based on the Ga-bilayer model. The calculated results are shown in Fig. 8.26. Here,

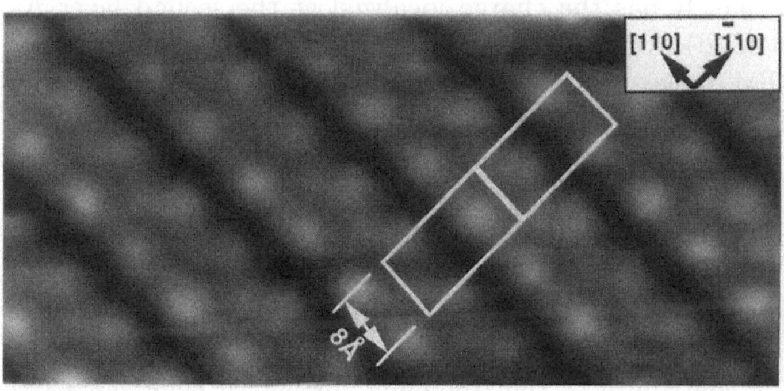

Fig. 8.25. Empty states image (at +2.2 V) of the GaAs(001) Ga-rich $4\times2/c(8\times2)$ reconstruction. The individual humps, evenly spaced by 8 Å along the [110] direction, are attributed to unoccupied states associated with first layer Ga dimers. One $c(8\times2)$ unit cell is indicated by the *white rectangle* [130]

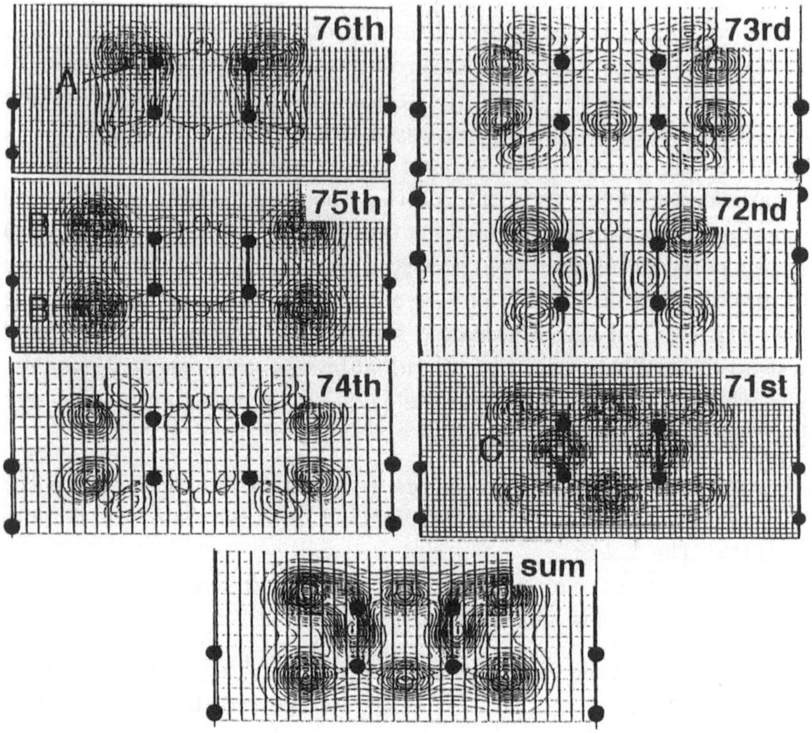

Fig. 8.26. Charge distributions of the local density of states calculated by Ohno on the basis of the Ga-bilayer model for the GaAs(001)-4 × 2 surface (Fig. 8.22b) at 0.9 Å above the first layer Ga-dimer position for the 71st to 76th bands. The *lowest panel* displays a sum of the 71st, 72nd, 73rd, 74th and 75th bands [126]

the 75th band is the HOMO-derived band, which should be dominant in the filled state STM image. It has the charge localized at the second layer As (peak B of the 75th band). As for the contribution by the top layer Ga dimer to the filled state STM image, its charge distribution becomes significant only at the deep 71st HOMO-derived filled band with its peaks in the middle of the dimer (peak C in Fig. 8.26). The charge distribution from the Ga dimer gradually disappears with increasing energy level (from 71st to 75th) and the As-derived charge becomes more dominant. The charge density for the 76th LUMO is located at one of the dangling bond positions (peak A) and should be empty for this ideal semiconducting surface. Under the filled states STM imaging condition at −1.8 V, all local densities of states between the 71st and 76th bands contribute to the tunneling current to form the STM image. Because of the smaller potential barrier height for tunneling from the 75th band, the 75th HOMO makes the most significant contribution to tunneling, in addition to those from the overlapping 74th, 73rd and 72nd bands, all of which are basically the contribution from second layer As atoms as individual brighter protrusions. On the other hand, the contribution for the top layer Ga dimer only becomes appreciable at the 71st band in the middle of the Ga dimer. Thus, the top layer Ga dimer is observed as a single faint hump (instead of a pair-like feature). However, there are significant amounts of surface defects, such as adsorbates and vacancies which may be charge-accumulated, as indicated by the arrows in the image (Fig. 8.24). As a result, the 76th LUMO-derived band is partially filled due to charge transfer from the charge-rich singularities on the surface, and contributes to the tunneling process, leading to a node-like image for the Ga dimer. The image shown in Fig. 8.24 basically corresponds to the total charge contribution shown in the bottom panel of Fig. 8.26, a sum of the bands from the 71st to 76th level. The 76th LUMO orbital is drastically buckled and its charge center moves further away from the geometric center of the dimer with increasing distance from the surface, due to the influence of third layer Ga dimers. Since the calculated densities of states shown in Fig. 8.26 are mappings of the charge distribution only 0.9 Å away from the surface, while the STM image reflects the charge distribution approximately 10 Å away, peaks A and B should move further away from the atom position. Taking all these effects into account, the STM image (Fig. 8.24) and the Ga-bilayer model (Fig. 8.22b) are in good agreement [126].

One important point here is that the high resolution image can only be obtained on a surface with a certain number of defects, implying that the surface in this case may be metallic. More convincing evidence supporting the Ga-bilayer model was obtained in STM observations of a genuine 4×6 structure where such an effect becomes more obvious. This will be discussed in Sect. 8.4.2.

In concluding this section, we would like to discuss an important implication of the conclusions drawn so far for the most stable As-rich 2×4 β phase and the Ga-rich 4×2. Figure 8.27 presents the ball-and-stick models of the As-rich 2×4 and the Ga-rich 4×2 reconstructions. The surface unit cell of the As-rich 2×4 phase has two As dimers on the top layer and one As dimer on the third layer [61]. It therefore contains 3/4 ML of As on the surface. The unit cell of the Ga-rich 4×2 phase has two Ga dimers on the top layer and one Ga dimer on the third layer, resulting in 3/4 ML of Ga on the surface [6]. Among all previously proposed models for both surfaces, the surface charge neutrality can be maintained quite simply only for the present model scheme. If one does not differentiate atom species, these two structures are mirror-symmetric: either can be obtained by 90°C rotation of the other along the (001) crystal axis. Therefore, addition of one ML of Ga atoms to the As-rich 2×4 phase results in the Ga-rich 4×2 phase and addition of one ML of As on that surface simply transfers it to the original As-rich 2×4 phase. This fact may have profound implications in understanding the physical processes involved in layer-by-layer MBE growth of the GaAs(001) surface.

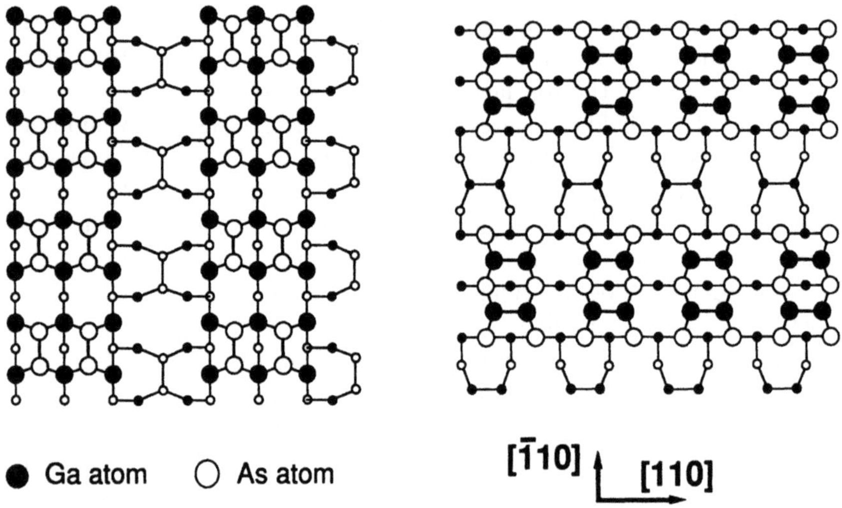

Fig. 8.27. Atomic models for the GaAs(001) As-rich 2×4 (*left panel* [21]) and Ga-rich 4×2 (*right panel* [6]) reconstructions, demonstrating that the two structures are essentially mirror-symmetric. One monolayer of Ga adsorption on the As-rich surface results in the Ga-rich 4×2 phase and one monolayer of As deposition on the Ga-rich 4×2 phase turns it back into the As-rich 2×4 phase

8.4.2 4×6 Phase

Successful structural determination of the 4×6 phase is another good illustration of the power of the STM technique with regard to spatial resolution and the resolution of local electronic structure. The earliest report of the 4×6 can be found in the RHEED study by Cho [114]. LEED studies of this surface include those by Massies et al. [33,131] and Drathen et al. [32]. Before the STM study, not a single structural model had been proposed, due to its large and complicated unit cell, too difficult to access by conventional diffraction techniques. Since none of the published diffraction patterns showed a complete set of spots with nonintegral coordinates in both [$\bar{1}$10] and [110] directions, it was believed that a 'genuine' 4×6 reconstruction might not exist and that the 4×6 was a mixture of the 4×1 and 1×6 reconstructed domains [32,35,121]. By UHV annealing at temperatures higher than that necessary to establish the $4 \times 2/c(8 \times 2)$ structure, Biegelsen et al. reported the observation by LEED and STM of the genuine 4×6 reconstruction [6]. However, noisy tunneling current prevented them from obtaining a high resolution STM image in order to document the atomic structure.

Creighton [119] proposed, on the basis of a TMGa adsorption study, that the 1×6, 4×6 and $c(8 \times 2)$ actually have the same stoichiometry since they all display the same chemistry for TMGa adsorption. He speculated that all Ga-rich surface phases can be constructed under a scheme of superposition based on the 4×2 three-Ga-dimer model. On the other hand, as regards relative surface coverage, many groups reported that the 4×6 phase is more As-rich than the 4 × 2 phase [32,34,35,70,108,110,111,114,121], while other experimental data suggested that the 4 × 6 phase can also be more Ga-rich [37,122]. RDS data for the 4 × 2 and 4 × 6 phases show significantly different features and imply that they are associated with different structures [111].

An attempt was made to establish a surface phase diagram using the vicinal surface, even though there is no reliable equilibrium surface phase diagram for GaAs(001) [15]. The result showed that two or more structures usually overlap in the Ga-rich regime [108,122]. These discrepancies in the geometric structure and stoichiometry remained puzzling for a long time. In the following, we will see how these discrepancies have been resolved by STM.

'Genuine' 4 × 6 Reconstruction. Xue et al. [126,129,130] and Behrend et al. [8] have identified two distinctly different 4 × 6 phases by direct STM observations. We first discuss the one which exhibits the 'genuine' 4 × 6 (hereafter referred to as G4 × 6) translational symmetry observed by Xue et al. [126]. The G4 × 6 phase can be prepared either by annealing the 2 × 6 surface in vacuum at 600°C for more than 20 min or by growth at 500°C using higher Ga flux (As_4/Ga = 6) than that needed for the 4 × 2 phase in MEE on the 2 × 6 surface. Large scale images from the As-deposited and annealed surfaces, respectively, are shown in Fig. 8.28a and b. A high resolution image for the filled states is shown in Fig. 8.29. The 'genuine' 4 × 6 phase is uniquely characterized by a regular array of large oval protrusions located at each corner of the unit cell. These oval protrusions always occupy the middle of the As rows of the 4×2 surface. By superimposing the 4×2 phase model with the 4 × 6 phase image, the 4× periodicity of the surface in the

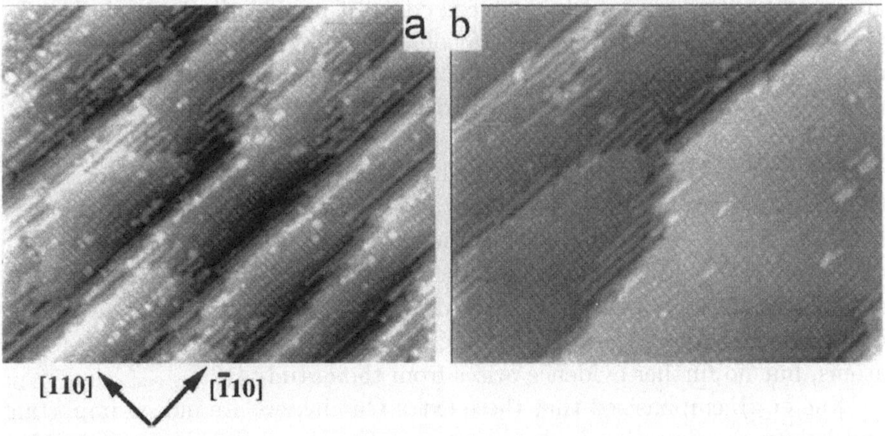

Fig. 8.28. Large scale filled states STM images (1 600 Å×1 200 Å) of a typical genuine 4 × 6 surface prepared (**a**) by MEE method and (**b**) by UHV annealing of the 1 × 6 substrate. On both surfaces, some regions of 1 × 6 always appear near step edges

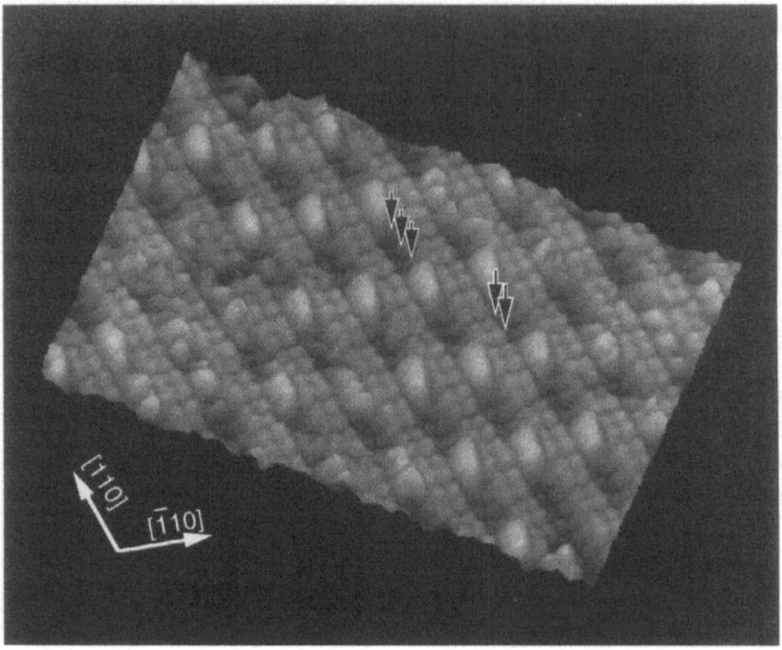

Fig. 8.29. Three-dimensionally displayed filled stated STM image of the GaAs(001)-G4 × 6 reconstruction. The image was recorded at a sample bias voltage of −1.8 V with a tunneling current of 40 pA [126]

direction [$\bar{1}10$] becomes immediately evident. In this image, the individual Ga atoms forming a dimer with enhanced contrast are clearly resolved. The 6× periodicity is seen in the regular array of large oval protrusions every three Ga dimers along the [110] direction. It can also be seen how the presence of the Ga-dimer vacancy defect (indicated by the white arrow) causes the $c(8 \times 2)$ symmetry by a 4 Å shift of the 4 × 2 subunit along the [110] direction. The oval features are ∼ 1.0 Å (∼ $H/2$) higher than the Ga dimers but lower than the 2 × 6 As rows by 0.9 Å. Their spatial distribution extends approximately from $3a_0$ to $4a_0$ (where $a_0 = 4$ Å is the surface lattice constant) in the [110] direction and ∼ $2a_0$ in the [$\bar{1}10$] direction. Xue et al. have identified these oval protrusions as Ga clusters formed at high temperature or in extremely heavy Ga flux, and suggested that each Ga cluster probably contains 6 to 8 atoms, but no further evidence arises from their study [129].

Xue et al. emphasized that these extra Ga clusters are indeed important and helpful in obtaining high resolution STM images of the surface [129]. The high contrast imaging of the Ga dimers is due to charge transfer from the clusters, leading to partial filling of previously empty Ga dangling bonds. The origin of this contrast enhancement effect is clearly demonstrated when a single oval protrusion adsorbs on the pristine 4 × 2 surface, as indicated by

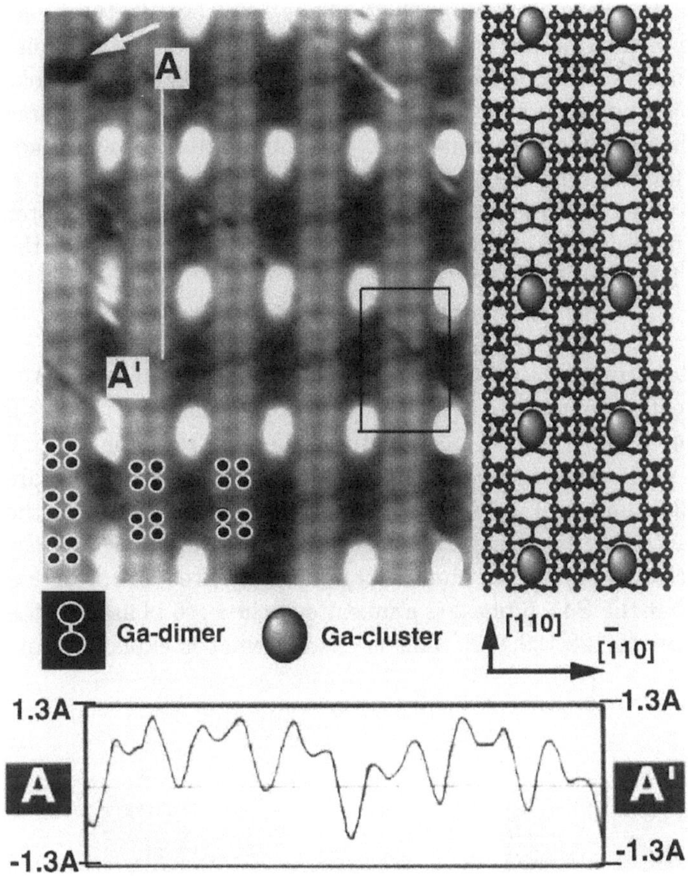

Fig. 8.30. Atomic resolution filled states STM image of the G4 × 6 surface taken at −1.8 V (the same image as Fig. 8.29, but displayed in gray scale). In the image, individual top-layer Ga atoms forming a dimer can be clearly resolved, as highlighted by *white open circles*. The line plot (A to A′ in the STM image) is shown in the *lower panel*. The height asymmetry within the dimers is clearly visible. This is due to charge transfer from the Ga clusters. The 4 × 6 unit cell is indicated by a *black rectangle*. The *white arrow* in the *upper left* indicates a missing-atom defect, which causes the transition between the out-of-phase and in-phase arrangements of the 4 × 2 subunit. The *schematic* shown to the *right* of the STM image is the proposed model

the arrow in Fig. 8.24. In the immediately surrounding region of the cluster (yellow), Ga dimers suddenly become brighter and As rows fade out. With increasing numbers of clusters, image contrast between the As and Ga atoms of the 4 × 2 surface shifts and eventually reverses completely, resulting in that of the G4 × 6 surface (Fig. 8.29). Figure 8.30 is the gray-scale G4 × 6 STM image with the line scan taken on the top of the Ga dimers along the [110] direction, which further corroborates the charge transfer effect.

The brightness asymmetry of individual dimers is evident in the line scan. It is attributed to the charge screening effect of the large oval feature. The closer the dimer is to the oval feature, the greater the degree of the asymmetry. The central dimer surrounded by equivalently distributed clusters looks symmetric due to an equal degree of charge transfer. Occurrence of charge transfer implies that the G4 × 6 is metallic. This observation provides additional evidence for the bilayer Ga-dimer model by Biegelsen et al. [6]. If we assume that Xue et al.'s model in Fig. 8.30 is correct and each cluster consists of approximately 6 (to 8) Ga atoms, the surface Ga coverage of the G4 × 6 should be 1 ML (to 1.2 ML).

Pseudo-4 × 6 Reconstruction. We discuss the second 4 × 6 phase, the pseudo-4 × 6 (hereafter referred to as P4 × 6), that also displays the 4 × 6 RHEED diffraction pattern. This is completely different from the G4 × 6 as regards atomic scale structure [8,126,129,130,132]. Typical examples are shown in Figs. 8.31a and b. Figure 8.31a demonstrates coexistence of the 1 × 6 phase (characterized by zigzag lines along the [$\bar{1}$10] direction) and the 4 × 2 phase (characterized by doublet lines along the [110] direction), whereas Fig. 8.31b shows that the P4 × 6 phase is a mixture of the 1 × 6 phase and the 'genuine' 4 × 6 phase [8,126,129,130]. This is exactly what is expected from

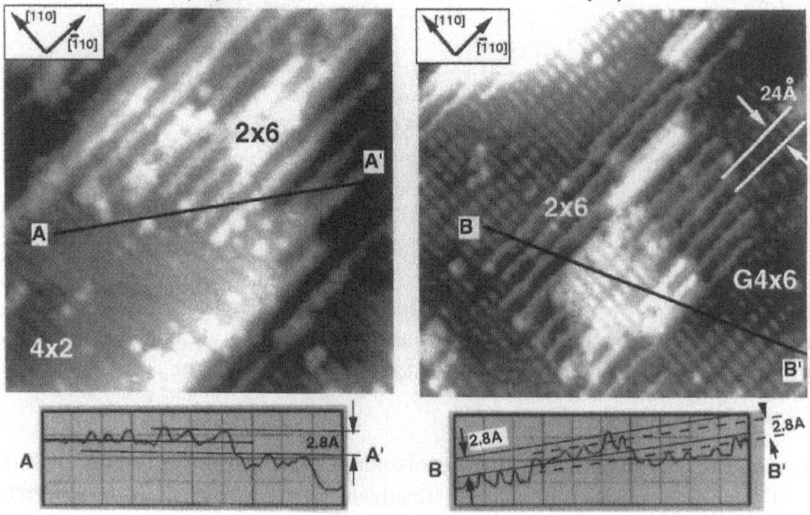

Fig. 8.31. The filled states STM images of the pseudo GaAs(001)-4 × 6 reconstruction. (**a**) (420 Å × 270 Å) The mixed 1 × 6 and 4 × 2 surfaces, and (**b**) (440 Å × 290 Å) the mixed 1 × 6 and 'genuine' 4 × 6 phases. The cross-section profiles AA′ in (**a**) and BB′ in (**b**) are shown just *below* the corresponding images. Both surfaces display a similar 4 × 6 RHEED pattern

previous LEED studies [32,122]. The cross-section profile (AA′) in Fig. 8.31a indicates a height difference of approximately 1.4 Å between the 1 × 6-As and the 4 × 2-Ga related features. Xue et al. and Behrend et al. further found that by increasing the Ga coverage and temperature, and extending the annealing time and/or increasing the Ga flux during MEE, the 4 × 2 domain becomes more abundant than the 1 × 6 domain on the surface and vice versa.

Behrend et al. [8] and Xue et al. [129] observed a new irregular structure formed in the annealed surface that coexists with the 4 × 2 or 1 × 6 phase at temperatures higher than that for forming the 1 × 6 phase. This is shown in Fig. 8.32. This new structure has no long range ordering and is thus called a disordered phase. It may be associated with the continuous (second-order) phase transition between the 2 × 4-As and 4 × 2-Ga [133]. This disordered phase is currently not well understood.

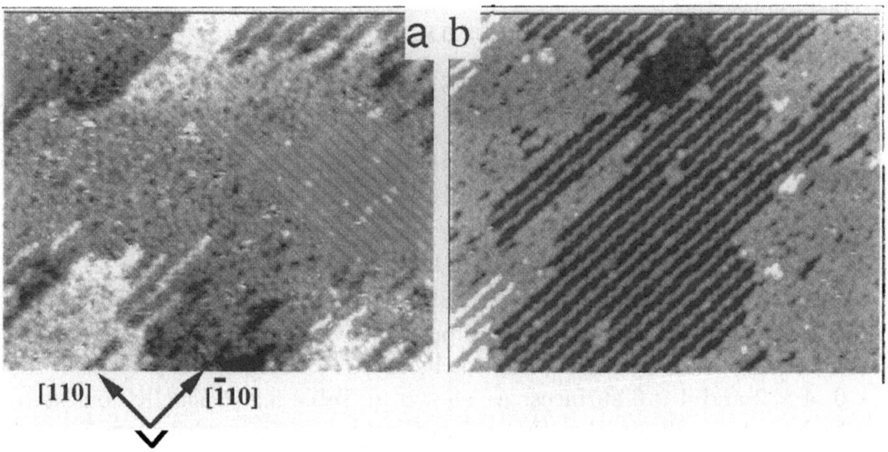

Fig. 8.32. Typical filled states STM images of (**a**) the coexisting 4 × 2 and disordered surface (420 Å×320 Å) and (**b**) coexisting 1 × 6 and disordered surface (750 Å×670 Å) at negative (−2.0 V) sample bias. Note that the contrast changes between the features for the disordered and 4 × 2, 2 × 6 regions [129]

Concluding this section, the STM studies by two groups [8,126,129,130] clearly demonstrated that the P4 × 6 reconstruction does indeed divide into two cases: the first one is a coexistence of the 1 × 6 and 4 × 2 phases with or without a structureless disordered phase (depending on the preparation methods), and the second one is a mixture of the 1 × 6 and 'genuine' 4 × 6 phases. Without such direct observations by STM, understanding of the 4 × 6 surface might never have been achieved.

Stoichiometry of the Ga-Rich 4 × 6 Reconstructions. STM cannot directly measure the stoichiometry of a surface. In this section, we present a qualitative discussion of the stoichiometry of Ga-rich reconstructions based

Table 8.2. Summary of surface Ga/As composition vs. various GaAs(001) surface phases reported by previous workers

reconstruction	4 × 6	4 × 2	2 × 6
Ref. 33 (Auger As/Ga)	x	2.0	2.5
Ref. 32 θ_{As} (Auger As/Ga)	0.27ML	0.22ML	0.52ML
Ref. 37 θ_{As} (core-level 3d As/Ga)	0.31ML	0.52ML	0.42ML
Ref. 39 (core-level 3d As/Ga)	0.91	0.78	x
Ref. 119 (Auger As/Ga)	1.27	1.27	1.27
Ref. 70 θ_{As} (average As coverage)	0.22ML	0.03ML	x
Ref. 122 (core-level 3d As/Ga)	overlag with the 4 × 2	0.8–1.0	x

on the structural diagram of various surfaces established by STM. We will see how discrepancies have been easily resolved using these STM studies. The main results in surface stoichiometry obtained by previous studies for the 1 × 6, 4 × 2 and 4 × 6 surfaces are summarized in Table 8.2.

Because of the different characterization techniques used, it is not possible to compare the absolute surface composition with universal standards. We shall therefore limit our discussion to the relative stoichiometry between the 1 × 6, 4 × 2 and 4 × 6 surfaces. As shown in Table 8.2, basically consistent results have been obtained in the case of the 1 × 6 surface, with the exception of Bachrach et al. [37], suggesting that the 2 × 6 is more Ga-rich than the 4 × 2 phase. There is, however, an obvious discrepancy in relative composition of the 4 × 6 and 4 × 2 surfaces [32–35,37,70,108,110,111,119,121,122].

The STM studies have shown that the 4 × 6 surface is not necessarily a single phase. It can be:

- a mixture of the 1 × 6 and 4 × 2 phases,
- a mixture of the 1 × 6 and the G4 × 6 phases,
- the single 'genuine' 4 × 6 phase.

Which domain appears and how large a given domain is in the course of the phase transition depend critically on the preparation process. This kind of local variation cannot be detected by conventional analysis techniques. According to the structural models established in Sects. 8.3.3, 8.4.1 and 8.4.2, the 1 × 6 surface is the most As-rich and the G4 × 6 surface is the most Ga-rich, while the 4 × 2 surface has an intermediate surface composition with a well defined narrow range (0.75 ML Ga). When the P4 × 6 is a mixture of the 2 × 6 and 4 × 2 phase, it should always be more As-rich than the 4 × 2 phase, as reported by most studies [32–35,70,108,110,111,121]. It will finally reach the composition value of the 4 × 2 surface with increased annealing temperature or higher MEE Ga flux [70,108,132]. On the other hand, when the P4 × 6 is formed with the 2 × 6 and G4 × 6 phases, the situation is rather complicated. The surface composition can be either more As-rich (when the 2 × 6 phase dominates on the surface) [32–35,70,108,110,111,121], or more Ga-rich (when the G4 × 6 dominates on the surface) [37,122] than that of the single phase 4 × 2, or the same as that of the 4 × 2 structure (when the 2 × 6 and 'genuine' 4 × 6 phases mix with an appropriate domain ratio [119]). The surface composition thus overlaps with that for the 4 × 2 surface [108,122]. In the decomposition region, the 'genuine' 4 × 6 phase dominates on the surface, and the composition is always more Ga-rich than the 4 × 2 phase. Under this scheme, the surface composition with the 4 × 6 diffraction symmetry can vary from 0.33 ML As (the theoretical value from the model for 1 × 6 proposed by Biegelsen et al., see Fig. 8.20a) on the extremely As-rich side to more than 1 ML Ga on the extremely Ga-rich side (for the G4 × 6 surface, see Fig. 8.29).

The above results are summarized in the schematic evolution diagram shown in Fig. 8.33, where various phases are plotted as a function of As/Ga surface coverage. Three different coverages: 0.3 ML As, 0.75 ML Ga and approximately 1 ML Ga are indicated on the horizontal axis, corresponding to the 2 × 6, 4 × 2 and 'genuine' 4 × 6 (G4 × 6) phases, respectively. All the phases discussed in this review are placed according to their surface composition. The lower path is the evolution of the surface phase during MEE growth with increasing Ga flux and the upper path is that due to annealing with increasing temperature. The pseudo-4 × 6 (P4 × 6) phase does indeed exhibit

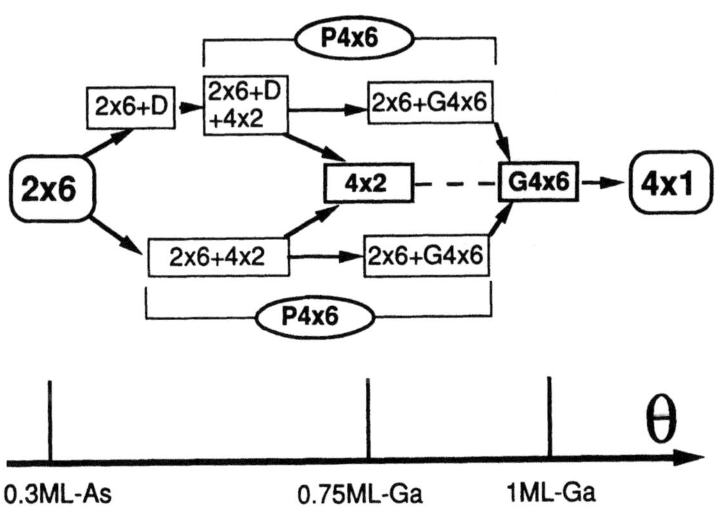

Fig. 8.33. The phase evolution diagram of the GaAs(001) Ga-rich surface as a function of Ga coverage. In the figure, D represents the disordered (structureless) region. P4 × 6 denotes the pseudo-4 × 6 phase and G4 × 6 the 'genuine' 4 × 6 phase

a varied surface composition: its value varies from 0.33 ML As (for the 2×6 phase) to 1 ML (for the G4×6 phase). According to this diagram, with increasing Ga concentration, the 2×6 phase undergoes a transition from the 2×6 and 4×2 mixed structure to the G4×6 phase via the 2×6 and G4×6 mixture. After that, the surface usually transforms into the Ga-rich 4×1 surface. In the case of annealing, the disordered Ga-rich phase (D) always exists, as shown by Behrend et al. [8] and Xue et al. [129]. This disordered phase may play an important role in continuous evolution. Evolution may also take place along a different transition path: the 2×6 phase may directly change into the 4×2 phase under certain conditions. For example, deposition of the right Ga density by MEE should produce the perfectly flat 4×2 surface without any 2×6 islands.

8.4.3 Ga-Rich 4×1 Phase

When the 1×6 As surface is deposited with about 2 ML Ga or heated at 650°C for 15 min, Xue et al. observed a new phase, the 4×1 phase [129]. Since this surface is too Ga-rich, it would not appear to correspond to the 4×1 phase reported previously by Massies et al. [35]. As shown in Fig. 8.34, the basic characteristics of this newly defined 4×1 phase are the same as those of the G4×6 surface.

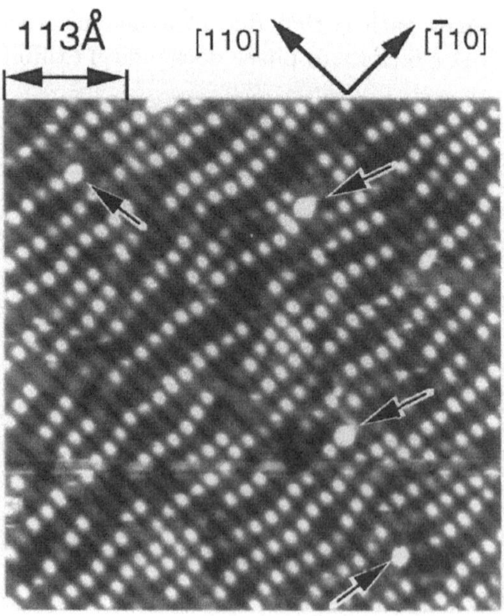

Fig. 8.34. The filled states STM image (460 Å×430 Å) of the newly defined 4×1 phase. *Arrows* indicate four Ga clusters (Ga droplet precursors) larger than the regular-sized ones

They consist of the Ga 4 × 2 structure as a subunit, which is modulated with randomly distributed oval-shaped Ga clusters along the [110] direction. But along the [$\bar{1}$10] direction, Ga clusters are uniformly separated by 16 Å, thus appearing as a 4 × 1 symmetry. Several larger clusters, i.e., Ga droplet precursors (indicated by arrows) have formed, since the desorption of As from the surface becomes more drastic due to annealing at higher temperature. The density of Ga clusters in the 4 × 1 surface was measured 15% higher than in the ideal G4 × 6 phase and it rises with increasing annealing time. Higher temperature annealing (for example at 700°C) causes decomposition of the material and results in extremely rough morphology. This structure might be the last stable single phase under the static Ga-rich condition for the GaAs (001) surface [129].

8.5 Other Arsenide (001) Surfaces

8.5.1 InAs(001) Surface

InAs is a low band-gap compound semiconductor material (E_g = 0.36 eV, compared to 1.42 eV for GaAs), promising for applications in high-speed transistors, infrared detectors and lasers. Compared with the complicated nature of the reconstructed GaAs(001) surface, reconstructions occurring on InAs(001) are rather simple. Two main reconstructions (As-covered 2 × 4 and In-covered 4 × 2) have been reported [30,136,137], and the phase transition between the 2 × 4 and 4 × 2 is found to be first order [136,137]. Monte Carlo simulation shows that this first-order transition occurs when the lateral interaction between the surface species is strong enough [133]. This simple situation makes the InAs(001) surface a good candidate for studying the microscopic mechanism of the surface phase transition by STM. The structure and stoichiometry of the InAs(001) surface have not been investigated much either experimentally or theoretically, since emphasis has been predominantly placed on GaAs. Despite relatively little work on the InAs(001) surface [138], STM has been the main technique employed for determining the structure of reconstructed surfaces so far [139–148].

As-Rich 2 × 4 Phase. The first STM study of the InAs(001) As-rich 2 × 4 phase was performed by Yamaguchi and Horikoshi in 1993 [139]. Their samples were prepared by the decapping technique, similar to that used by Pashley et al. [5] for preparing the GaAs(001) As-rich 2 × 4 surface. In their study, a non-doped InAs buffer layer (0.5 μm) was first grown on the InAs substrate at 480°C after the oxide film was desorbed. The samples were then cooled down to −10°C under the As flux to form a protective amorphous As layer, and transferred into the UHV STM chamber where the protective layer was desorbed at 260°C. As a result, an ordered and clean As-rich 2 × 4 surface was obtained [139,140]. The clean 2 × 4 InAs surface can also be prepared by a simple annealing process in an As flux without buffer layer growth, as described by Ohkouchi et al. in the first in situ MBE-STM study of the InAs(001) surface [142]. For preparation of defined surfaces on the misoriented substrates, Ohkouchi and Ikoma used in situ MBE growth followed by quenching of the sample in the STM chamber [145]. In each case, the STM images indicate a contaminant-free and well-ordered surface.

Figure 8.35 shows a typical STM image of the InAs(001)-2 × 4 surface prepared with the decapping technique by Yamaguchi and Horikoshi [139]. The dark rows running along the $[\bar{1}10]$ direction in the image are the missing-dimer row structures, a characteristic of the 2 × 4 reconstruction already established on GaAs(001). Here the dark missing-dimer rows are running very straight with far fewer kinks than on the GaAs surface, and the InAs 2 × 4 structure is thus highly uniform. It was suggested that it is due to a stronger lateral interaction on the InAs(001) surface that causes high coherence in the reconstruction [139,141].

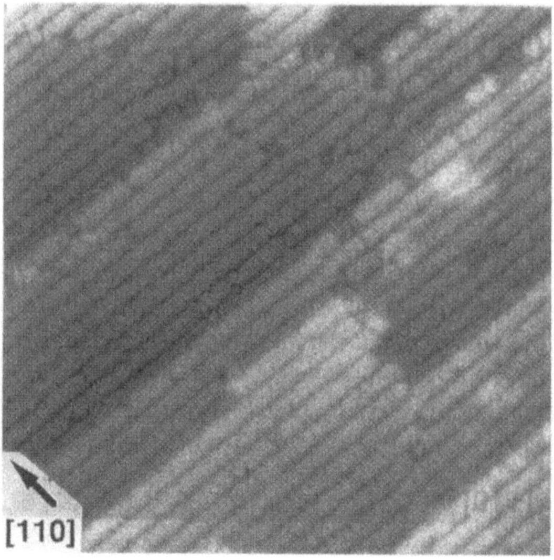

Fig. 8.35. STM image of an InAs(001)-2 × 4 surface obtained by annealing at 260°C. The image size is 500 Å×500 Å. The bias voltage was −3.0 V and the tip current 0.08 nA [140]

An atom-resolved image of the surface is shown in Fig. 8.36. The surface unit cell consisting of two As dimers with two missing dimers is clearly visible [146]. Among several models proposed for the 2 × 4 As reconstruction and shown in Fig. 8.37, the STM image observed here agrees with either Chadi's two-dimer model (Fig. 8.37b) or Frankel et al.'s two-dimer-model (Fig. 8.37c). The InAs(001) 2 × 4 surface prepared at higher temperatures has a different structure, as shown in Fig. 8.38. Due to As desorption, the surface now consists mainly of just one top-layer As dimer in a unit cell. Moreover, an In atomic layer was observed after the top-layer As desorbed (indicated by B in Fig. 8.38), but the previous As-vacancy row (indicated by

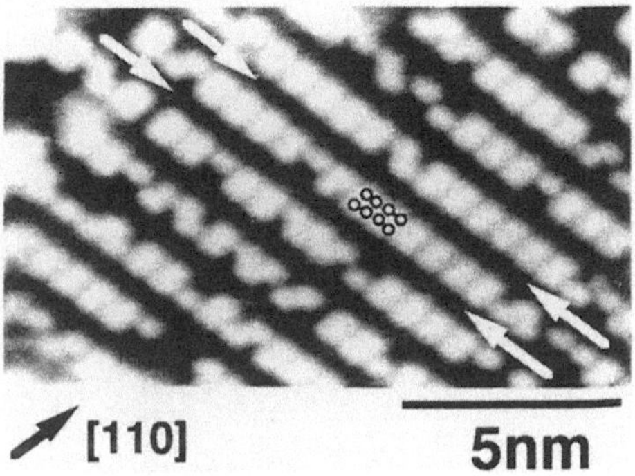

Fig. 8.36. High resolution STM image of the InAs(001) As-rich 2×4 surface just after in situ MBE growth, showing the two-dimer structure of the surface [146]

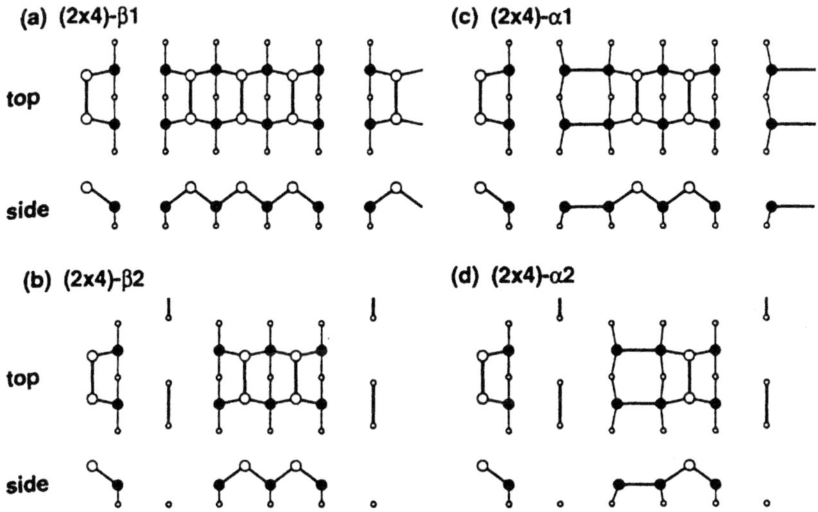

Fig. 8.37. Schematic representations of InAs(001) As-rich 2×4 structures with As coverages of 0.75 ML (**a**) and (**b**), and 0.5 ML (**c**) and (**d**). *Top* and *side views* are both represented [146]

A in Fig. 8.38) is lower than the In layer. With these observations, Yamaguchi and Horikoshi concluded that the InAs(001)-2×4 As-rich reconstruction can be best described by Chadi's two-As-dimer structure (Fig. 8.37b), rather than the 2×4 model (Fig. 8.37c), and the annealed surface containing single dimers (Fig. 8.38) corresponds to the structure depicted in Fig. 8.38d. Despite the occurrence of single dimers, which makes the dimer rows somewhat zigzag in appearance, the dimer-vacancy rows are perfectly straight, without formation of kink defects, over a wide range (more than 500 Å) along the $[\bar{1}10]$ direction. This is very different from what is observed for the GaAs(001) 2×4 surface [89,99]. The origin of the higher coherence on the InAs(001) 2×4 re-

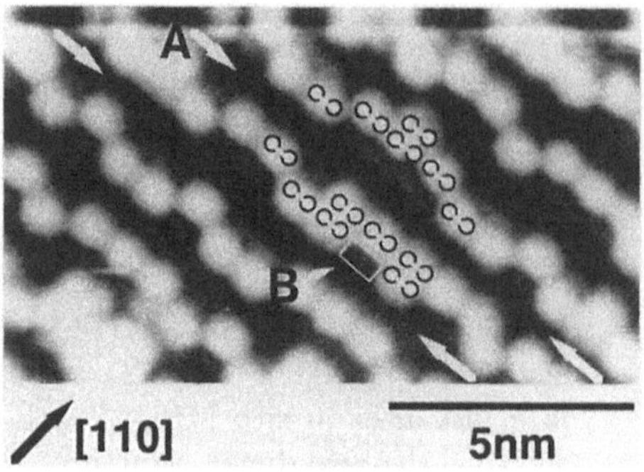

Fig. 8.38. STM image of an InAs(001)-2 × 4 surface annealed at 340°C for 10 min. The sample bias voltage was −2.0 V and the tip current was 0.10 nA. *Arrows* labeled A show the dimer vacancy rows, and As dimers are depicted on the image following the expected model. The *arrow* labeled B indicates the exposed In layer, one atomic layer below the top As layer [146]

construction, and thus stronger lateral interaction among the surface species, could be attributed to the different formation energies associated with the atomic structures of the 2 × 4-α2 for InAs (Fig. 8.37d) and 2 × 4-α1 for GaAs (Fig. 8.37c) [140,146]. It has been shown that the first-order character of the phase transition on InAs(001) was caused by such a strong interaction between surface species [146]. With their STM operating at high temperatures, Yamaguchi and Horikoshi were able to monitor the dynamic process in the course of phase transition from the As-rich 2 × 4 to the In-rich 4 × 2 phase. They observed a significant step motion that predominantly occurs in the dimer row direction [139,140,146], as shown in Fig. 8.39. This preferential motion agrees with the anisotropic step (kink) energy as observed on

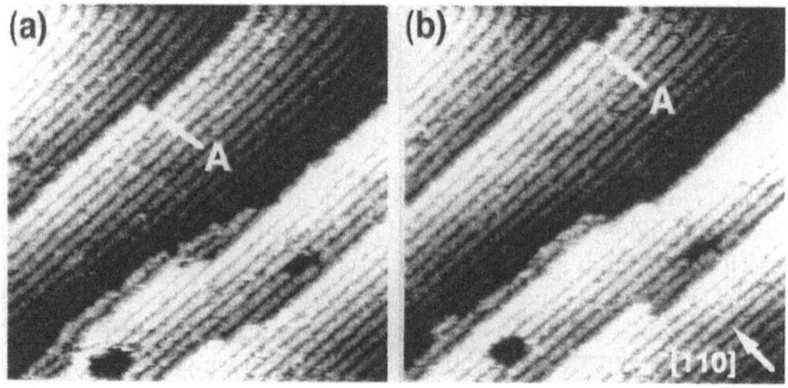

Fig. 8.39a, b. STM images of an InAs(001) surface obtained at 35 s intervals at 350°C. The image size is 500 Å×500 Å. The bias voltage was −3.5 V and the tip current 0.08 nA [146]

the GaAs(001)-2 × 4 surface [49,65], and Yamaguchi and Horikoshi speculated that the step motion on GaAs during MBE growth should resemble the present result, and that it might explain the anisotropic step flow growth of GaAs surfaces [106].

One interesting observation is that the surface can steadily maintain its 2 × 4 structure even though the diffusion of surface In and As atoms is appreciable at the temperature the images were taken, further documenting the strong lateral interaction on the InAs surface. They have used STM, together with RHEED, SEM and Monte Carlo simulations, to investigate the microscopic mechanism for the first-order phase transition between the 2 × 4 and 4 × 2 structures [139,141]. The transition was directly observed to be facilitated by coexistence of the As-terminated 2 × 4 and In-terminated 4 × 2 domains, as predicted by the two-dimensional lattice-gas model, and the discontinuous stoichiometry change in the first-order transition is immediately evident. The 4 × 2 domains appear in regions surrounded by the 2 × 4 domains, and form by direct As desorption. This observation is consistent with their Monte Carlo simulation and can be explained once again by strong lateral interaction of As units. In order to study the influence of finite size effects on the first-order transition between the 2 × 4 and 4 × 2 phases on the InAs(001) surface, they also performed STM observations of the misoriented InAs substrate. Their images show that the transition preferentially occurs at the step edge parallel to the [110] direction. No As desorption was observed from the step edges parallel to the [$\bar{1}$10] direction. Similar results were also observed by Ohkouchi and Ikoma [143]. These observations are consistent with predictions for the finite-size effect on the first-order phase transition on the InAs(001) surface [133], and give direct evidence for preferential As desorption at step edges.

Yamaguchi and Horikoshi also studied the kink defects formed on heavily doped InAs(001)-2 × 4 surfaces [148]. Figure 8.40 shows their measured kink densities as a function of doping concentration for InAs together with GaAs. As discussed in Sect. 8.3.1, the kink density measured by STM on the GaAs(001) surface coincides with the surface states density needed for Fermi level pinning at the midgap [64]. In the case of InAs, however, kink defects are very difficult to form and the observed density of kinks is smaller than the surface states density required to move the Fermi level to the midgap of InAs, for the entire range of the doped concentrations studied. They suggested that the Fermi level does not lie at the midgap [148].

In search of a possible explanation for the difference in kink density between InAs and GaAs, they performed self-consistent calculations, and found that kink formation is governed by two competing processes: energy loss by new kink formation and energy gain by the electron trap from the conduction band to the surface state formed by a newly created kink. Using a modeled value of 0.1 eV for kink formation energy, they were able to explain the lower density of kinks on the InAs(001) surface. As the band gap of InAs is only

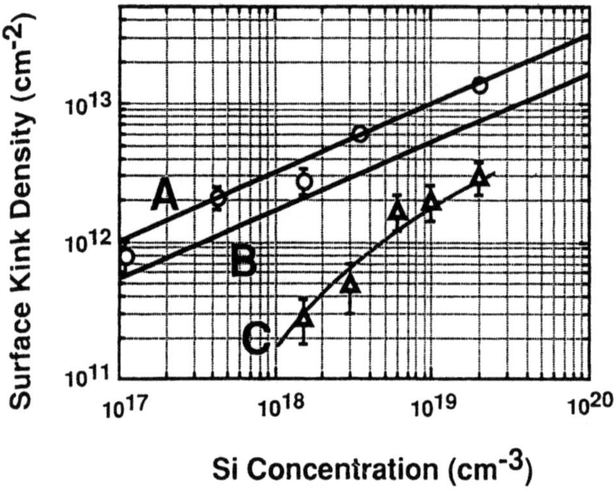

Fig. 8.40. Observed kink densities as a function of doping concentration for GaAs (*circles*) and InAs (*triangles*). *Solid lines* A and B show the surface-state densities needed to move the surface Fermi level to midgap for GaAs and InAs, respectively. *Solid line* C shows the calculated kink density with finite kink formation energy for an InAs(001) surface [148]

0.36 eV, band bending is very small, if we assume that the surface state lies at the midgap. Kinks are difficult to form since the energy gain from trapping an electron from the CBM to the surface states is almost comparable to the kink formation energy. However, in the case of GaAs, which has a large band gap (1.4 eV), such a process is favorable as the energy gain is much larger than the kink formation energy [148].

Step structures of the InAs(001) surface with different misorientations have been studied by Ohkouchi and Ikoma [143,145]. Figure 8.41 shows two STM images obtained from the vicinal InAs(001) surfaces misoriented by 1° toward (a) the [110] direction (A-type surface) and (b) the [$\bar{1}$10] direction (B-type surface), respectively [145]. Both surfaces are reconstructed into the 2×4. It is obvious that the A-type steps running parallel to the $2\times$ ([$\bar{1}$10]) direction are straight with very few kinks over a distance of 100 Å (Fig. 8.41a), whereas the B-type steps running along the $4\times$ direction are highly kinked and very ragged (Fig. 8.41b). This is basically the same as the observation of the GaAs(001) surface by Pashley et al. [64]. Ikoma and Ohkouchi found that in most cases the kink occurs in one or two units of complete 2×4 unit cells. This agrees with the energetically stable configuration of the 2×4 unit cell described by the model shown in Fig. 8.37b, because there are exactly the correct number of electrons to allow all As dangling bonds to be filled with two electrons, and all In dangling bonds to be completely empty [20,146]. The kink is difficult to form at the A-type step since it creates the B-type step, and that costs higher energy [65]. All the steps shown in Fig. 8.41 are monolayer steps with a height of 3 Å.

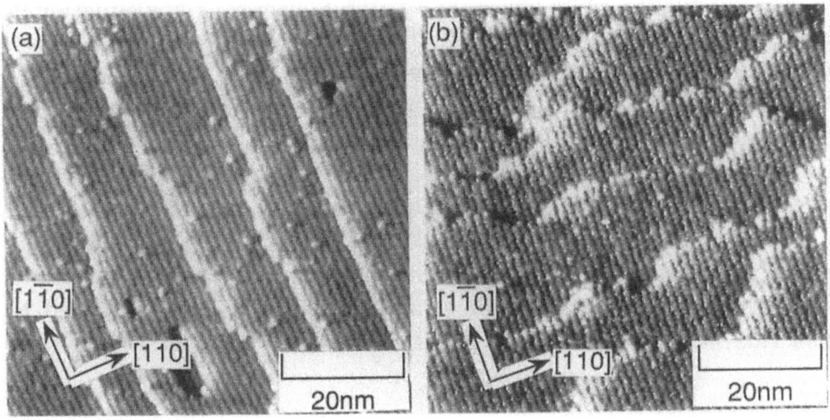

Fig. 8.41. STM images on the InAs(001) vicinal surfaces misoriented by 1 Å toward (**a**) the [110] direction and (**b**) the [$\bar{1}$10] direction of the initial 2 × 4 samples [145]

Ikoma and Ohkouchi [145] found that the step structures on the InAs(001) In-rich 4 × 2 surface are very different from those on the As-rich 2 × 4 surface mentioned above, and also different from what is observed for the GaAs(001) Ga-rich 4 × 2 surface [132]. The STM images for the case of In-rich 4 × 2 are shown in Fig. 8.42.

On the A-type surface, significant step bunching takes place. Figure 8.42a shows two bunched steps consisting respectively of 8 (on the right) and 10 (on the left) monolayer steps [145]. Additional AFM measurements on the same sample by Ikoma and Ohkouchi indicate an average height of approximately

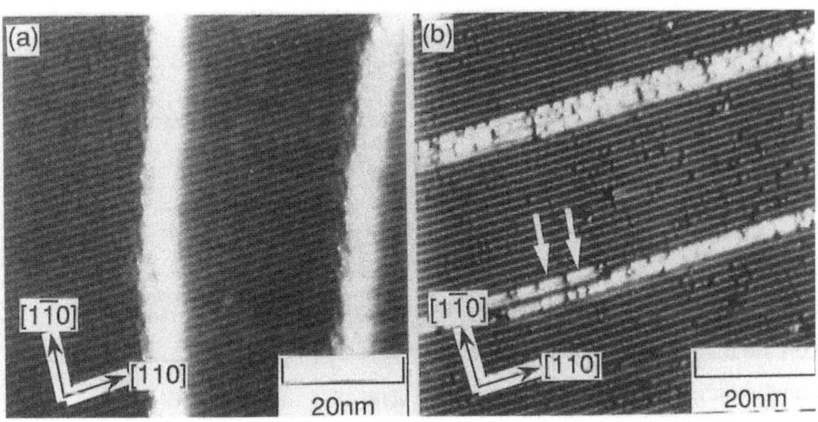

Fig. 8.42. STM images on the InAs(001) vicinal surfaces misoriented by 1 Å toward (**a**) the [110] direction and (**b**) the direction of the 4 × 2 samples. The *white arrows* in (**b**) indicate the white lines running parallel to the [110] direction and about 8 Å wide along the [1$\bar{1}$0] direction [145]

40 Å (corresponding to 13 monolayers) for step bunching. The inclination of the steps shown in Fig. 8.42a is about 20 Å suggesting the (114) facetting. However, the step-bunching behavior on the B-type surface is not as significant as on the A-type surface (Fig. 8.42b), the bunched steps typically containing only two or three levels of the monolayer step.

The image with improved resolution has allowed them to address the question of structure. Figure 8.43 shows a model they proposed for the bunched steps. Step bunching reduces the total surface tension [149], and should be energetically favorable. The proposed structure was thus believed to accommodate dangling bonds and reduce strain. Further evidence is needed to establish this model. In contrast, the step-bunching behavior becomes in-

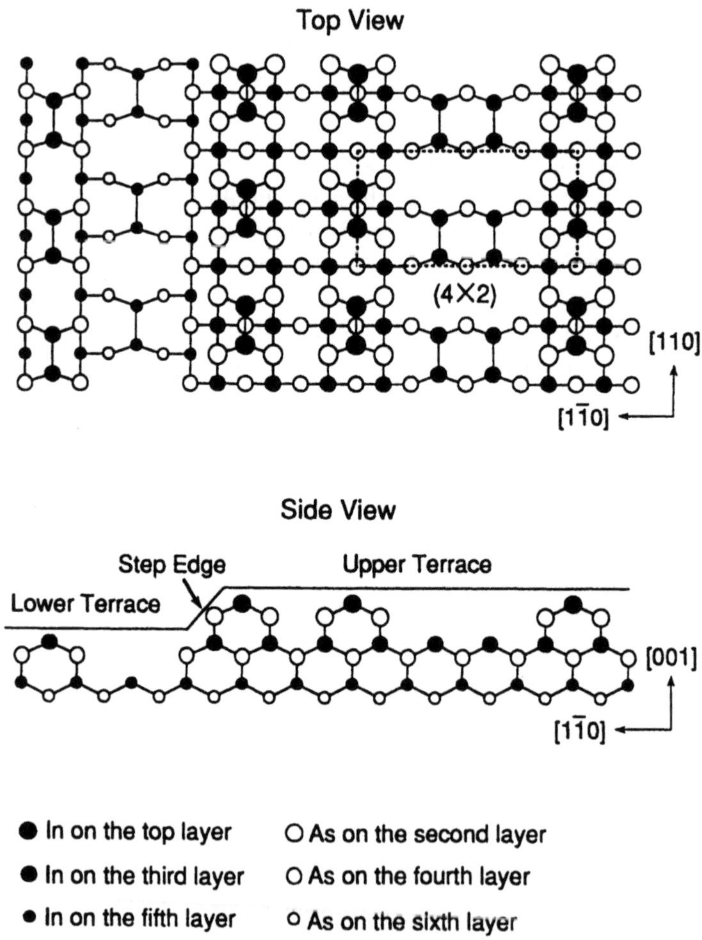

Fig. 8.43. Proposed ball-and-stick model for the step along the [110] direction on InAs(001) under the 4×2 reconstruction. To simplify the illustration, the fifth and sixth layers are not drawn under the upper terrace in the *top view*. The 4×2 unit cell is indicated by *dotted lines* [145]

significant on the B-type surface. Ikoma and Ohkouchi explained that it is due to limited kinetics, e.g., existence of anisotropic diffusion. Because of the anisotropic configuration associated with the missing-dimer structure, diffusion across the dimer row might be much slower than along the missing-dimer row direction [150], making it difficult for surface atoms to reach the step edge in the case of the B-type step.

In-Rich 4 × 2 Phase. The first atomic model for the In-rich 4 × 2 phase reconstruction was proposed by Ohkouchi and Ikoma based on a thermally cleaned surface (see Fig. 8.44) [144]. Since the STM images obtained from the As-grown InAs(001) surface [145] and the surface prepared by the IBA method [147] are essentially identical, the structure should be the same. As we see in Fig. 8.42, the surface is almost perfectly ordered and characterized with straight bright lines separated by 17 Å, corresponding to the 4× periodicity along the [$\bar{1}$10] direction. Hence, this reconstruction is also of the missing-row-dimer type [138]. With a decreased bias voltage, Kendrick et al. have revealed a ladder-like structure that exhibits 2× periodicity along the [110] direction in between these bright lines [147]. Since both the filled and empty states images obtained by Kendrick et al. show the same position for the bright lines, they were attributed to a combination of tunneling out of the first-layer In dimer bonds and the second-layer As filled dangling bonds [147]. They also studied some dislocation lines, and depression and protrusion defects on the IBA-prepared surface [147]. It was found that the defect for the transition from the 4 × 2 to the $c(8 \times 2)$ phase is negatively charged, similar to compensating

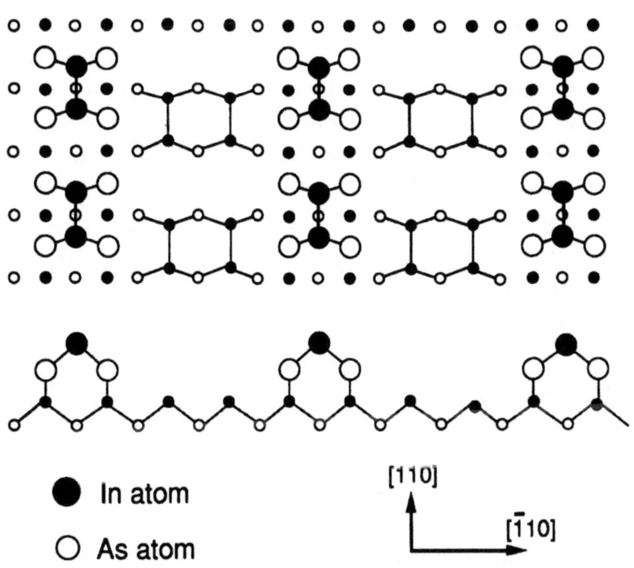

Fig. 8.44. Proposed ball-and-stick model for an InAs(001) In-stabilized 4 × 2 reconstruction. The 4 × 2 unit cell is indicated by *dotted lines* [145]

surface defects that have been reported for the GaAs(001) 2×4 phase, induced by a high density of Si doping [58]. Through bias-dependent atomic-resolution imaging, they were able to distinguish electronic features from topographic features. All observations are fully consistent with the model suggested by Ohkouchi and Ikoma, shown in Fig. 8.44 [144,151].

Strained In-Rich 4×2 and 6×2 Reconstruction. Semiconductor low-dimensional nanostructures, i.e., quantum dots, wires and wells, are gaining greater importance due to their potential in optoelectronic engineering and industry. In these sandwich-like structures, where electrons are confined to a single direction, two-dimensional (2-D) planar growth is required with both morphologically and compositionally abrupt interfaces, since a well defined interface is crucial for the high performance of strained-layer-based optoelectronic devices [152–158]. However, strain-induced morphology roughening usually occurs beyond a critical thickness, e.g., ~ 2 ML for InAs/GaAs, due to the intrinsic lattice mismatch. For example, for InAs/GaAs, Behrend et al. and Xue et al. recently reported two new structures: strained 4 × 2 (hereafter referred to as S4 × 2 phase) and 6 × 2 [159–161]. They have found that when the subsequent InAs growth is performed under special conditions that lead to S4 × 2 or 6 × 2 phase, a novel layer-by-layer InAs/GaAs(001) multilayer growth can be achieved. The S4 × 2 and 6 × 2 phases were prepared by deposition of approximately 0.6 ML In, at the substrate temperature 500°C, on the GaAs(001) As-rich 2 × 4 surface, miscut by 0.5° toward [111] and [$\bar{1}$11], respectively. The STM images (Fig. 8.45 and Fig. 8.46) of the S4 × 2 phase display a high degree of perfection. The contrast of the humps is significantly enhanced in the case of the empty states image. The STM images rule out the possibility of this 4 × 2 phase being the well known homoepitaxial InAs(001)-4 × 2 surface [141,143,146,162]. As the 2× direction agrees with the In dimerization direction, Xue et al. concluded that these humps represent the In dimers. According to the theoretical calculation by Chang and Aspnes [98], the empty states are located preferentially at the In sites. This contrast variation further supports the identification of the 2× humps as In. STM images with high resolution reveal the 1× periodicity along the straight lines, i.e., the lines in the [110] direction of Fig. 8.45 and Fig. 8.46 are made of small humps with 4 Å separation [159]. Since the small 1× humps always straddle the 2× humps along the [110] direction, they should correspond to a different species, namely, the As atoms. Tunneling becomes unstable with smaller bias than ±1.4 V, implying the semiconducting nature of the surface.

Based on these observations, Xue et al. proposed the model shown in Fig. 8.47. A variant of this model may be the more stable bilayer model (see Fig. 8.22b) [6] since the rebonding of second-layer As in the missing row in the present model may induce additional tensile strain. However, it may no longer be appropriate to the present case, because a symmetric double row is expected from the equivalent second-layer As dangling bonds for the filled

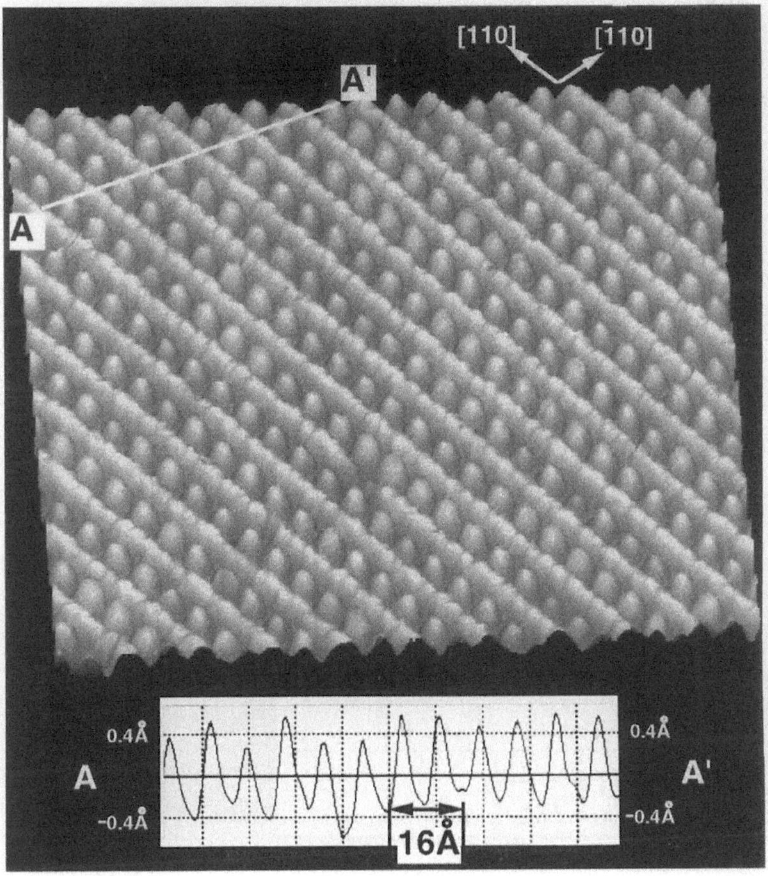

Fig. 8.45. Three-dimensionally color-coded filled states STM image (213 Å × 170 Å) of the new In-terminated 4×2 surface with a cross-section profile AA′ revealing the perfect ordering of the surface. We see no measurable contrast difference between the line and hump along the cross-section AA′. The imaging bias (V_s) used is −2.2 V [160]

states in this model [126]. The present model provides a natural explanation for the high surface tension of the In-rich reconstruction, needed for layer-by-layer growth [156]. In reality, structural relaxation must occur due to this asymmetric geometry [72,74], so that the rebonded As in the missing row moves laterally to the center of the missing row and vertically toward the surface, while the related Ga atoms shift inward. This tendency leaves the dangling bonds of Ga in a more *p*-like state, while the dangling bonds of the As are in a more *s*-like state. Therefore, relaxation shifts the Ga dangling bonds higher and the As dangling bonds lower in energy, widening the surface energy gap, and completes the charge transfer from the Ga to As dangling bonds, further stabilizing the model. The feature from another second-layer

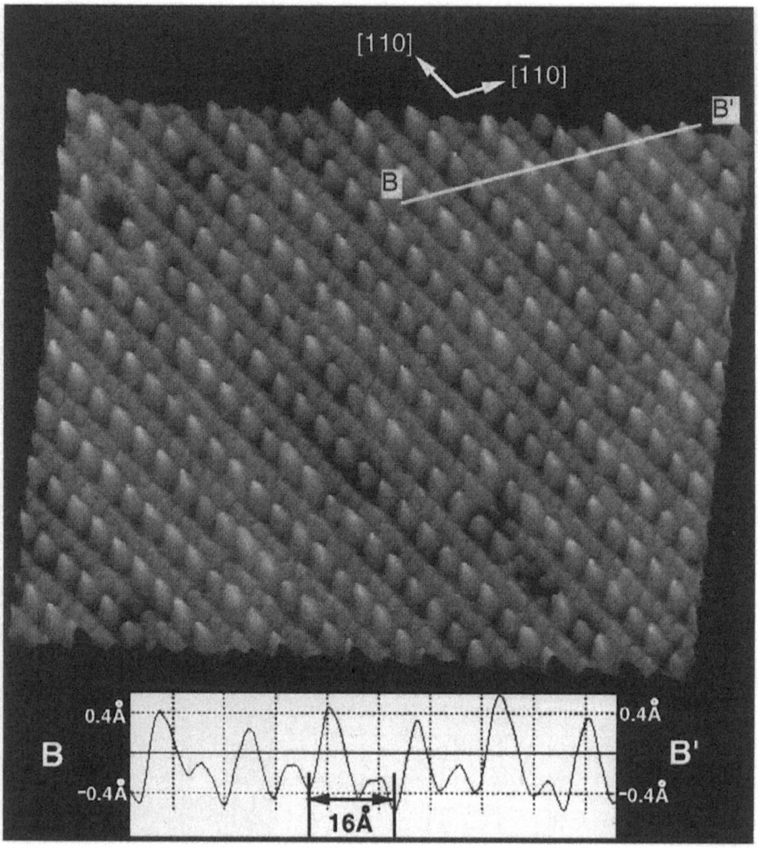

Fig. 8.46. Three-dimensionally color-coded empty states (V_s = +2.5 V) images (215 Å×194 Å) of the new In-terminated 4 × 2 surface. All basic features are very similar to those of the filled states (Fig. 8.11a), but the contrast of the humps is significantly enhanced, higher than the lines by as much as 0.6 Å on average (see the corresponding cross-section BB′). The image of this 4 × 2 surface is quite different from those observed for GaAs(100) As-rich 2×4, Ga-rich 4×2 and InAs(001)-4×2, as discussed in this review, and consists of straight line and evenly spaced humps along the [110] direction with a 16.0 Å×8.0 (±0.23) Å unit cell which thus completely accommodates the 7.2% compressed strain from the lattice mismatch. The 4× and 2× translational symmetries are easily identified by cross-section profiles (AA′ in Fig. 8.45 and BB′ in Fig. 8.46) [160]

threefold coordinated As whose dangling bonds are also doubly occupied is rather weak. This is most likely due to the fact they are too close to the In dimer and charge densities overlap with that of the In dimer.

Resch-Esser et al. and Behrend et al. have also studied the 4 × 2 surface formed by deposition of 0.5 ML In and proposed different models [151,161]. Since the STM images and their dependence on the bias polarity appear

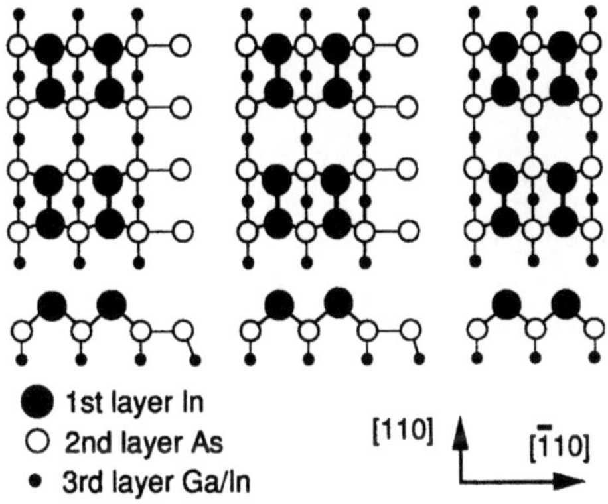

Fig. 8.47. The proposed ball-and-stick model of the In-terminated GaAs(001)-S4 × 2 surface. According to this model, the individual humps seen in STM images are attributed to tunneling from the first layer In dimers, and the 1× small humps to tunnelling from the doubly occupied dangling bonds of the second layer As, which rebond to the neighboring As in the missing row, leading to 0.5 ML In surface coverage, in good agreement with the experimental value (0.6 ML) [160]

quite different, occurrence of different structures is possible. All the models are tentative, and further study is necessary. The morphology improvement shown by these groups implies that the In termination may be energetically more stable than the Ga, As and Sb termination [147,148].

The STM images for the 6 × 2 surface by Xue et al. [160] are shown in Figs. 8.48 and 8.49. Figure 8.48 gives an overview of the surface morphology with a high degree of ordering. Atomic-resolution filled and empty states

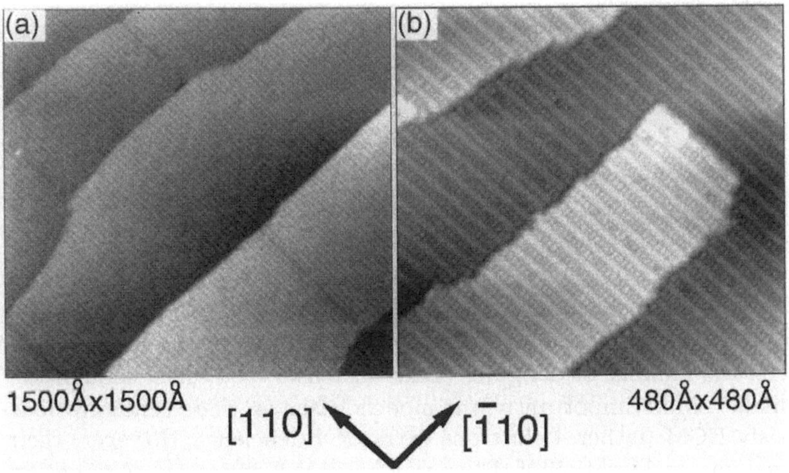

Fig. 8.48a, b. Large scale filled states (at −2.0 V) STM images of an In-rich 6×2/*c*(12×2) reconstruction formed on GaAs(001) [160]

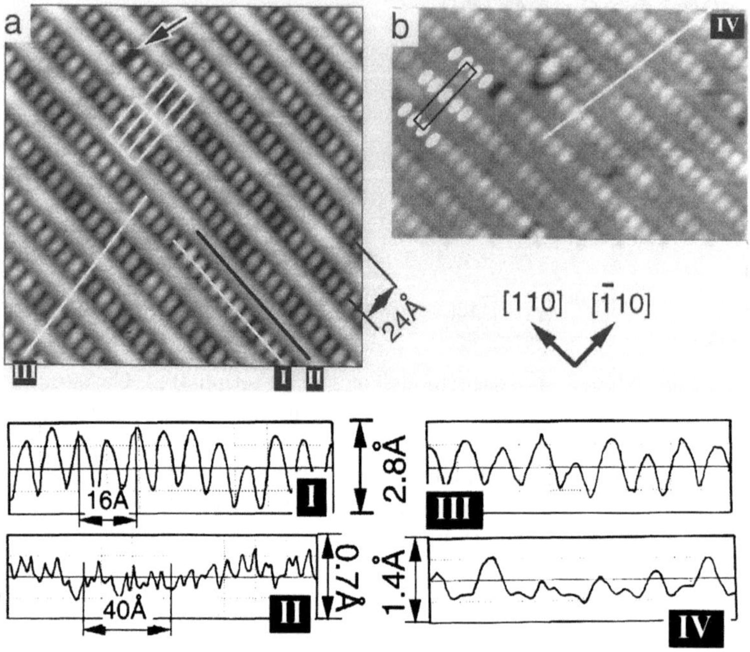

Fig. 8.49. Zoom-in STM images of an In-rich $6\times2/c(12\times2)$ reconstruction together with cross-sectional line plots showing the contrast variations over the image features. (**a**) Filled states at −2.0 V and (**b**) empty states at +2.0 V. Similar to the S4 × 2 case, this surface consists of characteristic oblongs sandwiched between straight lines along the [$\bar{1}$10] direction for both the filled and empty states. However, the line separation is now 24 Å instead of 16 Å. The oblongs extend 10 Å (about 2.5 times the surface lattice) along the [$\bar{1}$10] direction. By a comparison of (**a**) the filled and (**b**) empty states images, the higher contrast for humps in the empty states is evident [160]

images are shown in Fig. 8.49a and b. The observations are very similar to those for the S4 × 2 phase. Realizing that it has never been reported in any static or dynamic phase diagram of InAs/GaAs(001), Xue et al. speculated that the formation of the 6 × 2 phase is stabilized by a built-up strain energy. Their proposed model is shown in Fig. 8.50, which satisfies ECM. If all four As atoms in a unit cell in the missing row bond to the As atoms that bond to first-layer In, the number of dangling bonds reduces by 4, but a larger tensile strain may result. When all four second-layer As atoms in the missing row form dimers, the number of dangling bonds increases by 4, even though the strain is smaller. More importantly, the models based on these configurations do not satisfy ECM (either 4 electrons extra or 4 deficient). However, their model, based on consideration of the energy balance between the elimination of dangling bonds and the minimization of surface stress, cannot conclude the structure.

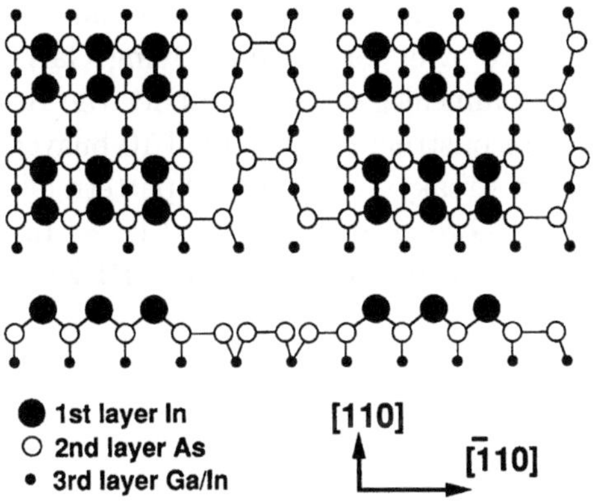

Fig. 8.50. The ball-and-stick model for the 6 × 2 In-rich surface. The oblong in the image is referred to as the three-In-dimer block, and the line as the 1× separated second-layer As. This model results in an In surface coverage of 0.5 ML [160]

8.5.2 AlAs(001) 3 × 2 Surface

AlAs and AlGaAs alloy semiconductors are usually grown on the GaAs(001) substrate because they have matching lattice parameters. AlGaAs/GaAs multilayer structures have been widely used to create various optoelectronic and electronic devices. The AlAs (001) surface, including AlGaAs alloy, exhibits a number of reconstructions, such as $c(4 \times 4)$, 2×4, 2×2 and 3×2, depending on the growth condition [163]. Recently, the first real-space image of the AlAs(001) surface grown on the GaAs(001) substrate using MBE was reported by Behrend et al. [164]. Their images clearly show that the AlAs(001)-2×4 surface also consists of two As dimers with two missing at the top layer, agreeing with the well established As-2×4 model. Due to the fact that Al atoms have the higher chemical reactivity and slower surface migration rate compared with that of Ga under the same growth conditions, high quality AlGaAs MBE growth on GaAs is difficult and often complicated by rough surface morphology, especially when the Al content is large. Based on the RHEED study of AlAs growth on GaAs(001), Dabiran et al. recently found that a layer-by-layer growth mode could coincide closely with the onsets of surface reconstruction transition at both high (2×4–3×2) and low ($c(4 \times 4)$–2×4) sample temperatures [165]. It is our experience that during the MBE growth of AlGaAs, the RHEED pattern becomes sharpest with the longest streaks only when the 3×2 reconstruction results. Realizing its importance, many groups have attempted to study this 3×2 structure by STM, without much success.

Gwo et al. in 1996 reported the first successful STM/STS study of the AlAs(001)-3×2 reconstruction [9]. This surface is found to be As-terminated, and is characterized with unusual double π-bonded As dimers along the [110] direction that distinguish it from almost all the surfaces discussed above. In this section, we will concentrate on their work. The samples were prepared by first growing a 10-bilayer AlAs buffer at a temperature of 600°C on top of a thick GaAs buffer layer as a protective layer for the underlying GaAs substrate. The AlAs buffer layer was then annealed at 700°C under the As flux for a few minutes to obtain the 3×2 reconstruction. The final 10-bilayer AlAs was grown at 690°C with a sharp and streaky 3×2 RHEED pattern [9].

The large area STM image obtained by Gwo et al. (shown in Fig. 8.51a) demonstrates a smooth surface morphology, as expected from the RHEED

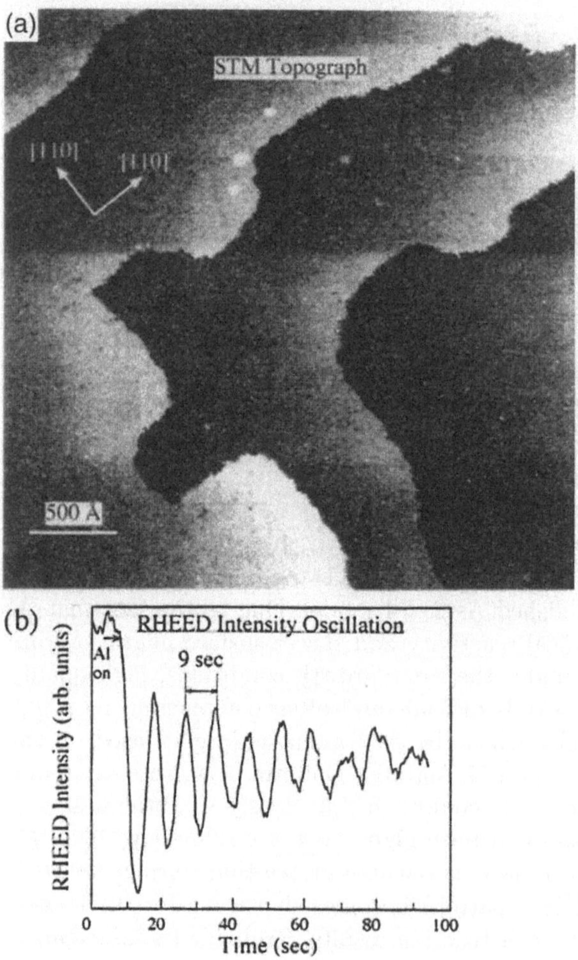

Fig. 8.51. (**a**) Large-scale (2500 Å×2500 Å) STM image of the AlAs(001)-3×2 surface grown on a GaAs substrate by MBE at 690°C. The steps in the image are all single-bilayer-height steps (2.83 Å). The image was recorded at a sample bias voltage (V_s) of −3.0 V with a tunneling current (I_t) of 0.25 nA. (**b**) RHEED spectrum beam intensity oscillation plot recorded along the $[\bar{1}10]$ azimuth during growth of the AlAs(001)-3×2 [9]

pattern. The layer-by-layer growth by which the smooth surface results is documented by a nice RHEED intensity oscillation shown in Fig. 8.17b. A high-resolution STM image (Fig. 8.52a) reveals two types of topographic maxim: type-I site (illustrated by yellowish dots) and type-II site (illustrated by bluish dots). Based on the bias-voltage and annealing-flux dependent STM

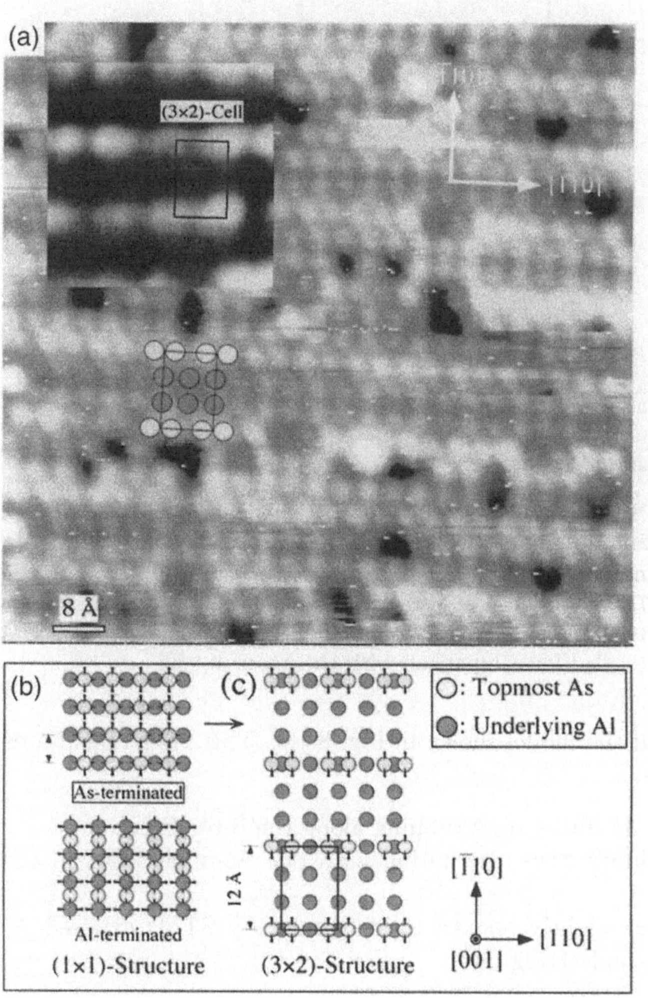

Fig. 8.52. (**a**) Color-code high resolution STM topography (V_s = −3.0 V, I_t = 0.25 nA) of the AlAs(001)-3 × 2 surface. The *inset* is a close-up view obtained at high bias conditions (V_s = −4.75 V, I_t = 1.0 nA), where the As dimers show a bean-shaped feature. (**b**) Schematic of the arrangement of dangling bonds on the As-terminated and Al-terminated (001) faces of AlAs before the surface reconstructs. (**c**) The proposed model for the As-terminated AlAs(001)-3 × 2 surface. For clarity, only the topmost As layer and the underlying Al layer are sketched. The dangling bonds are shown only for As atoms [9]

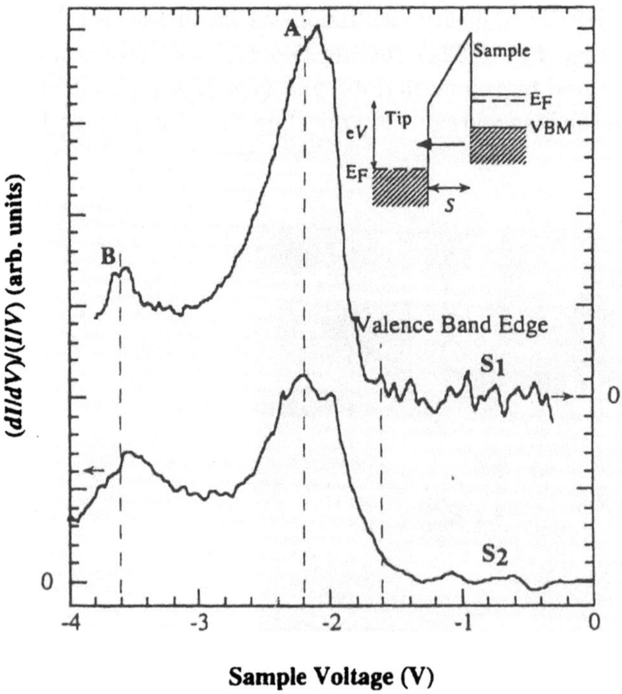

Fig. 8.53. Normalized tunneling conductance spectra of the AlAs(001)-3 × 2 surface obtained at constant tip–sample separations. The *upper* (S_1) and *lower* (S_2) *curves* correspond to measurements obtained at different initial tunneling conditions: −3.0 V and −4.75 V, and thus at varied tip–sample separations ($S_1 < S_2$). Two filled state structures (A and B) can be clearly seen from the spectra. The *upper curve* has been rescaled for comparison with the *lower one* [9]

images, they proposed the model shown in Fig. 8.18c. This model facilitates the following features:

- double π-bonded As-dimer rows running along the [110] direction,
- two missing As-dimer rows alternating with one As-dimer row, in the $[\bar{1}10]$ direction,
- unreconstructed or slightly relaxed (undetectable by STM) exposed As atoms of the first underlying layer.

In order to understand the surface electronic structure and observed bias-voltage-dependent behavior, Gwo et al. also recorded the STS spectra of the surface (see Fig. 8.53). The two most prominent normalized conductance structures located at −2.2±0.1 eV (A) and −3.6±0.1 eV (B) below the Fermi level are visible. They further found that the B states are related to type-I sites and correspond to the filled surface states of the surface dimer rows. Both type-I and type-II sites, located in the topmost surface bilayer and well resolved in the STM topographies, are the surface sites of the A states. The observed isotropic properties in diffusion and nucleation were explained to be closely related to the delocalized character of double π-bonded As dimers, lending additional support to their model. Since the model is associated with a significant number of dangling bonds, Gwo et al. pointed out that establishment of a complete model needs further theoretical considerations [9].

8.6 Phosphide, Antimonide and Nitride (001) Surfaces

8.6.1 Phosphide (001) Surfaces

InP(001)-2 × 4 and 4 × 2 Phases. Recently, two independent STM studies of the 2 × 4 reconstruction on the InP(001) surface were reported [11,166]. Typical STM images of the clean InP(001)-2 × 4 surface for both filled and empty states by Shimomura et al. are shown in Fig. 8.54 [166]. In the filled states image, the prominent feature is the bright rows containing nearly three-fold trimer-like units, which are separated by 8 Å along the $[\bar{1}10]$ direction, whereas on the empty states images only large singular protrusions were observed in the row. The images obtained by MacPherson et al. are essentially the same but with new features: faint zipper-like rows in the second layer on each side of the trimer rows [11]. The first atomic model proposed by MacPherson et al. is shown in Fig. 8.55, which satisfies ECM. The 2× periodicity of the surface is due to a repetition of the trimer unit. The trimer-like structure for the filled states and large protrusions for the empty states making the bright rows, can be associated with the P trimer atoms, and the zipper-like features seen in both the filled and empty states images are due to tunneling from or out of the surface states of the In dimers in the second layer. The most striking conclusion in this study is that the InP(001)-2 × 4 surface constructs itself, an exception amongst all other III–V (001) surfaces, and its structure cannot be described by the conventional dimer-missing dimer models [11]. Such a trimer-based structural model was also suggested by Shimomura et al. [166] and Sung et al. [167]. MacPherson et al. pointed out that a similar alternative model, such as one with two missing In dimers, is also possible [11]. Therefore, further total energy calculations are needed to establish a complete model.

As mentioned in Sect. 8.4.1, the first atomic model for the InP(001)-4 × 2 surface was proposed in 1987 by Hou et al., based on their LEED, UPS and HREELS studies, for the IBA cleaned samples [124]. According to this model, each unit cell is composed of three In dimers with one missing dimer (see Fig. 8.56a). This model is still widely accepted. In 1991, Ohkouchi and Tanaka investigated the InP(001)-4 × 2 surface by STM [162]. Their STM image convincingly demonstrated the dimer missing mechanism as suggested by Hou et al. [124]. The doublet features in the image do not seem to agree with a three-dimer unit, so they proposed a similar model, but containing two In dimers in a unit (Fig. 8.56b). Since the sample was prepared by annealing at 510°C in the As flux, an unknown amount of As may be incorporated

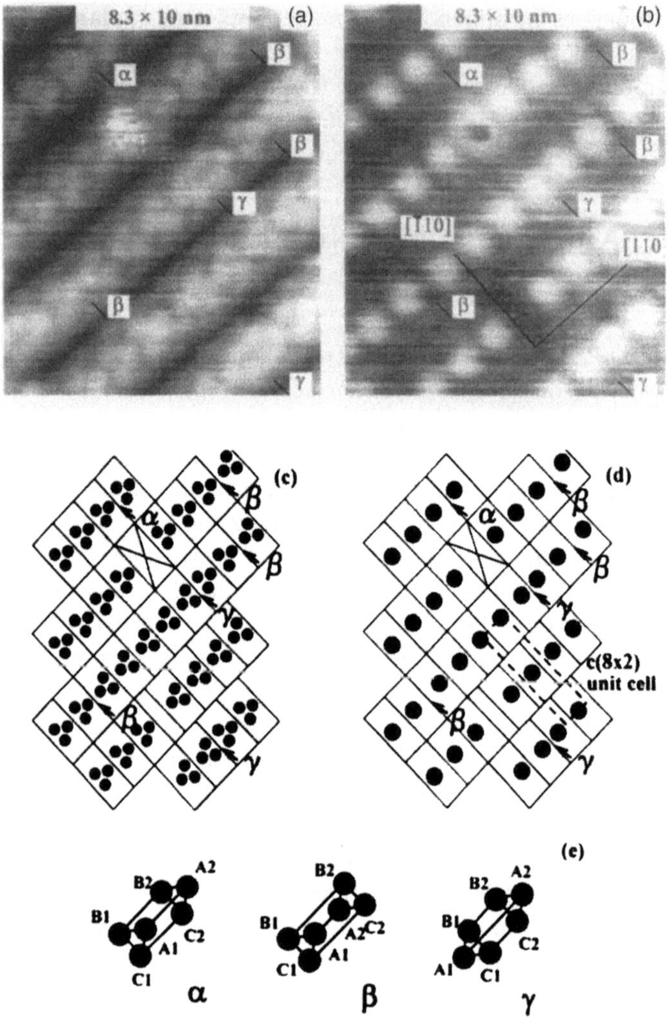

Fig. 8.54. (**a**) Filled states (−1.5 V, 0.3 nA) and (**b**) empty states (+1.5 V, 0.2 nA) STM images of the InP(001)-2 × 4 surface, where the measurements were carried out over the same area. *Schematics* for (**a**) and (**b**), and for three types (α, β and γ) of combination of the filled states protrusions are shown in (**c**), (**d**) and (**e**), respectively [166]

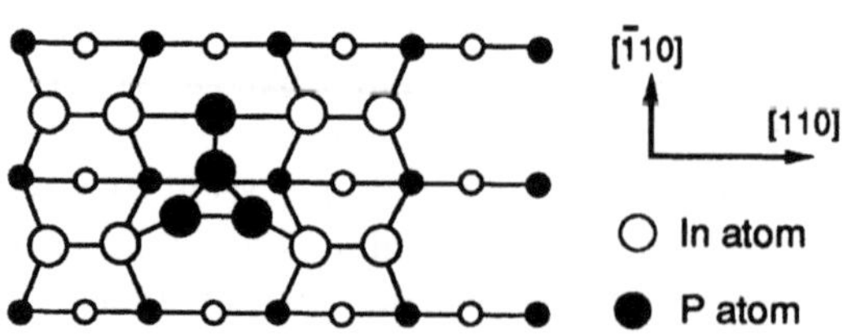

Fig. 8.55. Proposed P trimer adatom model for the InP(001)-2 × 4 reconstruction by MacPherson et al. [11]. It is noteworthy that this structure differs from the dimer-based structures usually proposed for III–V (001) surfaces

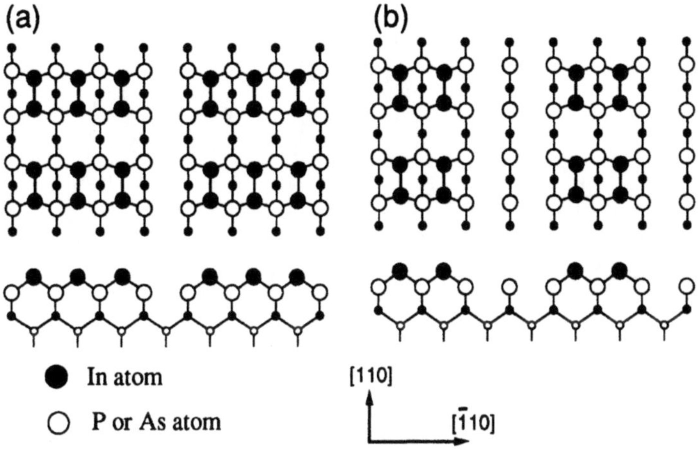

Fig. 8.56. Proposed models for the InP(001)-4 × 2 reconstruction. (**a**) Three-In-dimer model by Hou et al. [124]. (**b**) Two-In-dimer model by Ohkouchi and Tanaka [162]

into the surface layers, and the clean InP(001)-4 × 2 surface may actually be different. The present experimental data is not sufficient to determine the structure of this reconstruction from the available models.

GaP(001)-4×2 Phase. GaP is another 'wide-gap' material with an indirect band gap of 2.78 eV. The clean reconstructed surface exhibits the 2 × 4 and 4 × 2 symmetries. The structure of the P-rich GaP(001)-2 × 4 surface has not been studied extensively. It may resemble that of the GaAs(001) As-rich 2×4 reconstruction. So far, there exists only one STM study of the GaP(001) Ga-rich 4 × 2 reconstruction by Sanada et al. in Fukuda's group [168]. They found that the 4×2 phase is Ga-terminated and its structure can be described by Biegelsen et al.'s bilayer model [6]. The STM image of the GaP(001)-4×2 (Fig. 8.57a), together with cross-sectional profiles (Fig. 8.57b and c), obtained by Sanada et al. is shown in Fig. 8.57 [168]. The regular bright and dark rows running along the [110] direction document the dimer-missing character of the surface, while the 2× periodicity is evident from the evenly spaced rectangular blocks in the bright rows. Convincing evidence of its structure was achieved by analyzing the cross-sectional profiles. As can be seen in Fig. 8.57b, along the 4× ([1̄10]) direction, each bright block exhibits two features separated by 3.8 Å, which corresponds to 1× periodicity as suggested by the model (shown in Fig. 8.58). The vertical contrast difference for the bright and dark rows is 2.1–2.4 Å in good agreement with the value of 2.73 Å expected from the model. On the other hand, the profile taken in the 2× direction in Fig. 8.57c, only shows a vertical contrast change of approximately 0.7 Å which was attributed to the electronic cloud overlapping of the dimers, since they are very close along the [110] direction. All these observations are fully consistent with the bilayer two-Ga-dimer model [6].

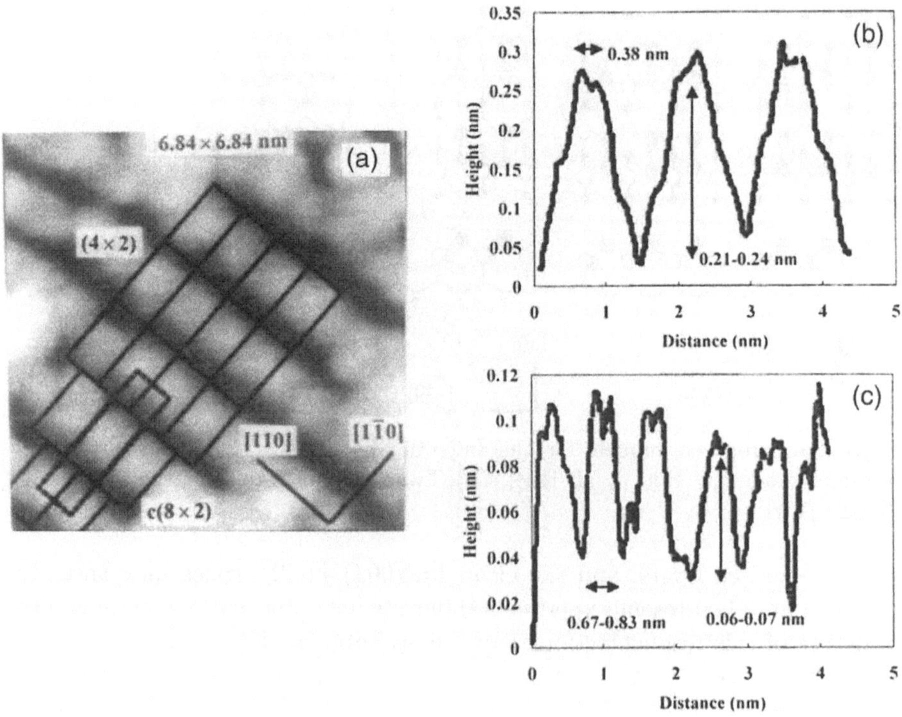

Fig. 8.57. (**a**) A filled states STM image (68.4 Å×68.4 Å) of the GaP(001)-4 × 2 Ga-rich surface measured at −3.00 V, 0.30 nA, and cross-sectional profiles along (**b**) the [$\bar{1}$10] and (**c**) the [110] directions. The 4 × 2 and *c*(8 × 2) unit cells are shown in (**a**) [168]

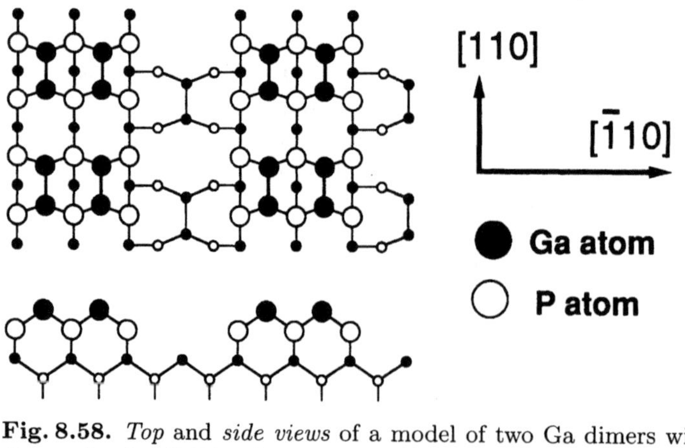

Fig. 8.58. *Top* and *side views* of a model of two Ga dimers with two missing Ga dimers for the GaP(001)-4 × 2 structure, proposed by Sanada et al. [168]

8.6.2 Antimonide (001) Surfaces

InSb(001) Surface. So far, InSb(001) is the only surface that does not display the commonly observed 2×4 reconstruction with respect to other III–V (001) discussed above. Previous RHEED studies showed that the InSb(001) surface reconstructs into the 1×1, $c(4 \times 4)$, 7×5, 4×3, 1×3 and $4 \times 2/c(8 \times 2)$ in order of increasing In surface coverage [169–171]. The Sb-rich $c(4 \times 4)$ and In-rich $c(8 \times 2)$ surfaces appear to be the two most ordered phases of all the reconstructions. The InSb (001) 1×1, $c(4 \times 4)$, 1×3 and $4 \times 2/c(8 \times 2)$ reconstructions have been systematically studied by STM [172–174]. Preparation of the samples involves both chemical etching treatment (with a mixture of lactic, nitric and hydrofluoric acids) and UHV IBA methods [172–174].

Atomically resolved STM images of the InSb(001)-$c(4 \times 4)$ for both the filled and empty states obtained by McConville et al. are shown in Figs. 8.59a and b, respectively [173]. The filled state image is similar to that of GaAs(001)

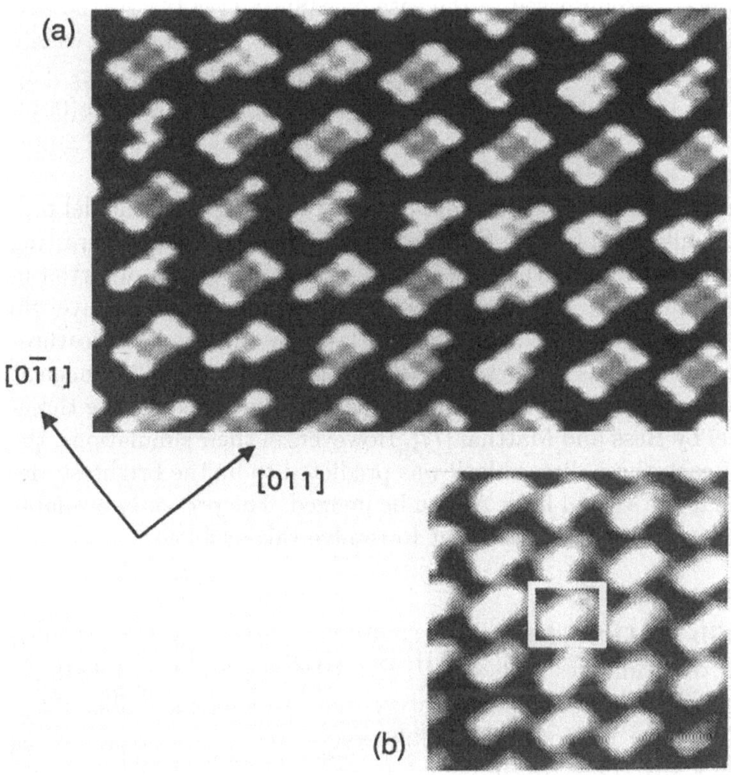

Fig. 8.59. High resolution STM images of the InSb(001)-$c(4 \times 4)$ surface showing individual Sb atoms arranged in blocks of three dimers separated by a missing dimer at the topmost layer of the surface. (**a**) (105 Å×70 Å) Filled states at −2.5 V, $I_t = 1$ nA. (**b**) Empty states at +2.5 V, $I_t = 1$ nA [173,174]

[6,90,113]. The larger lattice constant (6.5 Å) of InSb, compared with 5.6 Å for GaAs, results in a greater separation of electron density associated with each Sb lone-pair orbital, and each dimer atom is resolved more clearly [173]. The observation is consistent with the model shown in Fig. 8.17c. Interestingly, the center dimer in each block appears to have less electron density associated with it and is thus imaged darker, whereas in the empty states image, such features and also the internal structure of the dimer are absent. McConville et al. pointed out that both electronic and steric effects are probably involved [173].

According to the theoretical prediction by Bass and Matthai, the center dimer of the three-dimer block should be imaged brighter for the filled states and darker for the empty states [77]. The experimental results are somehow reversed here. However, the simulations were performed with the three-dimer model for the GaAs(001)-2×4 As-rich surface. As the bonding configurations for the topmost layer As are different for the 2×4 and $c(4 \times 4)$ surfaces, similar simulations on the InSb(001)-$c(4 \times 4)$, or at least the GaAs(001)-$c(4 \times 4)$, should be helpful to understanding these features and their differences. Such a phenomenon was also observed in the case of AlSb(001)-$c(4 \times 4)$ [12], and the underlying mechanism may be identical. It may also be due to the widely separated charge density distribution associated with the larger lattice constant of InSb. The high resolution image was also acquired for the InSb(001)-$4 \times 2/c(8 \times 2)$ surface by the same groups [172,173]. A typical filled states image is shown in Fig. 8.60.

The data have been interpreted in terms of the three-In-dimer model proposed by John, Miller and Chiang [175] (see Fig. 8.61). The most interesting point in this interpretation is that the brighter circular features observed in the image are attributed to the occupied lone-pair orbitals of second layer Sb exposed in the missing rows [172], and the first layer In-dimer-related features are completely absent. This seems to agree with the simulated STM image at low bias (−1 V) of the GaAs(001) Ga-rich 4×2 surface based on the three-Ga-dimer model by Bass and Matthai [77]. However, in their simulations, the center dimer of each three-dimer block was predicted to be the brightest, despite the fact that the second layer As can be imaged. The presently available experimental data are still not sufficient to resolve this problem.

GaSb and AlSb (001) Surfaces. The GaSb(001) surface has been studied using a number of techniques such as RHEED, PES and STM [176–181]. As a function of growth/annealing temperature and III/V element flux ratio, the GaSb(001) surface is shown to exhibit several reconstructions such as 1×3, $c(2 \times 6)$, $1 \times 5/c(2 \times 10)$ and 2×5 in order of increasing surface Sb coverage [179]. The atomic model for the $c(2 \times 6)$ phase proposed by Sieger et al. [179] is shown in Fig. 8.62a and it basically agrees with that by Resch-Esser et al. [181]. The model for the 1×3 phase also proposed by Sieger et al. is shown in Fig. 8.62b.

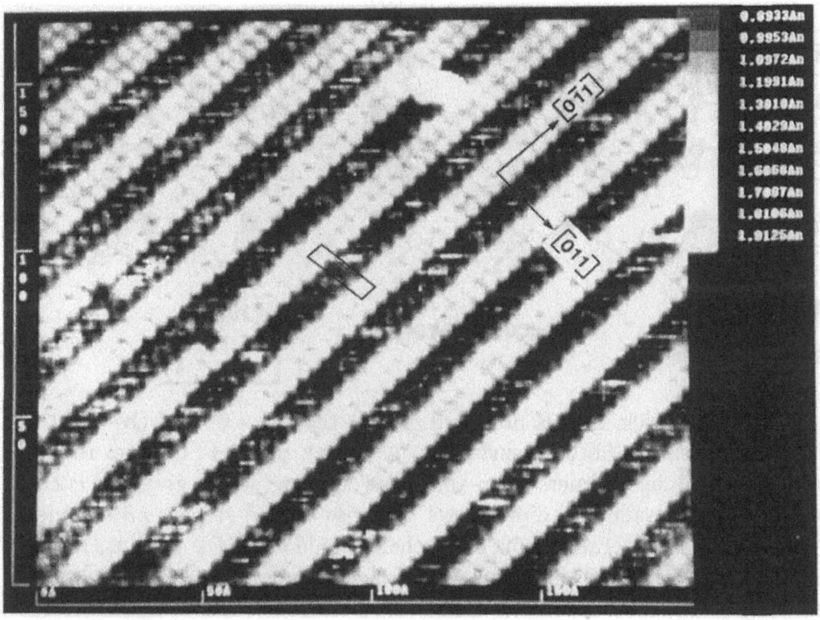

Fig. 8.60. Representative filled states STM image (200 Å×200 Å) of the InSb(001) In-rich $4 \times 2/c(8 \times 2)$ reconstruction prepared using the IBA method by Schweitzer et al. [172]. The visible double rows are the occupied lone-pair orbitals of the second layer Sb atoms. The 4×1 unit cell is indicated in the image, which was recorded with a sample bias of -3 V and a tunneling current of 1 nA

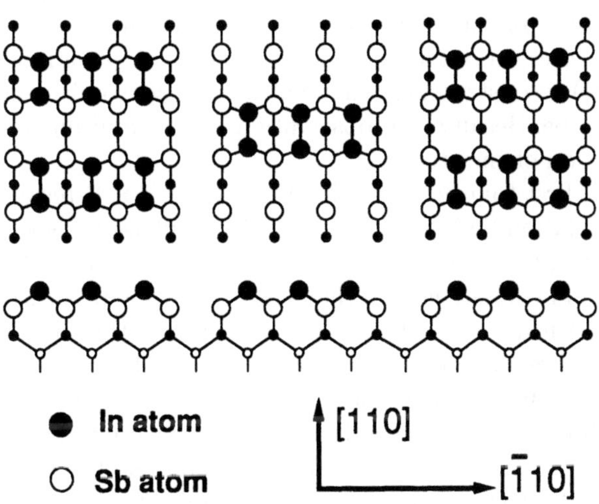

Fig. 8.61. Proposed three-In-dimer model for the InSb(001)-$4 \times 2/c(8 \times 2)$ reconstruction [172,175]

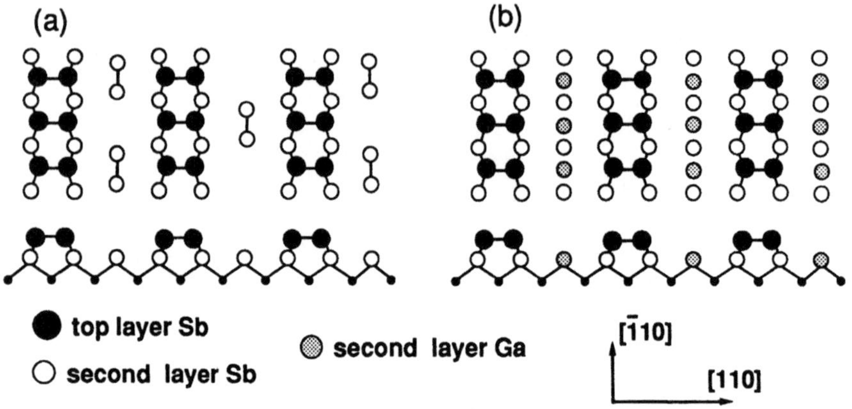

Fig. 8.62. Ball-and-stick models proposed for (**a**) the $c(2 \times 6)$ and (**b**) the 1×3 reconstructions of the GaSb(001) surface [179]. In (**a**), parallel Sb dimer rows are separated by second layer dimers in an antiphase ordering, which gives the $c(2 \times 6)$ symmetry. In (**b**), parallel Sb dimer rows give rise to the strong $\times 3$ symmetry observed by RHEED. Lateral bucking of Ga chains could cause the weak $\times 2$ features in the RHEED pattern, and contribute to surface disordering. Both *top* and *side views* are shown

In the remaining part of this section, we discuss two 'odd' structures: the 1×5 and 2×5 reconstructions observed on the GaSb(001) surface. These two structures are unique and represent an interesting exception to all other reconstructions found on III–V (001) surfaces. With STM/STS observations and first-principles electronic structure calculations, Whitman et al. found that, the GaSb(001)-$n \times 5$ reconstructed surfaces are weakly metallic and violate ECM [12]. Hence, the electron counting model, the basic rule governing the structures of III–V (001) reconstructions, may not be nearly as universal as it is currently assumed to be. Because this observation is so significant, we will give a detailed description of the study.

Whitman et al. were the first to investigate the AlSb(001)-$c(4 \times 4)$ reconstruction [12]. The high resolution STM image of the $c(4 \times 4)$ phase is shown in Fig. 8.63a, similar to the $c(4 \times 4)$ phase observed on the InSb and GaAs surfaces mentioned above, and consistent with the model originally proposed for the GaAs(001)-$c(4 \times 4)$ surface (see Fig. 8.63b) [6]. On the other hand, Whitman et al. did not observe the formation of the $c(4 \times 4)$ phase on the GaSb(001) surface under Sb-rich and low substrate temperature preparation conditions; instead, they observed the 1×5 and 2×5 phases with increasing Sb coverage. Moreover, the 1×5 and 2×5 phases show an abnormal behavior with respect to tunneling. The STS spectra for the 1×5 and 2×5 phases together with that of the AlSb(001)-$c(4 \times 4)$ are shown in Fig. 8.64.

Despite the existence of a small pseudo-gap, observation of midgap conductivity is a direct demonstration that the surfaces are metallic [12]. On both surfaces, Whitman et al. obtained atomically resolved STM images:

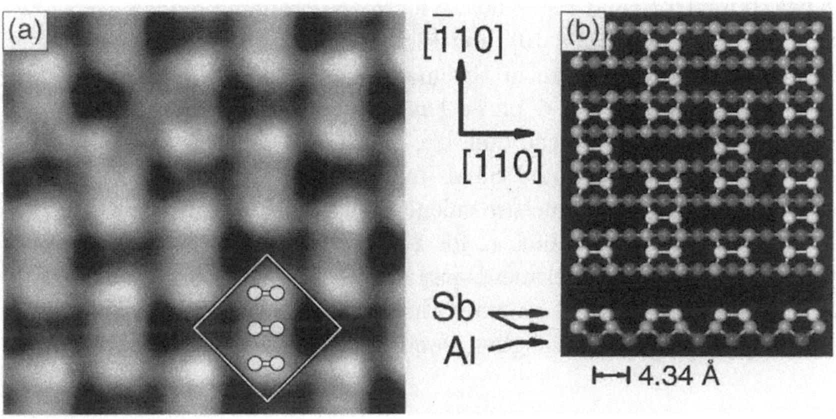

Fig. 8.63. (**a**) Filled state STM image of AlSb(001)-$c(4 \times 4)$, 48 Å×48 Å acquired at 3.3 V. A unit cell is indicated along with some of the surface Sb dimers. (**b**) Structural model of the surface, consisting of an Sb double layer [12]

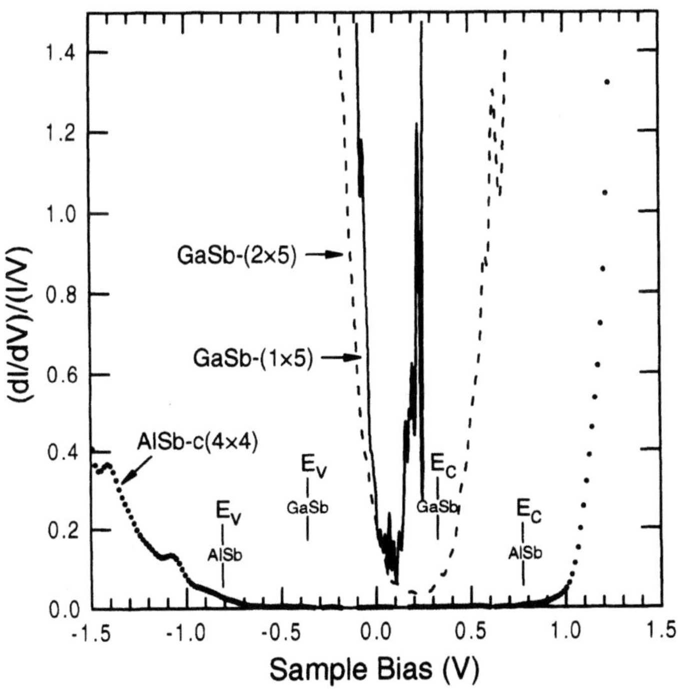

Fig. 8.64. Tunneling spectra acquired on AlSb(001)-$c(4 \times 4)$ (*dotted line*), GaSb(001)-1×5 (*solid line*), and GaSb(001)-2×5 (*dashed line*). The spectra are displayed as normalized conductivity $(\mathrm{d}I/\mathrm{d}V)/(\overline{I/V})$ (approximately proportional to the density of states), calculated from the measured I vs. V, where $(\overline{I/V})$ is exponentially broadened with a width of 0.7 V. The bulk band edges for both substrates are indicated. Note that the Fermi level is approximately in the middle of the bulk band gap on both surfaces. Whereas the AlSb surface is insulating, the GaSb surfaces are weakly metallic, with nonzero conductivity for all voltages [12]

Fig. 8.65a, b for 1×5, and Fig. 8.66a, b for 2×5. Proposed models are shown in Fig. 8.65c, for $1 \times 5/c(2 \times 10)$, and Fig. 8.66c, for 2×5.

The most striking feature of the models is that they have more than one monolayer Sb. The 1×5 phase has a double layer whereas the 2×5 phase has a triple layer, on the bulk-like GaSb (001) surface (see Fig. 8.65c and Fig. 8.66c). The simulated STM images based on these models using first-principles electronic structure calculations are shown in Fig. 8.65d, for $1 \times 5/c(2 \times 10)$, and Fig. 8.66d, e, for $2 \times 5/c(2 \times 10)$. These are in excellent agreement with the experiments and hence justify their models. Electron counting of the models gives an extra three valence electrons per 2×5 unit cell after every surface Sb dangling bond is fully filled with two electrons,

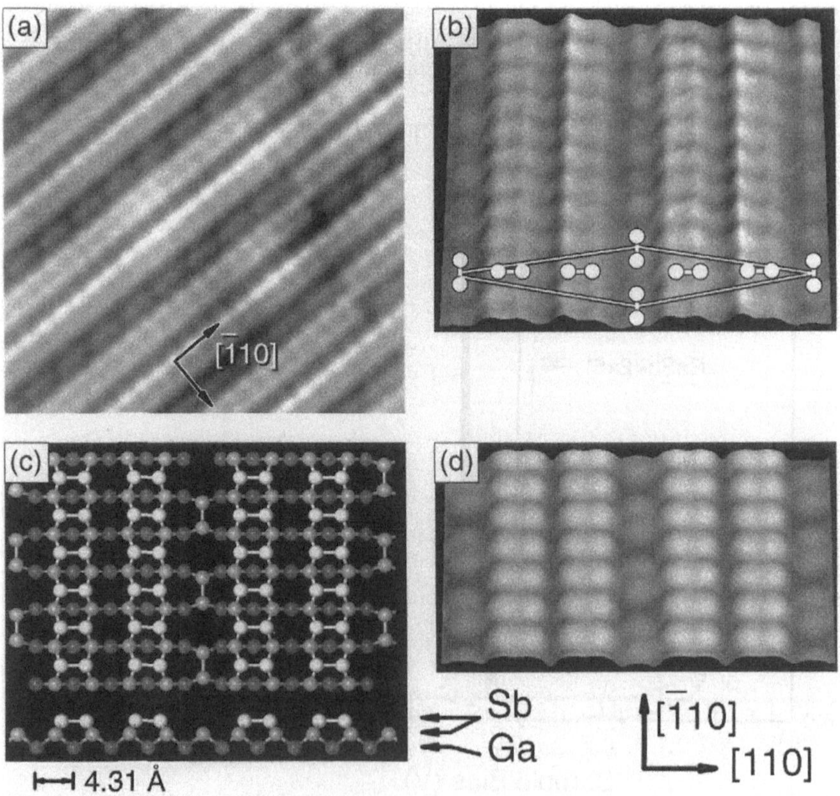

Fig. 8.65. Filled state STM images of GaSb(001)-1×5 acquired at 1.8 V: (**a**) 80 Å ×80 Å and (**b**) 48 Å×48 Å. The lower dimers give the surface a 2× periodicity along the [$\bar{1}$10] direction. A $c(2 \times 10)$ unit cell is indicated. Note that the slight asymmetry of the right dimer row of each row pair is a feedback artefact resulting from the large gain required to resolve the internal structure of the dimers. (**c**) Proposed $c(2 \times 10)$ structural model for this surface. (**d**) Simulated STM image of this structure calculated for filled states within 1.8 eV of the Fermi level [12]

and the surface is therefore also expected to be metallic. The mechanism has been analyzed in terms of heteroepitaxy of Sb on the III-Sb(001) surface [12] and results from the ideal lattice match (between the GaAs(001)-1 × 1 and Sb tetrahedral) and stronger Sb–Sb homo-bond with respect to the soft Sb–Ga hetero-bond, which is unique to this surface. The 1 × 5 surface compromises structural energy reduction and electronic energy increase associated with occupying high energy conduction-like states, and in this way violates ECM. They further pointed out that the III-Sb(001)-1 × 3 may also fall into this category. Some existing results seem to support such a speculation [176,180].

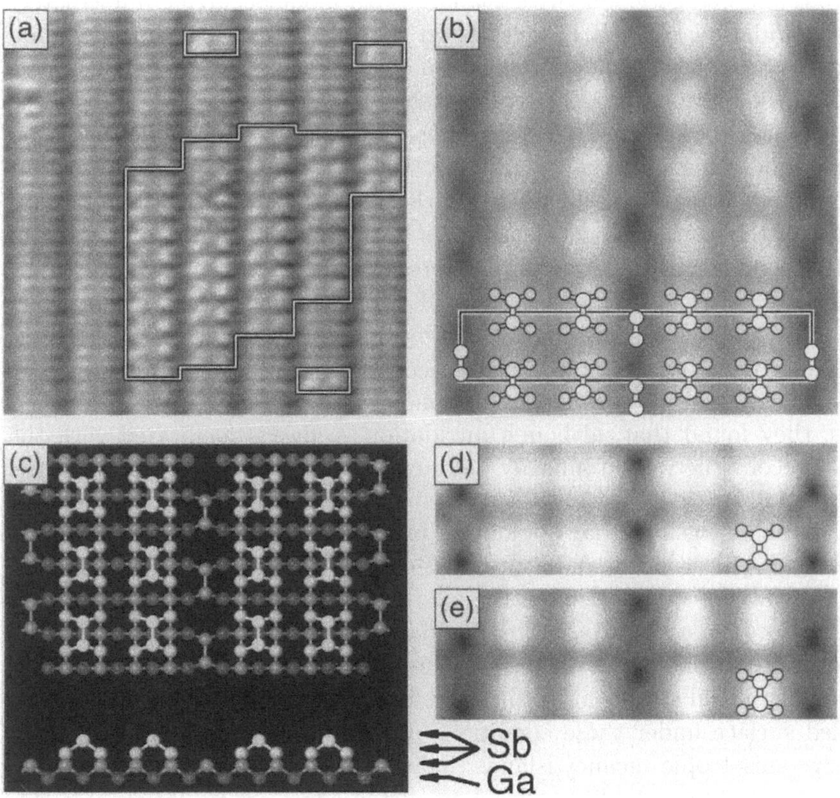

Fig. 8.66. (**a**) Filled state STM images of the GaSb(001) surface with both 1 × 5-like and 2 × 5-like regions (2.4 V, 150 Å×150 Å). The 2 × 5-like regions are *outlined.* (**b**) Higher magnification image of a well-ordered 2 × 5 (by RHEED) surface (0.4 V, 52 Å×52 Å). A 2 × 10 unit cell is indicated. (**c**) Proposed 2 × 10 structural model for this surface, consisting of the simple addition of Sb dimers on top of the dimer rows of the structure shown in Fig. 8.65. Simulated filled state STM images of this structure calculated for (**d**) 0.4 eV and (**e**) 2.4 eV [12]

8.6.3 GaN(001) 3 × 3 Surface

Along with other Group III nitrides, AlN and InN, GaN is useful for the fabrication of short wavelength optoelectronic devices, particularly high brightness Light-Emitting Diodes (LED) and laser diodes [13,14]. GaN (III-nitrides) can be synthesized in either the hexagonal wurtzite (α-GaN) or cubic zinc blende (β-GaN) form, and metastable cubic GaN would have superior electronic properties [182–184]. The GaAs(001) is one of the most frequently adopted substrates [185–194]. An understanding of the initial growth stage is of the utmost importance since it directly influences the crystal structure and quality of the grown film [186,187,190]. Previous studies with ECR nitrogen sources have shown that smooth nitride layers are very difficult to achieve, due to surface damage and preferential etching by excited radicals, high energy electrons and energetic ions with island growth mode occurring immediately under nitridation [195]. It is well known that the GaAs(001) surface, upon exposure to nitrogen plasma, displays a well defined (3 × 3) RHEED pattern [192,196]. Gwo et al. [197] reported the first successful STM study of the nitrided GaAs(001)-(2 × 4) As-rich surface and proposed a model based on N-dimer-and-missing-dimer for the (3 × 3) reconstruction. It is shown that the N-dimer formation is unstable on the reconstructed GaN(001) surface because of the very small covalent radius of the N atom compared to that of Ga [198]. Further justification is therefore necessary.

Recently, Xue et al. have used MBE-STM to study the nitridation of various GaAs(001) surfaces by an N-atom RF plasma source [199]. By examination of the effects of substrate reconstruction, plasma exposure time and substrate temperature on the morphology and structure of the nitrided layers, they found that, although nitridation is always facilitated with the (3 × 3) reconstruction, atomically flat surfaces can only be obtained on the As-rich (2 × 4) and (2 × 6) reconstructed surfaces. With these observations, they were able to show that nitridation proceeds via competing mechanisms between (3 × 3) ordering and continuous etching. The main results obtained by Xue et al. are summarized in the following [199].

The (2×4) As-rich surface after nitridation at 500°C is shown in Fig. 8.67. The overall morphology with terrace-plus-step structure is basically similar to that of the substrate, indicating the formation of an atomically smooth nitrided surface under these conditions. Despite this smooth surface, there are large anisotropic vacancy islands (indicated by arrows in Fig. 8.67a) populating the terraces. Additionally, A-type step edges are no longer straight, but appear somewhat ragged. It is speculated that both changes are due to preferential etching of the steps by N atoms, and etching of B-type steps is more favorable than etching of A-type steps. Figure 8.67b is a typical zoom-in STM image of the surface, showing a well defined 3× periodicity along the [$\bar{1}$10] direction, as highlighted by an array of solid lines. The irregular distribution of spots along the [110] direction is probably a manifestation of strain relief by local bond deformation.

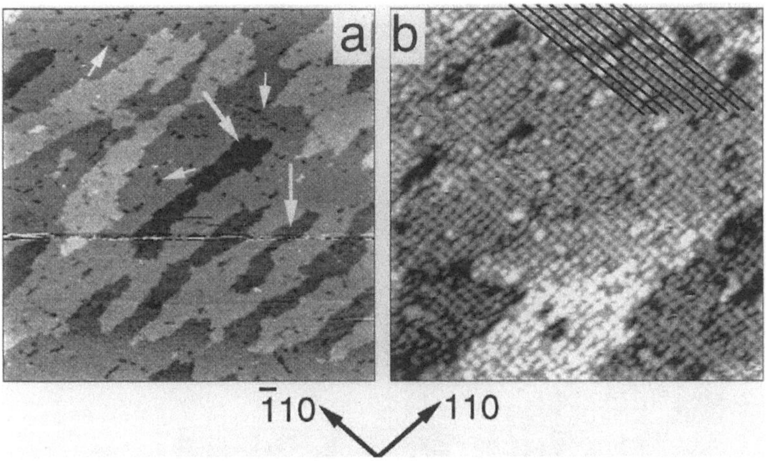

Fig. 8.67. STM images of the nitrided GaAs(001) (2×4) As-rich surface after 10 s plasma exposure at a power of 100 W, displaying (**a**) an overview of the surface morphology (2 000 Å×2 000 Å) and (**b**) the 'distorted' (3×3) reconstruction (500 Å×500 Å). Substrate temperature during exposure was 500°C. Sample bias voltage $V_s = -3.5$ V and tunneling current $I_t = 40$ pA [199]

STM images recorded from the nitrided 2×6 surfaces are summarized in Fig. 8.68. A preferential etching of the steps is again evident as shown by the series of images in Fig. 8.68. A different observation is that now the (3×3) domains are well-ordered (see the inset of Fig. 8.68c). On the Ga-rich (4×2) surface, nitridation is characterized by the immediate formation of small 3-D GaN crystallites (see Fig. 8.69). In this case, the apparent Ga diffusion on the surface begins and the local exchange mechanism no longer operates, thereby providing a kinetic pathway for GaN islanding.

8.7 Conclusions

In this review, we have overviewed the STM studies of almost all reconstructions found on polar III–V compound semiconductor (001) surfaces. Its capability for high spatial resolution in real space combined with chemical sensitivity, make STM an extremely powerful tool for the structural determination of complicated surfaces. We are able to show how the puzzling features reported by various investigators can be sorted out using STM, thus demonstrating the unique role it plays in structural determination. Bringing together the results of experimental and theoretical investigations, we can make the following conclusions.

- Most reconstructions occurring on III–V (001) surfaces can be understood in terms of dimer formation with periodic missing-dimer defects.

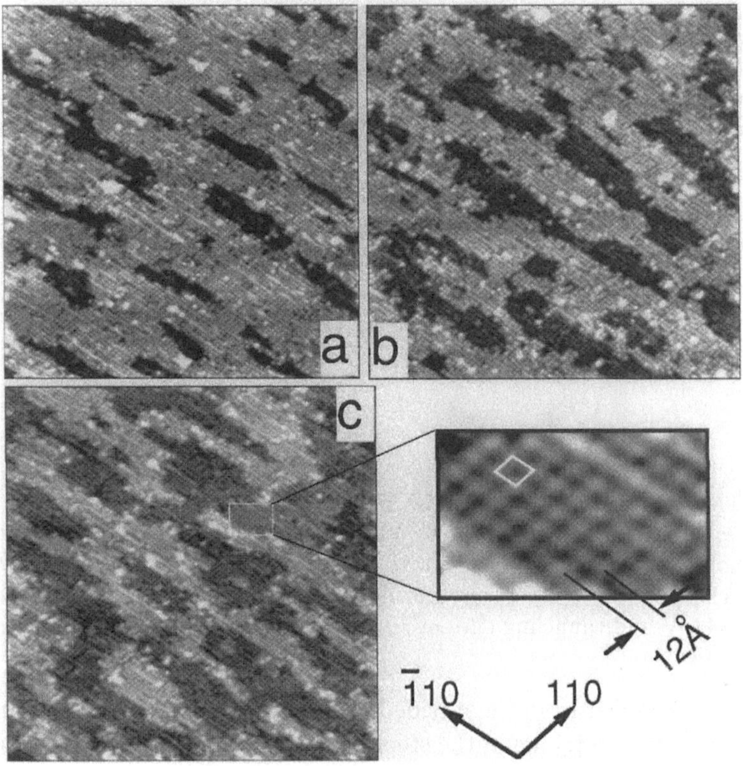

Fig. 8.68. A series of STM images of the nitrided (2 × 6) As-rich surface at 550°C showing surface morphology evolution and etching process with different plasma exposures: (**a**) 5 s (1 000 Å×1 000 Å) , (**b**) 10 s (1 000 Å×1 000 Å, (**c**) 17 s (1 000 Å×1 000 Å). The *inset* of (**c**) indicates the well defined (3 × 3) ordering. $V_s = -3.3$ V, $I_t = 40$ pA [199]

Fig. 8.69. Surface morphology (3 000 Å×3 000 Å) of the nitrided (4 × 2) Ga-rich surface showing the formation of small 3-D GaN islands near step edges. Plasma exposure time was 10 s and substrate temperature was 610°C. $V_s = -2.5$ V, $I_t = 40$ pA [199]

Dimerization reduces the number of surface dangling bonds, while dimer-missing maintains charge neutrality. The stable surface is generally semiconducting.

- Depending on the chemical potential of the III (or V) element, a number of reconstructions may occur. ECM narrows the range of possible candidates for each reconstruction phase, and STM/STS observation further reduces the number of candidates. Bias-dependent images are most helpful for the correct interpretation of data.
- STM is the most powerful technique available for documenting the situation when several phases coexist, as is amply demonstrated in the case of two 4×6 phases on the GaAs(001) surface.
- AlAs(001)-3×2 is unique with double π-bonded dimers, and InP(001)-2×4 (P-rich) is an exception, where trimerization of the dimer-missing-dimer rows takes place. GaSb(001)-'$n \times 5$' (Sb-rich) is another exception that violates ECM and is weakly metallic.

In addition, this review describes the potential STM has for studying growth phenomena. Much progress has been made in III–V related homo- and hetero-epitaxy, for a range of surfaces, such as (111) [200] and high-index (114) [201], which lie beyond the scope of this review. Breakthroughs in synthesizing group III-nitrides, the most promising material for short wavelength opto-electronic device applications, have brought STM into the realm of III-nitride surfaces. Some excellent results are now being reported [197,202]. STM has played, and will continue to play, a crucial role in this important field.

Acknowledgements

We would like to thank Professor J.M. Zhou, Institute of Physics, Academia Sinica, Professor A. Ichimiya, Nagoya University, Dr. T. Ohno, National Research Institute of Metals, and Professors I.S.T. Tsong, Arizona State University, and Y. Hasegawa, Tohoku University for fruitful collaborations. We would also like to acknowledge Professors Y. Fukuda, Shizuoka University, Max. G. Lagally, University of Wisconsin at Madison, and Roy F. Willis, Pennsylvania State University, and Drs. H. Yamaguchi, NTT, Tim S. Jones, Imperial College, W.G. Schmidt, Friedrich-Schiller University, L.J. Whitman, Naval Research Laboratories, S. Ohkouchi, Sumitomo Electric. Corp., and H. Tokumoto, JRCAR-NAIR, for generously providing us with figures for this review.

We wish to thank Messrs. T. Ogino, H. Kiyama, and Q.Z. Xue for invaluable assistance during MBE-STM experiments. One of the authors (Q.K.X) would like to thank Prof. D.E. Aspnes, North Carolina State University, for continuous encouragement to him.

The work was partially supported by Special Promoted Research Project on BEEM (No. 08102001) and Priority Area Project (Single-Electron-Devices Project), Department of Education.

References

1. G. Binnig and H. Rohrer, Helv. Phys. Acta. **55**, 726 (1982).
2. G. Binnig, H. Rohrer, Ch. Gerber and E. Weibel, Phys. Rev. Lett. **50**, 120 (1982).
3. H. Neddermeyer, Scanning Tunneling Microscopy of Semiconductor Surfaces, Rep. Prog. Phys. **59**, 701 (1996).
4. J.A. Kubby and J.J. Boland, Scanning Tunneling Microscopy of Semiconductor Surfaces, Surf. Sci. Rep. **26**, 61 (1996).
5. M.D. Pashley, K.W. Haberern, W. Friday, J.M. Woodall and P.D. Kirchner, Phys. Rev. Lett. **60**, 2176 (1988).
6. D.K. Biegelsen, R.D. Bringans, J.E. Northrup and L.-E. Swartz, Phys. Rev. **B41**, 5701 (1990).
7. R.M. Feenstra, J.A. Stroscio, J. Tersoff and A.P. Fein, Phys. Rev. Lett. **58**, 1192 (1987).
8. J. Behrend, M. Wassermeier, L. Daweritz and K.H. Ploog, Surf. Sci. **342**, 63 (1996).
9. S. Gwo, H. Ogino and H. Tokumoto, Phys. Rev. **B55**, R1962 (1997).
10. I. Chizhov, G. Lee, R.F. Willis, D. Lubyshev and D.L. Miller, Phys. Rev. **B56**, 1013 (1997).
11. C.D. MacPherson, R.A. Wolkow, C.E.J. Mitchell and A.B. Mclean, Phys. Rev. Lett. **77**, 691 (1996).
12. L.J. Whitman, P.M. Thibado, S.C. Erwin, B.R. Bennett and B.V. Shanabrook, Phys. Rev. Lett. **79**, 693 (1997).
13. S. Nakamura, T. Mukai and M. Senoh, Appl. Phys. Lett. **64**, 1687 (1994).
14. I. Akasaki, H. Amano, S. Sota, H. Sakai, T. Tanaka and M. Koike, Jpn. J. Appl. Phys. **34**, L1517 (1995).
15. L. Daeritz and R. Hey, Surf. Sci. **236**, 15 (1990).
16. D.J. Chadi, in: The Structures of Surfaces III, S.Y. Tong, M.A. van Hove, K. Takayanagi and X.D. Xie (Eds.), Springer Series in Surface Sciences Vol. 24, 532. Springer, Berlin (1991).
17. C.B. Duke, Chem. Rev. **96** , 1237 (1996).
18. J.P. LaFemina, Surf. Sci. Rep. **16**, 133 (1992).
19. C.B. Duke, in: Physical Structure, Handbook of Surface Science, W.N. Unertl (Ed.), Vol. 1, 229. Elsevier, North Holland (1996).
20. M.D. Pashley, Phys. Rev. **B40**, 10481 (1990).
21. D.J. Chadi, J. Vac. Sci. Technol. **A5**, 834 (1987).
22. H.H. Farrell, J.P. Harbison and L.D. Peterson, J. Vac. Sci. Technol. **B5**, 1482 (1987).
23. W.A. Harrison, J. Vac. Sci. Technol. **16**, 1492 (1979).
24. M.C. Gallagher, R.H. Prince and R.F. Willis, Surf. Sci. **275**, 31 (1992).
25. J.R. Arthur and J.J. LePore, J. Vac. Sci. Technol. **6**, 545 (1969).
26. F. Jona, IBM J. Res. Develop. **9**, 375 (1965).
27. A.Y. Cho, J. Appl. Phys. **42**, 2074 (1971).
28. R.E. Schlier and H.E. Farnsworth, Semiconductor Surface Physics, 3. U. Pa. Press, Philadelphia, Pennsylvania (1957).
29. A.Y. Cho, J. Appl. Phys. **47**, 2842 (1976).
30. A.J. van Bommel and J.E. Crombeen, Surf. Sci. **57**, 437 (1976).
31. J.R. Arthur, Surf. Sci. **43**, 449 (1974).

32. P. Drathen, W. Ranke and K. Jacobi, Surf. Sci. **77**, L162 (1978).
33. A.J. van Bommel, J.E. Crombeen and T.G.J. van Oirschot, Surf. Sci. **72**, 95 (1978).
34. J.H. Neave and B.A. Joyce, J. Crystal Growth **44**, 387 (1978).
35. J. Massies, P. Etienne, F. Dezaly and N.T. Linh, Surf. Sci. **99**, 121 (1980).
36. P.K. Larsen, J.H. Neave and B.A. Joyce, J. Phys. C: Solid State Phys. **14**, 167 (1981).
37. R.Z. Bachrach, R.S. Bauer, P. Chiaradia and G.V. Hansson, J. Vac. Sci. Technol. **19**, 335 (1981).
38. P.K. Larsen, J.F. van der Veen, A. Mazur, J. Pollman, J.H. Neave, and B.A. Joyce, Phys. Rev. **B26**, 3222 (1982).
39. T.-C. Chiang, R. Ludeke, M. Aono, G. Landgren, F.J. Himpsel and D.E. Eastman, Phys. Rev. **B27**, 4770 (1983).
40. B.A. Joyce, J.H. Neave, P.J. Dobson and P.K. Larsen, Phys. Rev. **B29**, 814 (1984).
41. P.K. Larsen, P.J. Dobson, J.H. Neave, B.A. Joyce, B. Bolger and J. Zhang, Surf. Sci. **169**, 176 (1986).
42. D.J. Chadi, C. Tanner and J. Ihm, Surf. Sci. **120**, L425 (1982).
43. I. Ihm, D.J. Chadi and J.D. Joannopoulos, Phys. Rev. **B27**, 5119 (1983).
44. R.I.G. Uhrberg, R.D. Bringans, R.Z. Bachrach and J.E. Northrup, Phys. Rev. Lett. **56**, 520 (1986).
45. Guo-xin Qian, R.M. Martin and D.J. Chadi, Phys. Rev. **B38**, 7649 (1988).
46. D.J. Frankel, C. Yu, J.P. Harbison and H.H. Farrell, J. Vac. Sci. Technol. **B5**, 1113 (1987).
47. D.K. Biegelsen, R.D. Bringans and L.E. Swartz, SPIE, **1186**, 136 (1990).
48. M.D. Pashley and K.W. Haberern, Ultramicroscopy **42-44**, 1281 (1992).
49. E.J. Heller and M.G. Lagally, Appl. Phys. Lett. **60**, 2675 (1992).
50. M. Wassermeier, V. Bressler-Hill, R. Maboudian, K. Pond, X.S. Wang, W.H. Weinberg and P.M. Petroff, Surf. Sci. Lett. **278**, L147 (1992).
51. V. Bressler-Hill, M. Wassermeier, K. Pond, R. Maboudian, G.A.D. Briggs, P.M. Petroff and W.H. Weinberg, J. Vac. Sci. Technol. **B10**, 1881 (1992).
52. H. Xu, T. Hashizume and T. Sakurai, Jpn. J. Appl. Phys. **32**, 1511 (1993).
53. S. Miwa, Y. Haga, E. Morita, S. Arakawa, T. Hashizume and T. Sakurai, Jpn. J. Appl. Phys. **32**, 1508 (1993).
54. I. Tanaka and S. Ohkouchi, Jpn. J. Appl. Phys. **32**, 2152 (1993).
55. M.D. Pashley, Phys. Rev. Lett. **70**, 3171 (1993).
56. R. Maboudian, V. Bressler-Hill and W.H. Weinberg, Phys. Rev. Lett. **70**, 3172 (1993) .
57. J. Falta, R.M. Tromp, M. Copel, G.D. Pettit and P.D. Kirchner, Phys. Rev. Lett. **70**, 3173 (1993).
58. M.D. Pashley and K.W. Haberern, Phys. Rev. Lett. **67**, 2679 (1991).
59. M.D. Pashley, K.W. Haberern and R.M. Feenstra, J. Vac. Sci. Technol. **B10**, 1874 (1992).
60. S. Ohkouchi, N. Ikoma and I. Tanaka, J. Vac. Sci. Technol. **B12**, 2030 (1994).
61. M.D. Pashley, K.W. Haberern, R.M. Feenstra and P.D. Kirchner, Phys. Rev. **B48**, 4612 (1993).
62. I. Tanaka, S. Ohkouchi and A. Hashimoto, Jpn. J. Appl. Phys. **31**, 2216 (1992).
63. J. Sudijono, M.D. Johnson, M.B. Elowitz, C.W. Snyder and B.G. Orr, Surf. Sci. **280**, 247 (1993).

64. M.D. Pashley, K.W. Haberern and J.M. Gaines, Appl. Phys. Lett. **58**, 406 (1991).
65. E.J. Heller, Z.Y. Zhang and M.G. Lagally, Phys. Rev. Lett. **71**, 743 (1993).
66. J. Falta, R.M. Tromp, M. Copel, G.D. Pettit and P.D. Kirchner, Phys. Rev. Lett. **69**, 3068 (1992)
67. P.K. Larsen, J.H. Neave, J.F. van der Veen, P.J. Dobson and B.A. Joyce, Phys. Rev. **B27**, 4966 (1983).
68. P.K. Larsen and D.J. Chadi, Phys. Rev. **B37**, 8282 (1988).
69. K. Kanisawa, J. Osaka, S. Hirono and N. Inoue, J. Cryst. Growth **115**, 348 (1991).
70. C. Deparis and J. Massies, J. Cryst. Growth, **108**, 157 (1991).
71. H.H. Farrell and C.J. Palmstrom, J. Vac. Sci. Technol. **B8**, 903 (1990).
72. J.E. Northrup and S. Froyen, Phys. Rev. Lett. **71**, 2276 (1993).
73. J.E. Northrup and S. Froyen, Phys. Rev. **B50**, 2015 (1994).
74. T. Ohno, Phys. Rev. Lett. **70**, 631(1993).
75. T. Ohno, Phys. Rev. Lett. **73**, 460 (1994).
76. T. Ohno, Thin Solid Films **272**, 331 (1996).
77. J.M. Bass and C.C. Matthai, Phys. Rev. **B50**, 11212 (1994).
78. W.G. Schmidt and F. Bechstedt, Phys. Rev. **B54**, 16742 (1996).
79. S.B. Zhang and A. Zunger, Phys. Rev. **B53**, 1343 (1996).
80. T. Ohno and K. Shiraishi, Phys. Rev. **B42**, 11194 (1990).
81. W.G. Schmidt and F. Bechstedt, Surf. Sci. **360**, L473 (1996).
82. N.N. Greenwood and A. Earnshaw, Chemistry of the Elements, 643. Pergamon, New York (1984).
83. S.A. Chambers, Surf. Sci. **261**, 48 (1992).
84. H. Li and S.Y. Tong, Surf. Sci. **282**, 380 (1993).
85. C. Xu, K.P. Caffey, J.S. Burnham, S.H. Goss, B.J. Garrison and N. Winograd, Phys. Rev. **B45**, 6776 (1992).
86. K. Sumitomo, H. Yamaguchi, Y. Hirota, T. Nishioka and T. Ogino, Surf. Sci. **355**, L361 (1996).
87. G.P. Srivastava and S.J. Jenkins, Phys. Rev. **B53**, 12589 (1996).
88. K. Shiraishi and T. Ito, J. Cryst. Growth **150**, 158 (1995).
89. T. Hashizume, Q.K. Xue, J.M. Zhou, A. Ichimiya and T. Sakurai, Phys. Rev. **B51**, 4200 (1995).
90. T. Hashizume, Q.-K. Xue, J.M. Zhou, A. Ichimiya and T. Sakurai, Phys. Rev. Lett. **73**, 2208 (1994).
91. P. Moriaty, P.H. Beton, Y.-R. Ma, A.W. Dunn, M. Henini and D.A. Woolf, J. Vac. Sci. Technol. **B14**, 943 (1996).
92. A.R. Avery, D.M. Holmes, T.S. Tones, B.A. Joyce and G.A.D. Briggs, Phys. Rev. **B50**, 8098 (1994).
93. M. Wassermeier, J. Behrend, L. Daweritz and K. Ploog, Phys. Rev. **B52**, R2269 (1995).
94. A.R. Avery, C.M. Goringe, D.M. Holmes, J.L. Sudijono and T.S. Jones, Phys. Rev. Lett. **76**, 3344 (1996).
95. M.D. Jackson, J.M.C. Thornton, D.A. Woolf and P. Weightman, Surf. Sci. **352-354**, 724 (1996).
96. L. Li, B.-K. Han, S. Gan, H. Qi and R.F. Hicks, Sci. Rep. RITU (Research Institutes of Tohoku University) **A44**, 149 (1997).
97. L.D. Broekman, R.C.G. Leckey, J.D. Riley, A. Stampfl, B.F. Usher and B.A. Sexton, Phys. Rev. **B51**, 17795 (1995).

98. Y.C. Chang and D.E. Aspnes, Phys. Rev. **B41**, 12002 (1990).
99. J.M. Zhou, Q.-K. Xue, H. Chaya, T. Hashizume and T. Sakurai, Appl. Phys. Lett. **64**, 583 (1994).
100. A. Ichimiya, Jpn. J. Appl. Phys. **22**, 176 (1983).
101. A. Ichimiya, Jpn. J. Appl. Phys. **24**, 1365 (1985).
102. A. Ichimiya, Surf. Sci. **235**, 75 (1990).
103. T. Hashizume, Q.K. Xue, A. Ichimiya and T. Sakurai, Appl. Surf. Sci. **87–88**, 373 (1995).
104. A. Ichimiya, Q.K. Xue, T. Hashizume and T. Sakurai, J. Cryst. Growth **150**, 136 (1995).
105. Y. Garreau, M. Sauvage-Simkin, N. Jedrecy, R. Pinchaux and M.B. Veron, Phys. Rev. B54, 17638 (1996).
106. P.R. Pukite, G.S. Petrich, S. Batra and P.I. Cohen, J. Cryst. Growth **95**, 269 (1989).
107. H. Norenberg and N. Koguchi, Surf. Sci. **296**, 199 (1993).
108. R. Duszak, C.J. Palmstrom, L.T. Florez, Y.-N. Yang and J.H. Weaver, J. Vac. Sci. Technol. **B10**, 1891 (1992).
109. M. Sauvage-Simkin, R. Pinchaux, J. Massies, P. Calverie, N. Jedrecy, J. Bonnet and I.K. Robinson, Phys. Rev. Lett. **62**, 563 (1989).
110. I. Kamiya, D.E. Aspnes, H. Tanaka, L.T. Florez, J.P. Harbison and R. Bhat, Phys. Rev. Lett. **68**, 627 (1992).
111. I. Kamiya, D.E. Aspnes, L.T. Florez and J.P. Harbison, Phys. Rev. **B46**, 15894 (1992).
112. I. Tanaka, S. Ohkouchi, T. Kato and F. Osaka, J. Vac. Sci. Technol. **B9**, 2277 (1991).
113. A.R. Avery, D.M. Holmes, J. Sudijono, T.S. Jones and B.A. Joyce, Surf. Sci. **323**, 91 (1995).
114. A.Y. Cho, J. Appl. Phys. **47**, 2841 (1976).
115. M. Kuball, D.T. Wang, N. Esser, M. Cardona, J. Zegenhagen and B.O. Fimland, Phys. Rev. **B51**, 13880 (1995).
116. Q.-K. Xue, T. Ogino, Y. Hasegawa, H. Shinohara and T. Sakurai, Phys. Rev. **B53**, 1985 (1996).
117. C.W. Snyder, J. Sudijono, Chi-Hang Lam, M.D. Johnson and B.G. Orr, Phys. Rev. **B50**, 18194 (1994).
118. H. Qi, P.E. Gee and R.F. Hicks, Phys. Rev. Lett. **72**, 250 (1994).
119. J.R. Creighton, Surf. Sci. **234**, 287 (1990).
120. R. Ludeke and L. Esaki, Phys. Rev. Lett. **33**, 653(1974).
121. W. Ranke and K. Jacobi, Prog. Surf. Sci. **10**, 1 (1981).
122. I.M. Vitomirov, A.D. Raisanen, A.C. Finnefrock, R.E. Viturro, L.J. Brillson, P.D. Kirchner, G.D. Pettit and J.M. Woodall, J. Vac. Sci. Technol. **B10**, 1898 (1992).
123. J. Cerda, F.J. Palomares and F. Soria, Phys. Rev. Lett. **75**, 665 (1995).
124. X. Hou, G. Dong, X. Ding and X. Wang, J. Phys. **C20**, L121 (1987).
125. S.L. Skala. J.S. Hubacek, J.R. Tucker, J.W. Lyding, S.T. Chou, and K.Y. Cheng, Phys. Rev. **B48**, 9138 (1993).
126. Q.K. Xue, T. Hashizume, J.M. Zhou, T. Sakata, T. Sakurai and T. Ohno, Phys. Rev. Lett. **74**, 3177 (1995).
127. P. Moriaty, private communications.
128. D.E. Aspnes, Y.C. Chang, A.A. Studna, L.T. Florez, H.H. Farrell and J.P. Harbison, Phys. Rev. Lett. **64**, 192 (1990).

129. Q.-K. Xue, T. Hashizume, T. Ohno, Y. Hasegawa, T. Sakurai, Sci. Rep. RITU **A44**, 113 (1997).
130. Q.K. Xue, T. Hashizume, T. Sakata, Y. Hasegawa, A. Ichimiya, T. Ohno and T. Sakurai, Thin Solid Films **281–282**, 556 (1996).
131. J. Massies, P. Devoldere, P. Etienne and N.T. Linh, in: Proc. 7th Intern. Vacuum Congr. and 3rd Intern. Conf. on Solid Surfaces, 639. Vienna (1977).
132. S.L. Skala, S.T. Chou, K.-Y. Cheng, J.R. Tucker and J.W. Lyding, Appl. Phys. Lett. **65**, 722 (1994).
133. H. Yamaguchi and Y. Horikoshi, Phys. Rev. Lett. **70**, 1299 (1993).
134. T. Sakurai, X.-D. Wang, Q.-K. Xue, Y. Hasegawa, T. Hashizume and H. Shinohara, Scanning Tunneling Microscopy Study of Fullerenes, Prog. Surf. Sci. **51**, 263 (1996).
135. Q.-K. Xue, T. Ogino, Y. Hasegawa, T. Hashizume, H. Shinohara and T. Sakurai, Mater. Sci. Eng. **A217/218**, 27 (1996).
136. J.M. Moison, C. Guille and M. Bensoussan, Phys. Rev. Lett. **58**, 2555 (1987).
137. H. Yamaguchi and Y. Horikoshi, Phys. Rev. **B45**, 1511 (1992).
138. M.M. Sung and J.W. Rabalais, Surf. Sci. **356**, 161 (1996).
139. H. Yamaguchi and Y. Horikoshi, Phys. Rev. **B48**, 2807 (1993).
140. H. Yamaguchi and Y. Horikoshi, Jpn. J. Appl. Phys. **33**, 716 (1994).
141. H. Yamaguchi and Y. Horikoshi, Jpn. J. Appl. Phys. **33**, L1423 (1994).
142. S. Ohkouchi, N. Ikoma and I. Tanaka, J. Vac. Sci. Technol. **B12**, 2033 (1994).
143. S. Ohkouchi and N. Ikoma, Jpn. J. Appl. Phys. **33**, L1700 (1994).
144. S. Ohkouchi and N. Ikoma, Jpn. J. Appl. Phys. **33**, 3710 (1994).
145. N. Ikoma and S. Ohkouchi, Jpn. J. Appl. Phys. **34**, 5763 (1995).
146. H. Yamaguchi and Y. Horikoshi, Phys. Rev. **B51**, 9836 (1995).
147. C. Kendrick, G. Lelay and A. Khan, Phys. Rev. **B54**, 17877 (1996).
148. H. Yamaguchi and Y. Horikoshi, Phys. Rev. **B53**, 4565 (1996).
149. C. Herring, Phys. Rev. **82**, 87 (1952).
150. K. Shiraishi, Appl. Phys. Lett. **60**, 1363 (1992).
151. U. Resch-Esser, N. Esser, C. Springer, J. Zegenhagen, W. Richter, M. Cardona and B.O. Firmland, J. Vac. Sci. Technol. **B13**, 1672 (1995).
152. W.J. Schaffer, M.D. Lind, S.P. Kowalczyk and R.W. Grant, J. Vac. Sci. Technol. **B1**, 688 (1983).
153. M.Y. Yen, A. Madhukar, B.F. Lewis, R. Fernandez, L. Eng and F.J. Grunthaner, Surf. Sci. **174**, 606 (1986).
154. O. Brandt, H. Lage and K. Ploog, Phys. Rev. **B43**, 14285 (1991).
155. E. Tournie, O. Brandt and K. Ploog, Appl. Phys. Lett. **60**, 2877 (1992).
156. C.W. Snyder, B.G. Orr and H. Munekata, Appl. Phys. Lett. **62**, 46 (1993).
157. A. Trampert, E. Tournie and K. Ploog, Appl. Phys. Lett. **66**, 2265 (1995); V. Bressler-Hill, S. Varma, A. Lorke, B.Z. Nosho, P.M. Petroff and W.H. Weinberg, Phys. Rev. Lett. **74**, 3209 (1995).
158. C.W. Snyder, B.G. Orr, D. Kessler and L.M. Sander, Phys. Rev. Lett. **66**, 3032 (1991).
159. Q.K. Xue, Y. Hasegawa, T. Ogino, H. Kiyama and T. Sakurai, J. Vac. Sci. Technol. **B15**, 1270 (1997).
160. Q.K. Xue, Y. Hasegawa, T. Ogino, H. Kiyama and T. Sakurai, Phys. Rev. **B57**, R6862-R6865 (1998).
161. J. Behrend, M. Wassermeier, K.H. Ploog, J. Cryst. Growth **167**, 440 (1996).
162. S. Ohkouchi and I. Tanaka, Appl. Phys. Lett. **59**, 1588 (1991).

163. L.L. Chang, L. Esaki, W.E. Howard and R. Ludeke, J. Vac. Sci. Technol. **10**, 11 (1973).
164. J. Behrend, M. Wassermeier, W. Braun, P. Krispin and K.H. Ploog, J. Cryst. Growth **175/176**, 178 (1997).
165. A.M. Dabiran and P.I. Cohen, J. Cryst. Growth **150**, 23 (1995).
166. M. Shimomura, N. Sanada, Y. Fukuda and P.J. Moller, Surf. Sci. **359**, L451 (1996).
167. M.M. Sung, C. Kim, H. Bu, D.S. Karpuzov and J.W. Radalais, Surf. Sci. **322**, 116 (1995).
168. N. Sanada, M. Shimomura, Y. Fukuda and T. Sato, Appl. Phys. Lett. **67**, 1432 (1995).
169. K. Oe, S. Ando and K. Sigiyama, Jpn. J. Appl. Phys. **19**, L417 (1980).
170. A.J. Noreika, M.H. Francombe and C.E.C. Wood, J. Appl. Phys. **52**, 7416 (1981).
171. A.G. de Oliveira, S.D. Parker, R. Droopad and B.A. Joyce, Surf. Sci. **227**, 150 (1990).
172. M.O. Schweitzer, F.M. Leibsle, T.S. Jones, C.F. McConville and N.V. Richardson, Surf. Sci. **280**, 63 (1993).
173. C.F. McConville, T.S. Jones, F.M. Leibsle, S.M. Driver, T.C.Q. Noakes, M.O. Schweitzer and N.V. Richardson, Phys. Rev. **B50**, 14965 (1994).
174. C.F. McConville, T.S. Jones, F.M. Leibsle and N.V. Richardson, Surf. Sci. **303**, L373 (1994).
175. P. John, T. Miller and T.-C. Chiang, Phys. Rev. **B39**, 1730 (1989).
176. G.E. Franklin, D.H. Rich, A. Samsava, E.S. Hirschorn, F.M. Leibsle, T. Miller and T.-C. Chiang, Phys. Rev. **B41**, 12619 (1990).
177. J. Piao, R. Beresford and W.I. Wang, J. Vac. Sci. Technol. **B8**, 276 (1990).
178. M. Yano et al., J. Cryst. Growth **111**, 609 (1991).
179. M.T. Sieger, T. Miller and T.-C. Chiang, Phys. Rev. **B52**, 8256 (1995).
180. P.M. Thibado, B.R. Bennett, B.V. Shanabrook and L.J. Whitman, J. Cryst. Growth **175/176**, 317 (1997).
181. U. Resch-Esser, N. Esser, B. Brar and H. Kroemer, Phys. Rev. **B55**, 15401 (1997).
182. M. Fanciulli, T. Lei and T.D. Moustakas, Phys. Rev. **B48**, 15144 (1993).
183. T. Lei, T.D. Moustakas, R.J. Graham, Y. He and S.J. Berkowitz, J. Appl. Phys. **71**, 1 (1992).
184. C.-Y. Yeh, Z.W. Lu, S. Froyen and A. Zunger, Phys. Rev. **B46**, 10086 (1992).
185. S. Strite, J. Ruan, Z. Li, N. Manning, A. Salvador, H. Chen, D.J. Smith, W.J. Choyke and H. Morkoc, J. Vac. Sci. Technol. **B9**, 1924 (1991).
186. H. Okumura, S. Misawa and S. Yoshida, Appl. Phys. Lett. **59**, 1058 (1991).
187. S. Fujieda and Y. Matsumoto, Jpn. J. Appl. Phys. **30**, L1665 (1991).
188. S. Miyoshi, K. Onabe, N. Ohkouchi, H. Yaguchi, R. Ito, S. Fukatsu, Y. Shiraki, J. Crystal Growth **124**, 439 (1992).
189. M.E. Lin, G. Xue, G.L. Zhou, J.E. Greene and H. Morkoc, Appl. Phys. Lett. **63**, 932 (1993).
190. A. Kikuchi, H. Hoshi and K. Kishino, Jpn. J. Appl. Phys. **33**, 688 (1994).
191. Z.Q. He, X.M. Ding, X.Y. Hou and X. Wang, Appl. Phys. Lett. **64**, 315 (1994).
192. R.J. Hauenstein, D.A. Collins, X.P. Cai, M.L. O'Steen and T.C. McGill, Appl. Phys. Lett. **66**, 2861 (1995).
193. O. Brandt, H. Yang, A. Trampert, M. Wassermeier and K. Ploog, Appl. Phys. Lett. **71**, 473 (1997).

194. D. Lederman, Z. Yu, T.H. Myers, M.R. Richards-Babb, Appl. Phys. Lett. **71**, 368 (1997).
195. R.S. Goldman, R.M. Feenstra, B.G. Briner, M.L. O'Steen and R.J. Hauenstein, Appl. Phys. Lett. **69**, 3698 (1996)
196. H. Okumura, K. Ohta, G. Feuillet, K. Balakrishnan, S. Chichibu, H. Hamaguchi, P. Hacke and S. Yoshida, J. Crystal Growth **178**, 113 (1997).
197. S. Gwo, H. Tokumoto and S. Miwa, Appl. Phys. Lett. **71**, 362 (1997).
198. A.D. Bykhovski and M.S. Shur, Appl. Phys. Lett. **69**, 2397 (1996).
199. Q.K. Xue, Q.Z. Xue, Y. Hasegawa, I.S.T. Tsong and T. Sakurai, Jpn. J. Appl. Phys. **36**, L1486 (1997).
200. H. Yamaguchi, J.G. Belk, X.M. Zhang, J.L. Sudijono, M.R. Fahy, T.S. Tones, D.W. Pashley and B.A. Joyce, Phys. Rev. **B55**, 1337 (1997).
201. H. Yamaguchi, T. Yamada and Y. Horikoshi, Jpn. J. Appl. Phys. **34**, L1490 (1995).
202. A.R. Smith, R.M. Feenstra, D.W. Greve, J. Neugebauer and J.E. Northrup, Phys. Rev. Lett. **79**, 3934 (1997).

9 Adsorption of Fullerenes on Semiconductor and Metal Surfaces Investigated by Field-Ion Scanning Tunneling Microscopy

T. Hashizume and T. Sakurai

Summary. We review our systematic investigation of adsorption and film growth of fullerenes on semiconductor and metal substrates using Field-Ion Scanning Tunneling Microscopy (FI-STM). The fullerenes investigated include pristine fullerenes (C_{60}, C_{70} and C_{84}) and metallofullerenes ($Sc@C_{74}/Sc_2@C_{74}$, $Sc_2@C_{84}$, $Y@C_{82}$ and $Gd@C_{82}$) and the substrates include Si(100)-2×1, Si(111)-7×7, GaAs (001)-2×4-As, Cu(111)-1×1 and Ag(111)-1×1 surfaces. Fullerene molecules are stable on the Si(100)-2×1 and Si(111)-7×7 surfaces at room temperature because of the strong bonding to substrate dangling bonds. As a result, unique intramolecular structures of the C_{60} molecule are observed. STM images of the C_{84}, $Sc@C_{74}/Sc_2@C_{74}$ and $Sc_2@C_{84}$ molecules are used to obtain information about molecular structures. In contrast to Si surfaces, fullerene molecules are mobile on the terrace of metal surfaces and initially segregate to the step. A well-ordered two dimensional overlayer forms with a close-packed arrangement upon annealing fullerene-covered surfaces. On the Cu(111) surface, the commensurate (4×4) phase forms for the cases of C_{60}, C_{70} and $C_{60x}C_{70(1-x)}$ adsorption, indicating a strong interaction between the Cu substrate and fullerenes. Beautiful intramolecular structures of C_{60} and C_{70} molecules are observed and interpreted as local mappings of the electron density of states. One-dimensional cluster formation of C_{60}, $Y@C_{82}$ and $Gd@C_{82}$ on the step of the Cu(111)-1×1 surface is analyzed and preferential dimer formation is only observed for the case of $Y@C_{82}$. This is attributed to the interaction between unpaired electrons of the $Y@C_{82}$ molecules. On the Ag(111) surface, the C_{60} and $Sc_2@C_{84}$ monolayer films exhibit several phases that have an almost identical nearest neighbour distance but are rotated relative to one another. Among the systems studied, C_{60} on the GaAs(001)-2×4-As surface shows a unique highly-strained fcc(110)-oriented multi-layer film growth, while all other systems show the well known fcc(111)-oriented multi-layer growth.

9.1 Introduction

After the discovery of C_{60} by Kroto et al. [1] and development of the method of extracting and purifying C_{60} and other fullerenes by Krätschmer et al. [2,3], there have been intensive studies on the properties of fullerene adsorption on semiconductor and metal surfaces. The discovery of superconductivity in K_3C_{60} crystals by Haddon et al. [4] further accelerated research in this field. In addition to the abundant C_{60} and C_{70}, other higher fullerenes such as C_{76} [5,6], C_{78} [6–8], C_{82} [6,8], C_{84} [5,6,8], C_{90} [5,6], C_{94} [5], C_{96} [6] and higher fullerenes as large as C_{300} [9] have also been successfully synthesized. Endohedral metallofullerenes, first proposed by Heath et al. [10], are gaining a wide interest because of their unique and noble structures and properties. The endohedral metallofullerenes so far produced, purified and isolated can encage La [11–15], Y [14,16,17] , Sc [14,18–20], U [21], Fe [22], Co [23], Gd [24,25] and most of the lanthanide elements [25,26].

Scanning Tunneling Microscopy (STM) [27] has been used to investigate morphology and electronic structure of fullerene films on various substrates, such as the Si(100) [28–37] and (111) [38–43], Ge(100) [44] and (111) [45], GaAs(110) [46–48] and (001) [49–51], Au(111) [52–60], (110) [55,61] and (100) [62], Cu(111) [63–70], Ag(111) [58,71,72] and HOPG (Highly-Oriented Pyrolytic Graphite) (0001) [73] surfaces. Among other important aspects, these studies document the adsorption and film-growth processes and elucidate the interaction between fullerene molecules and substrates. While interactions among fullerene molecules can generally be described well by the van der Waals force [74], interactions between fullerene molecules and substrates depend strongly on the nature of the substrates. In the case of fullerene adsorption on Si surfaces, fullerene molecules interact strongly with the substrate forming chemical bonds with surface dangling bonds [28–36,38,39]. For the case of the GaAs(110) and GaAs(001) surfaces, in contrast, a dangling bond of As is filled with two electrons whereas a dangling bond of Ga is empty, because of charge transfer from the Ga dangling bond to the As dangling bond [75]. Metal surfaces have less corrugation compared with semiconductor surfaces and the migration energy of the fullerene molecules is expected to be small.

There have been several reports on the different STM images of individual fullerene molecules as a result of different interactions between adsorbates and substrates. On relatively inert surfaces such as HOPG(0001) [73] and GaAs(110) [46,47] where the interaction between the substrate and adsorbates is not strong enough, round-shaped STM images of C_{60} molecules were observed, and interpreted as rotating C_{60} molecules. Kuk et al. examined C_{60} adsorption on the Au(100) surface and reported a squeezed electronic density of states because of a relatively strong interaction between C_{60} molecules and the Au(100) substrate [62]. However, the intramolecular structures were not observed and it was concluded that the C_{60} molecules are rotating on this surface. Zhang et al. observed several patterns of intramolecular structures of C_{60} molecules, such as

- one central dark ring surrounded by six rings,
- one central ring surrounded by five rings,

for the case of the Au(110) surface imaged in the 0.1 M $HClO_4$ solution [55]. Chen et al. observed several types of intramolecular structures on an evaporated Au substrate and interpreted them as *p*-electrons of every other carbon atom [53]. Altman and Colton reported intramolecular structures of C_{60} molecules adsorbed on the atomic step of the Au(111) surface and in a close-packed monolayer film [57]. They also observed bias-dependent sixfold filled state and fivefold empty state images in the C_{60} monolayer film on the Au(111) and Ag(111) surfaces [58]. They claimed that the symmetry of the adsorption site resulted in the sixfold filled state image and the symmetry of the molecular states resulted in the fivefold empty state image.

In this review, we report on our recent results [76,77] concerning the adsorption and film growth of fullerenes, including C_{60}, C_{70}, C_{84}, $Sc_2@C_{84}$, $Sc@C_{74}/Sc_2@C_{74}$ (a mixture of $Sc@C_{74}$ and $Sc_2@C_{74}$), $Y@C_{82}$ and $Gd@C_{82}$, on the Si(100)-2×1, Si(111)-7×7, Cu(111)-1×1, Ag(111)-1×1 and GaAs(001)-2×4-As surfaces using Field-Ion Scanning Tunneling Microscopy (hereafter abbreviated to FI-STM) [78,79]. Depending on the interaction of adsorbed fullerene molecules with substrates and its balance with the interaction between fullerene molecules, the latter exhibit random adsorption [Si(100)-2×1 and Si(111)-7×7], migration on the terrace and step-edge decoration [Cu(111) and Ag(111)], or site-specific pairing [GaAs(001)-2×4-As] during initial stage adsorption. Intramolecular structures are observed in STM images for the cases of C_{60} adsorption on the Si(100)-2×1 and Si(111)-7×7 surfaces. A well-ordered monolayer film forms with a close-packed arrangement upon annealing the fullerene-covered Cu(111) and Ag(111) surfaces and the intramolecular structures of C_{60} and C_{70} molecules are observed. They are interpreted as mappings of the local electron density of states and detailed information on adsorption structures is derived. $Y@C_{82}$ molecules adsorbed on step edges of the Cu(111) surface show preferential dimer formation as a result of spin–spin interactions between them. Most of the adsorption systems of fullerenes show fcc(111)-oriented multi-layer film growth, although C_{60} on the GaAs(001)-2×4-As surface shows a unique highly-strained fcc(110)-oriented multi-layer film growth.

9.2 Experiment

The FI-STM (Fig. 9.1) [78,79] used in this study was a high-performance STM equipped with a Room Temperature Field Ion Microscope (RT-FIM).

A scanning tip was electrochemically etched from a [111] oriented single crystal tungsten wire. By using the RT-FIM routinely, the atomic arrangement of the cap of the scanning tip can be monitored and the tip end can be fabricated by field evaporation to ensure a reliable atomic resolution in the STM imaging. The main UHV (Ultra High Vacuum) STM chamber was pumped by a 1 200 l/s ion pump and a 3 000 l/s titanium sublimation pump, and the operating pressure was 3×10^{-11} torr. The Si(100) and Si(111) samples were cut into 4×19 mm^2 pieces from commercial Si wafers. Prior to introducing the sample into the UHV chamber, the sample surface was cleaned simply by washing with methyl alcohol. A series of heatings up to 1200°C,

Fig. 9.1. FI-STM apparatus used in this study

maintaining the pressure below 1×10^{-10} torr, followed by an overnight degassing at 700°C, was used to clean the sample in the sample preparation chamber attached to the main STM chamber [78,79]. The experiment using GaAs(001)-2×4-As substrates was performed in a combined MBE (Molecular Beam Epitaxy)-STM system (Fig. 9.2) [80–82]. The GaAs(001) samples were cut from on-axis wafers (dopant Si, 1×10^{18} cm^{-3}), chemically etched in a standard 4:1:1 mixture of H_2SO_4:H_2O_2:DI H_2O and mounted on the sample holder with indium. The sample was outgassed overnight at 400°C in the MBE chamber. Surface oxides were removed by annealing at 600°C in an As_4 flux. The buffer layer was then grown with growth temperature ranging from 540 to 630°C, a typical As_4/Ga flux ratio of 30, and growth rate of 0.15 mm/h monitored by the RHEED (Reflection High-Energy Electron Diffraction) intensity oscillation. The Si doping level was kept below 1.5×10^{18} cm^{-3} during growth. The GaAs(001) As-rich 2×4-β phase was prepared by annealing the substrate at the growth temperature (540–630°C) whilst maintaining the As_4 flux [80–82].

The Cu(111) and Ag(111) samples were cut from single crystal rods of 99.9999% pure copper and 99.999% pure silver with a miscut angle of less than 0.5°. The surface was polished using a standard technique and was cleaned in situ by a combination of Ar sputtering (500 eV) and annealing (typically at 480°C for Cu and 420°C for Ag) [63,72]. The C_{60}, C_{70} and C_{84} samples were synthesized, extracted and purified in a similar way to the method

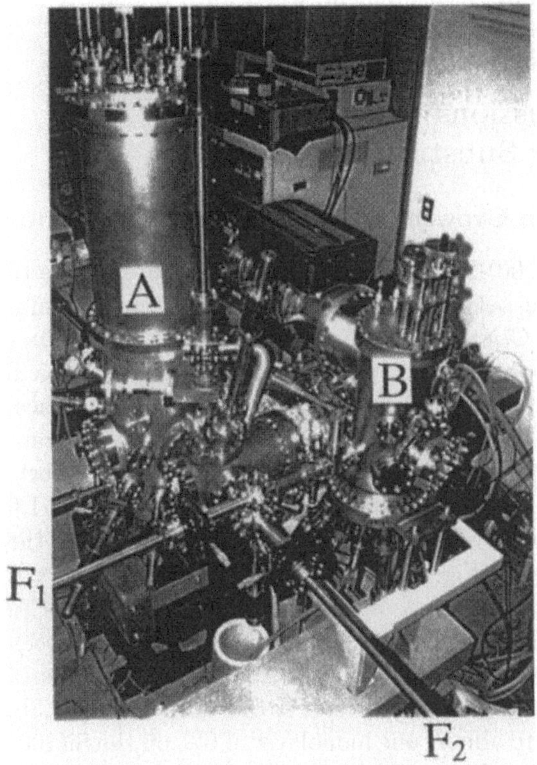

Fig. 9.2. The MBE-STM apparatus used in this study. A denotes the STM chamber, B the MBE chamber, and F1 and F2 are feedthroughs for sample transfer

used by Krätschmer et al. [2,3], described in detail elsewhere [9,83–86]. The composite-rod arc burning method [11,16,17,20,83,87,88] was used to produce metallofullerenes ($Sc@C_{74}/Sc_2@C_{74}$, $Sc_2@C_{84}$, $Y@C_{82}$ and $Gd@C_{82}$) and separation and isolation were performed using two-stage High-Performance Liquid Chromatography (HPLC) [88,89]. The purities of fullerene samples relative to other fullerenes were 99.95% (C_{60}), 99.5% (C_{70}), 98% (C_{84}), 97% ($Sc@C_{74}/Sc_2@C_{74}$), 99% ($Sc_2@C_{84}$), > 99.8% ($Y@C_{82}$), and > 99.8% ($Gd@C_{82}$), these values being confirmed by mass spectrometry. Each fullerene sample was placed in a small Ta dispenser, which was resistibly heated to temperatures ranging from 320°C to 500°C, depending on the vapor pressure of each fullerene, in order to deposit the fullerene molecules on the substrate placed 1 cm away from the dispenser. Deposition rates ranged from 0.01 ML (monolayer) to 1 ML per hour. (1 ML is defined here as the number of molecules in the fcc(111) face of the van der Waals bulk crystal of the fullerene: 1–2$\times 10^{14}$ molecules/cm^2 for C_{60} and 8.3×10^{13} molecules/cm^2 for $Sc_2@C_{84}$, for example.) The pressure during fullerene adsorption was kept below 1×10^{-10} torr after thoroughly outgassing the doser in UHV. For STM imaging, a sample bias V_s ranging from −4.0 to +4.0 V and a constant tunneling current I_t ranging from 20 to 50 pA were used, except for the STS

(Scanning Tunneling Spectroscopy) measurement, where an I_t of 300 pA was used for STM scanning.

9.3 Results and Discussions on Semiconductor Substrates

9.3.1 Adsorption and Film Growth of C_{60} on Si(100)-2×1

The dangling bond on the Si(100)-2×1 surface shows a typical example of the strong chemical interaction between a fullerene molecule and the substrate [28,29]. As a result, the C_{60} molecules are stable on the Si(100)-2×1 surfaces without rotation at room temperature. Figure 9.3a shows a typical STM image of the initial stage adsorption of C_{60} on the Si(100)-2×1 surface. The spherical protrusions are individual C_{60} molecules. They adsorb randomly and stably, without any specific interaction with step edge or defects on the surface. When we scanned the surface at bias voltages between −1.6 and +1.6 V, we occasionally observed C_{60} molecules being dragged by the tip. This is indicated by arrows in Fig. 9.3a. The lateral diameter observed for C_{60} molecules was approximately 20 Å, although the real size of the C_{60} molecule is 7.1 Å. This can be explained by the convolution effect between tip and C_{60} molecule, shown schematically in Fig. 9.3b. Note that even a single-atom tip exhibits a large convolution effect, as is shown in Fig. 9.3b. When we increase the coverage to almost one monolayer of C_{60} on the Si(100) surface, C_{60} forms a locally ordered layer, which is shown in Fig. 9.4. In this image, substrate Si dimer rows run along the $[0\bar{1}1]$ direction, indicated by an arrow in Fig. 9.4. The dark regions are defects, where C_{60} molecules are

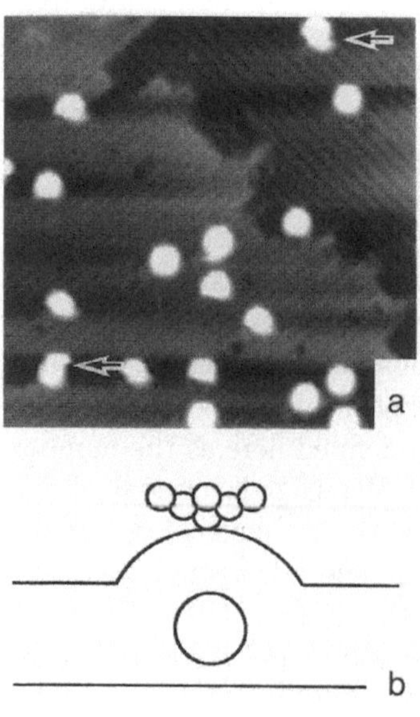
a

b

Fig. 9.3. (**a**) STM image of the Si(100)-2×1 surface with C_{60} coverage of 0.02 ML. Area 280 Å×280 Å, $V_s = -2.0$ V. (**b**) A model of the convolution effect between tip and C_{60} molecule

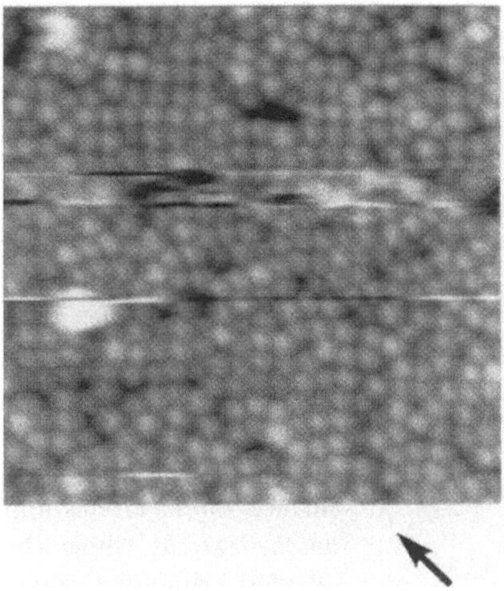

Fig. 9.4. STM image of the surface with C_{60} coverage of approximately 1 ML. Area 250 Å× 250 Å, $V_s = -4.0$ V. The *arrow* indicates the dimer row direction of the substrate

missing. We see a square-type local ordering, around top middle for example, and hexagonal-type local ordering in the lower region of the center part, for example. By analyzing these STM images, we obtained the model for adsorption up to one monolayer, as shown in Fig. 9.5. In this model, solid and hatched circles are buckled Si dimers [90]. Solid circles are up Si atoms of the buckled Si dimers, and large blank circles are C_{60} molecules.

We observed that buckling of Si dimers is stabilized by C_{60} adsorption, similar to the case for alkali metal adsorption on the Si(100)-2×1 surface [78,91]. At the initial stage, a C_{60} resides at position A, in the center of four Si dimers, surrounded by eight up Si atoms. When coverage is slightly increased, the second C_{60} molecule may occupy B or B′ positions in this model. STM data showed that the nearest neighbour distance at initial adsorption is 12 Å, larger than the schematic dimension of 10.9 Å for the B case and 9.6 Å for the B′ case. This suggests a repulsive interaction between neighbouring C_{60} molecules. Considering that the nearest neighbour distance in the bulk C_{60} fcc crystal is reported as 10.0 Å, the larger nearest neighbour distance implies that charge transfer is taking place from the Si to the C_{60} molecule. The local square arrangement is analyzed as the $c(4\times4)$ structure, and the nearest neighbour distance is measured to be 10.9 ± 0.3 Å. The local hexagonal arrangement turned out to be quasi-hexagonal and the adsorption model is shown in Fig. 9.5b, which is the (4×3) structure. The distance between A and A in Fig. 9.5b is measured to be 11.5 ± 0.4 Å, and between A and B′ is 9.6 ± 0.4 Å from the STM images. If we compare the number density of C_{60} molecules in these two arrangements with the close packed fcc(111) plane of the bulk C_{60} crystal, the $c(4\times4)$ arrangement is about 30% less dense and the (4×3) is only 2% less dense. The monolayer C_{60} film on the Si(100) surface is stretched slightly, compared with the bulk crystal.

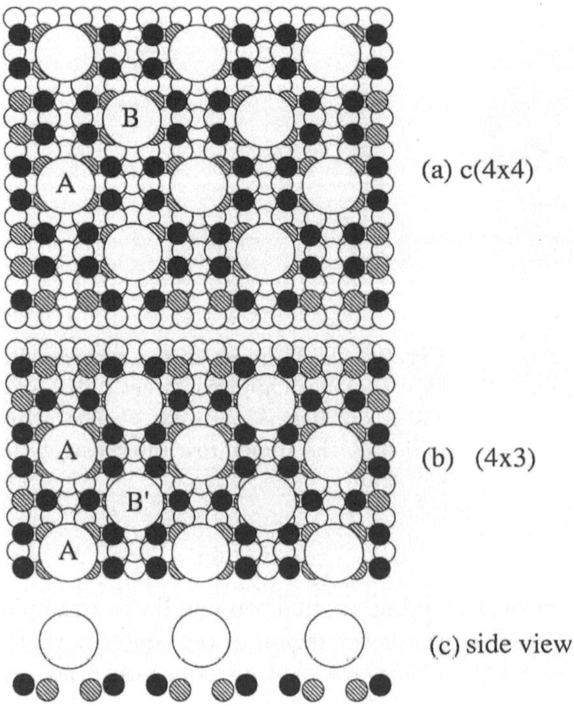

Fig. 9.5. Adsorption model for C_{60} adsorption on the Si(100)-2×1 surface. *Solid circles* and *hatched circles* show the buckled dimers, where *solid circles* are up Si atoms. At the initial stage, C_{60} adsorbs at positions A, B and B′. At monolayer coverage, C_{60} molecules form (**a**) *c*(4×4) and (**b**) (4×3) local orderings

When the coverage is increased, the second C_{60} layer formed as islands, which showed Stranski–Krastanov mode growth (Fig. 9.6). Figure 9.6 is a large scale STM image for about 2 ML C_{60} adsorption. We can observe the 1st, 2nd, 3rd and multi-layer islands. When we zoomed in on one of these multi-layer islands, we obtained the STM image shown in Fig. 9.7. Figure 9.7a shows an STM image of a 5-layer island. The 1st and 2nd layers have only local ordering, similar to the monolayer adsorption, shown in Fig. 9.4. For the 3rd, 4th and 5th layers, we observe the ordered hexagonal pattern, which can be clearly seen in the zoom-in image (Fig. 9.7b). The nearest neighbour distance is 10.4 Å in this case, slightly larger than the reported value for the bulk phase, which is 10.0 Å. Hence, even the 5th layer on this island is also stretched slightly, compared with the bulk crystal. If we carefully study the stacking of the 3rd, 4th and 5th layers, the crystallographic orientation of individual layers is slightly different and rotated through about 5° relative to each other. As described above, the 1st and 2nd layers are stretched more than the 3rd layer. Therefore, these observations imply the gradual relaxation of the atomic structure from the 1st layer to the bulk for multi-layer film growth of C_{60} on the Si(100) surface [92].

We used Scanning Tunneling Spectroscopy (STS) to analyze the electronic structure of C_{60} on the Si(100) surface. The solid lines in Fig. 9.8 are the STS results. Figure 9.8a shows the result for the monolayer of pure C_{60} on the

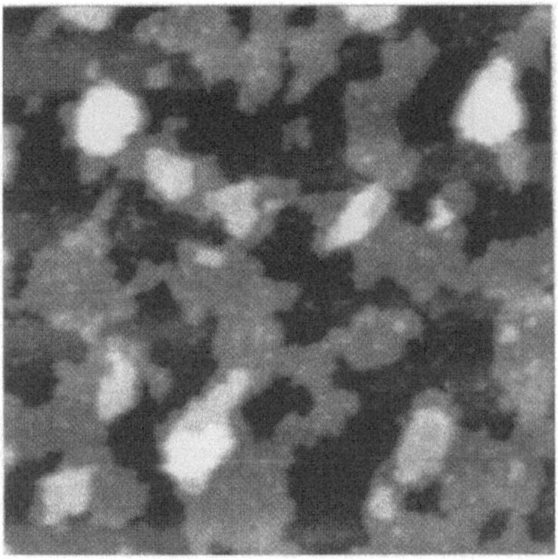

Fig. 9.6. A large scale STM image of the C_{60} multi-layers on the Si(100)-2×1 surface (1600 Å × 1600 Å, $V_s = -3.5$ V)

Si(100) surface. The dotted line is the density of states calculated theoretically by Kawazoe et al. [93]. The Fermi level position of the theoretical result is shifted by 0.4 eV downward to fit the STS results. We find good agreement of peak positions between theory and experiment, if we assume a charge transfer of 0.1 to 0.3 electrons per C_{60} molecule. Figure 9.8b shows the results for the monolayer of C_{60} on the Si(100) surface with 1:1 potassium coadsorption. A charge transfer of one electron per potassium atom is expected and we note a good agreement between theoretical calculation and STS results.

These results clearly indicate that the dangling bonds of Si interact with C_{60} molecules and there is charge transfer from Si to C_{60} molecules adsorbed on the Si(100) surface. It is known that C_{60} molecules rotate in the bulk

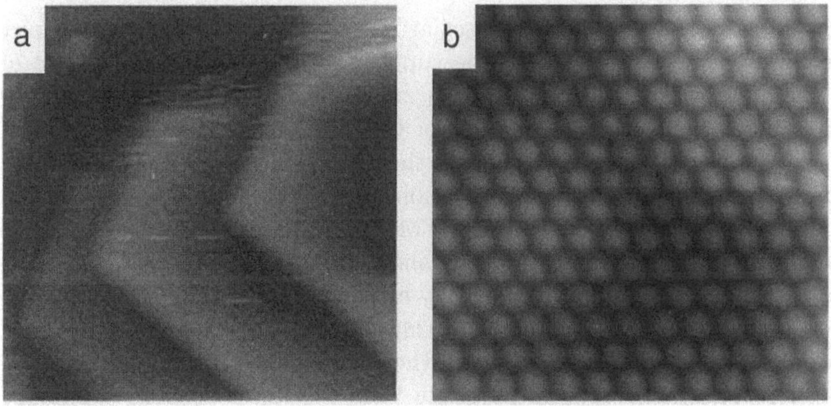

Fig. 9.7. (**a**) STM image of a C_{60} multi-layer island formed on the Si(100)-2×1 surface (350 Å×350 Å, $V_s = -3.5$ V). (**b**) Zoom-in image of the top layer (140 Å×140 Å, $V_s = -3.0$ V)

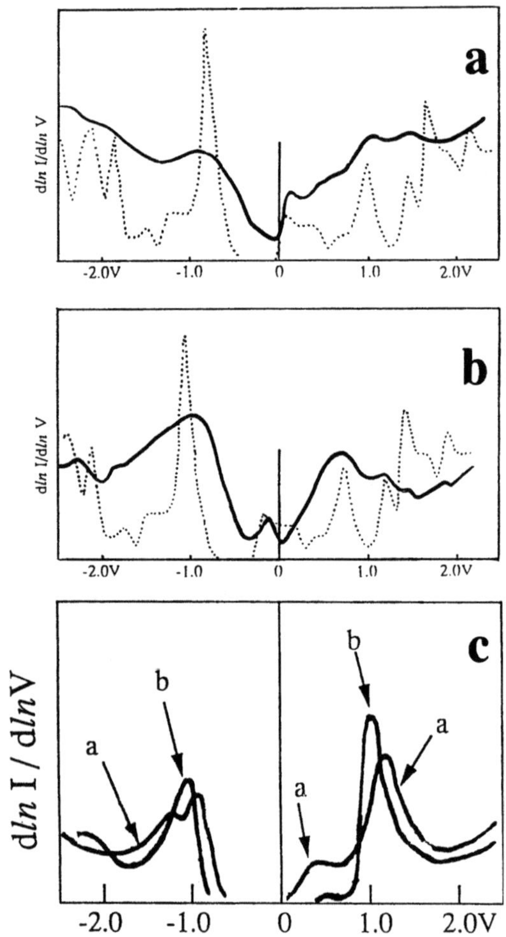

Fig. 9.8. STS results (*solid lines*) and calculated density of states [93] (*dotted lines*) for (**a**) the monolayer of pure C_{60} on the Si(100)-2×1 surface and (**b**) the monolayer of C_{60} on the Si(100)-2×1 surface with 1:1 potassium coadsorption. (**c**) Site-specific STS results obtained from a similar sample to (**a**) for areas with 'a' dark and 'b' bright contrast in the intramolecular structure, discussed below and in Fig. 9.9. Fermi level positions of the theoretical results are shifted by (**a**) 0.4 eV or (**b**) 0.6 eV downward to fit the STS results

fcc crystal at the rate of 10^9 turns/sec at room temperature [94]. We expect this rotation of C_{60} molecules to be suppressed or terminated, because of the strong bonding between Si and C_{60} molecules. Indeed, we were able to observe the intramolecular structures of C_{60} molecules. Figure 9.9 is taken for the C_{60} monolayer film adsorbed on the Si(100) surface. We can see three or four stripes in the individual C_{60} molecules. This kind of image can be observed only with high resolution STM. The important feature in these intramolecular structures is that they show twofold symmetry. In order to reveal the twofold symmetry of the C_{60} molecule, one of the carbon double bonds of a C_{60} molecule, interconnecting two hexagon rings, should face down on the Si(100) surface. If one of the pentagonal rings in the molecule faced the Si(100) surface, the molecule would reveal a fivefold symmetry instead of the observed twofold symmetry. If one of the hexagonal rings in the molecule faced down on the surface, the molecule would reveal a threefold symmetry. From this symmetry analysis of the intramolecular structure of

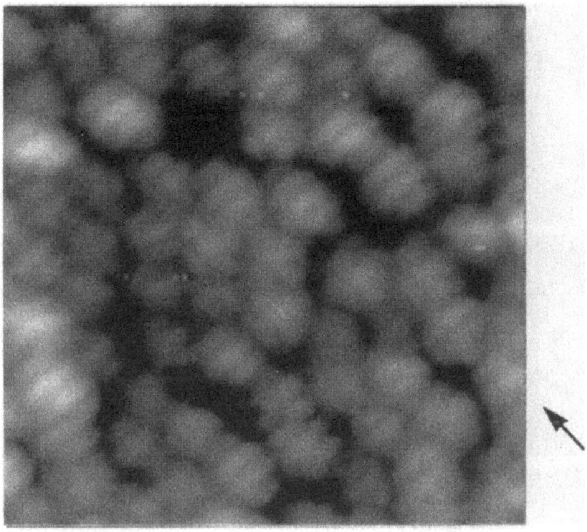

Fig. 9.9. Intramolecular structures observed in the STM image of the first layer of C_{60} adsorbed on the Si(100)-2×1 surface (90 Å×90 Å, $V_s = -1.6$ V). The *arrow* indicates the dimer row direction of the substrate

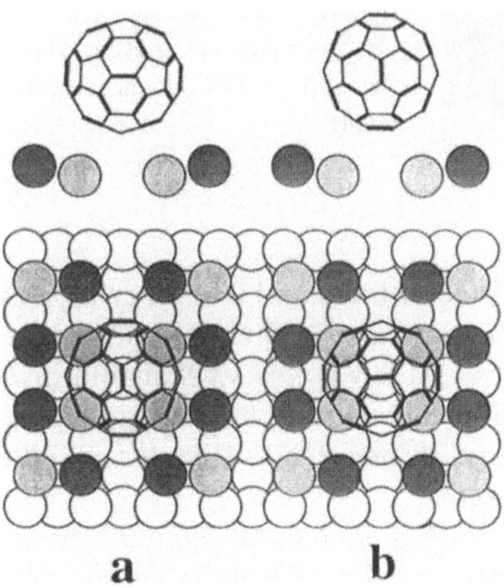

Fig. 9.10a, b. Adsorption model of C_{60} on the Si(100)-2×1 surface

the C_{60} molecule, together with the adsorption model (Fig. 9.5), we propose two possible adsorption orientations of C_{60} on the Si(100) surface, which are shown in Fig. 9.10a and b. In the STM images, we are observing the three-dimensional map of the local electron density of states near the Fermi level. We were able to calculate the simulated image of intermolecular structures by theoretical calculation [93], taking the adsorption geometry determined experimentally. Figure 9.11 shows the results of the theoretical calculation of the local density of states for each band of the C_{60} monolayer, assuming the adsorption model shown in Fig. 9.5b and the adsorption orientation shown in Fig. 9.10a. Figure 9.11a shows the total density of states. Solid lines show the position of the two hexagonal rings and two pentagonal rings. Figures 9.11b through f show the local density of states mapping for 180 through 184 level,

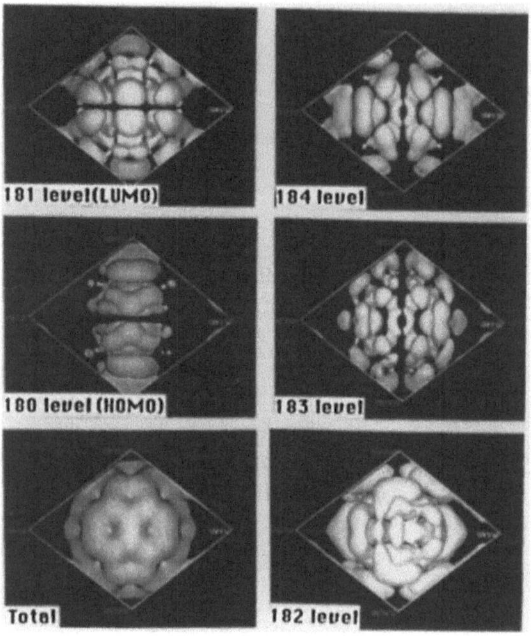

Fig. 9.11. Calculated distribution of charge density of states of a C_{60} molecule adsorbed on the Si(100)-2×1 surface. (**a**) is the total charge density. (**b**) to (**f**) are the charge density distributions for 180 to 184 levels, respectively

respectively. This 180 level is the original HOMO band and its energy level is located at −0.9 eV from the Fermi level (Fig. 9.8a). From the scanning bias condition of the STM image (Fig. 9.9, $V_s = -1.6$ V), this level should give the largest contribution to the STM imaging. The density of states mapping of the 180 level shows the peaks in the shape of four stripes (Fig. 9.11b) and gives a good agreement with the STM image.

Site specific STS results obtained from the C_{60} monolayer film and shown in Fig. 9.8c are also consistent with the theoretical simulations. Yajima and Tsukada recently performed first-principles theoretical calculations and simulated the Local Density of States (LDOS) of a C_{60} molecule at the STM tip trajectory using the Discrete-Variational Xα (DV-Xα) method with LCAO basis [95–97]. They included the interaction of a C_{60} molecule with the Si(100)-2×1 substrate as a chemical bonding [95,96] and also calculated the result for the physisorption case [97] or by neglecting the substrate interaction [95,96]. Their results showed that the type of bonding does affect the local density of states of the C_{60} molecule and the STM image can only be understood by assuming chemical bonding between the molecule and the Si substrate (Fig. 9.12) [95–97]. The results clearly disagree with a recent report by Chen and Salid [37], claiming that the C_{60} molecule physisorbs on the Si(100)-2×1 surface by dipole-induced interaction at room temperature, and conclusively support the chemisorption nature of C_{60} adsorption on the Si(100)-2×1 surface.

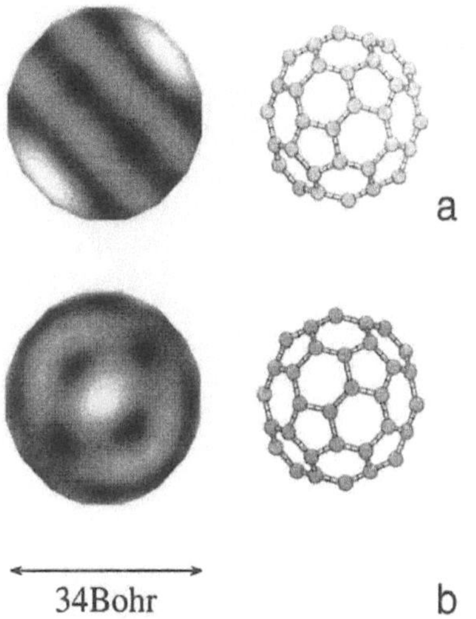

Fig. 9.12. Mappings of the local density of states of a C_{60} molecule at an STM tip trajectory simulated by the first-principles calculation. (**a**) Including the chemical bonding nature of the Si(100)-2×1 substrate and (**b**) neglecting interaction with the substrate [96]

9.3.2 Film Growth of C_{70} and C_{84} on Si(100)-2×1

Initial stage and monolayer/multi-layer adsorption of C_{70} [36] and C_{84} [29,30] fullerenes on the Si(100)-2×1 surface are also characterized by the chemical bonding of the fullerene molecules to Si dangling bonds, similar to the case of C_{60} [28,29]. At room temperature, C_{70} and C_{84} molecules reside stably in the trough formed by Si dimer rows. Figure 9.13a is the STM image of C_{70} molecules deposited on the Si(100)-2×1 surface at room temperature with a coverage of approximately 0.02 ML.

Figure 9.13b is the STM image of a C_{84} molecule (arrow) together with C_{60} molecules deposited on the Si(100)-2×1 surface at room temperature with a coverage of approximately 0.003 ML of C_{84}. They reside in the trough separated by the Si dimer rows and they are distributed randomly on the surface. One can observe that the C_{84} molecule is imaged larger than the C_{60} by 10 to 20%. Our data show that large bias voltages (> 3 V) are needed to obtain stable STM images, consistent with the fact that the C_{84} molecule has a large energy gap between the HOMO (Highest Occupied Molecular Orbital) and LUMO (Lowest Unoccupied Molecular Orbital) state levels. The C_{84} molecules adsorb on the surface with a minimum separation of 14 Å, when the C_{84} number density is small. Under these conditions, the STM image of C_{84} molecules shows very little deviation (within 10%) from a perfectly circular shape. At a coverage of nearly 1 ML, the C_{70} and C_{84} molecules form a flat first layer without three-dimensional island formation (Fig. 9.14). However, they only produce a short-range local ordering, while the long-range ordering is less evident than in the case of C_{60} [28,29], because of the larger size misfit of the molecules with the Si(100)-2×1 surface structure. When the first layer is completed, the second layer begins and islands form in the Stranski–Krastanov mode. The islands often show a hexagonal pattern.

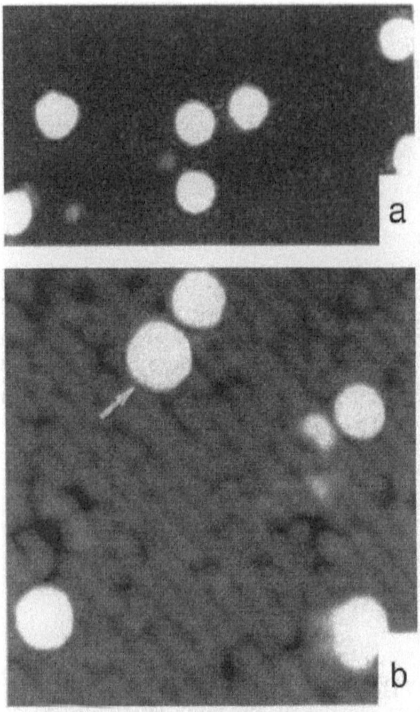

Fig. 9.13. STM images of the Si(100)-2×1 surface with (**a**) 0.01 ML of C_{70} (300 Å×150 Å, V_s = −2.5 V) and (**b**) 0.003 ML of C_{84} (140 Å×140 Å, V_s = −3.0 V). The *arrow* in (**b**) indicates a C_{84} molecule

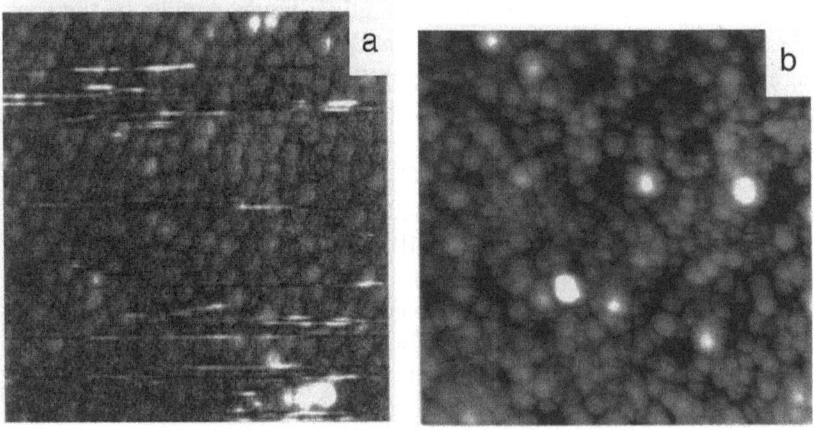

Fig. 9.14. STM image of the surface with approximately 1 ML of (**a**) C_{70} (230 Å×230 Å, V_s = −3.5 V) and (**b**) C_{84} (320 Å×320 Å, V_s = −4.0 V) molecules, showing only a short-range local ordering

When the monolayer-covered surface is exposed to C_{70} vapor while keeping the substrate at 120°C, a well-ordered C_{70} film grows on the surface, forming islands (Fig. 9.15). The second layer is still somewhat irregular. However, the C_{70} molecule overlayers (the third layer and up) grown on the second layer are well-ordered and almost perfectly close-packed (Fig. 9.15b), suggesting that the overlayers are formed purely by the van der Waals interaction

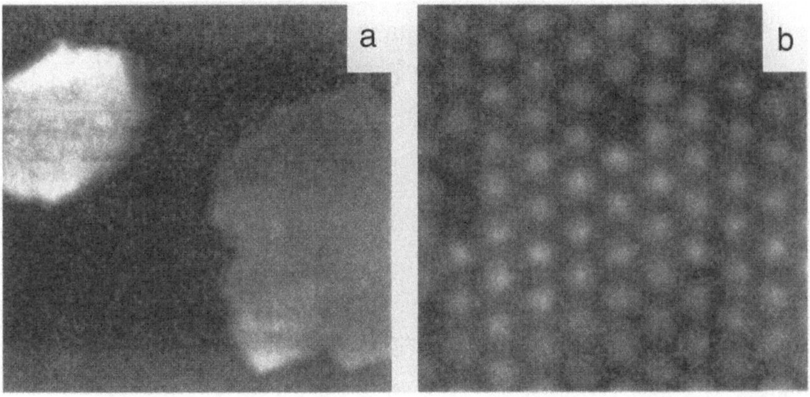

Fig. 9.15. (**a**) Large scale STM image of the Si(100)-2×1 surface after depositing C_{70} molecules while keeping the surface at approximately 120°C (1800 Å×1800 Å, V_s = +4.0 V) and (**b**) an enlarged area of the crystalline island shown in (**a**) (100 Å×100 Å, V_s = +4.0 V)

without interference from the substrate, similar to the case of C_{60} deposition on the Si(100)-2×1 surface [28,29]. The nearest neighbour distance in this C_{70} layer is 10.6 ± 1 Å, consistent with the nearest neighbour distance in the bulk C_{70} crystal, which is 10.6 Å. When the monolayer-covered surface is exposed to C_{84} vapor while keeping the substrate at 100 to 150°C, the C_{84} film grows on the surface, forming islands (Fig. 9.16). Similar to the case of C_{70}, the second layer is still irregular. The step height for the first and second layers is 11.4 Å, slightly larger than the rest of the film, where the step height is 9.9 Å. The C_{84} overlayers (the third layer and up) grown on the second layer are well-ordered (Fig. 9.16b). In Fig. 9.16a, one can observe a terrace (left side of line AA) with a facet of the (100) plane (upper right side of line AA). The perfectly ordered hexagonal packing is clearly observed in the enlarged image of the terrace (Fig. 9.16b).

The nearest neighbour distance in this C_{84} layer is 12.1 Å. These STM data allow us to conclude that the C_{84} overlayer is the fcc crystal, rather than the hcp crystal. An interesting and important observation from Fig. 9.16b is that individual C_{84} molecules are not imaged equally in size. Indeed there appear to be three distinctly different sizes in the STM image, shown by arrows I, II and III. Type I is the smallest with average height 9.3 Å. The observed heights of type II and III are 9.9 Å and 10.5 Å, respectively. This observation strongly suggests that there are several types of stable C_{84} molecule existing on the Si(100)-2×1 surface and that they are not completely spherical.

Our preliminary study shows that the present STM data agree best with D_{2d} and D_2 symmetry for the C_{84} structure, as proposed by the Iowa group [98]. When the fullerene layers on the Si(100)-2×1 surface are heated, the first

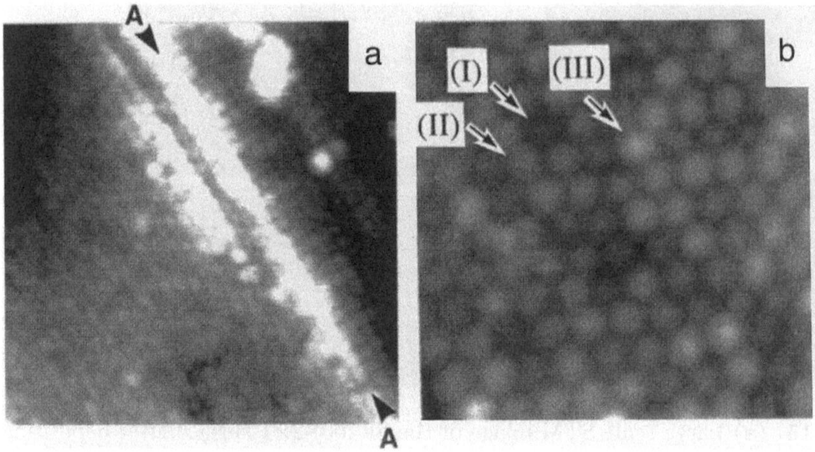

Fig. 9.16. (**a**) Large scale (320 Å×320 Å) and (**b**) small scale (120 Å×120 Å) STM images ($V_s = -4.0$ V) of the C_{84} multi-layer film formed on the Si(100)-2×1 surface while keeping the surface at a temperature ranging from 100 to 150°C. In (**a**), one can observe a terrace with hexagonal ordering (*left side of line* AA) and a facet with square ordering (*upper right side of line* AA). In (**b**), I, II and III show representative C_{84} molecules with three distinctly different sizes

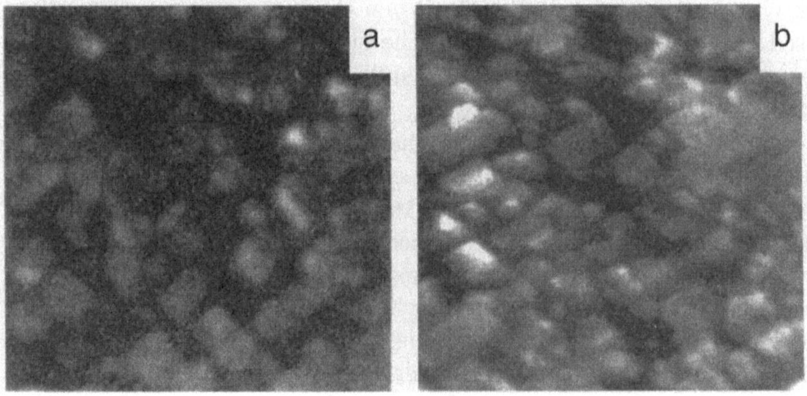

Fig. 9.17. Large-scale STM images showing SiC island formation after annealing (**a**) the C_{70} covered surface at 1 000°C (1 200 Å×1 200 Å, $V_s = -2.5$ V) and (**b**) the C_{84} covered surface at 1 000°C (2 000 Å×2 700 Å, $V_s = -2.5$ V)

layer adsorption stays intact until the temperature exceeds 1 000°C, although the layers above the first layer desorb or migrate away at moderate temperatures (250°C for C_{70} and 400°C for C_{84}). Above 1 000°C, fullerene molecules react with the Si substrate and silicon carbide islands can be observed to form using STM (Fig. 9.17).

9.3.3 Shape of Metallofullerenes ($Sc@C_{74}/Sc_2@C_{74}$ and $Sc_2@C_{84}$)

One of the most important properties of the metallofullerenes (MFs) is the encapsulation of metal atoms inside carbon cages. However, due to the difficulties involved in isolating MFs and the lack of techniques characterizing the properties of complex molecules, previous studies could not conclusively determine the geometry and structure of the MFs. As for the location of the metal atom with respect to the fullerene cage, the majority of experimental results appear to support the model in which metal atoms are encapsulated inside the carbon cage [10]. In contrast, Soderholm et al. suggested formation of a dimer of ($Y@C_{82}$)-X-($Y@C_{82}$), where -X- is a bridging carbon or oxygen, based on the EXAFS result of $Y@C_{82}$ [19]. Alvarez et al. also suggested that La atoms may replace two carbon atoms to form part of the cage, instead of being trapped inside the cage, based on the discovery that the Lan C_{84}-n series of complexes were most abundant in the synthesized soot [12].

In this section, we describe our analysis of the STM images of fullerene molecules adsorbed on the Si(100)-2×1 surface, which gives us important information on the size and shape of metallofullerenes (MFs), because of the random and stable adsorption and the relative uniformity of the adsorption geometry [32–34]. The STM observations show that the $Sc@C_{74}/Sc_2@C_{74}$ (a mixture of $Sc@C_{74}$ and $Sc_2@C_{74}$) and $Sc_2@C_{84}$ metallofullerene (MF) molecules are imaged spherically, similarly to the cases of C_{60}, C_{70} and C_{84}, suggesting that the Sc atoms are trapped inside the carbon cages of C_{74} and C_{84} [32–34]. Figure 9.18 is a large scale STM image showing randomly dispersed $Sc_2@C_{84}$ metallofullerene molecules adsorbed on the Si(100)-2×1 surface. The coverage in this case is approximately 0.001 ML. Large bright spots are $Sc_2@C_{84}$ molecules, which are imaged spherically. A similar result was obtained for the $Sc@C_{74}/Sc_2@C_{74}$ molecules. The MF molecules on the Si(100)-2×1 surface exhibit strong bonding with the Si substrate and do not show nucleation or segregation to step edges, similar to the cases of C_{60}, C_{70} and C_{84}. MF molecules are separated from each other without forming dimers or trimers. Impinging MF molecules have little possibility of migrating after they hit the silicon surface since bonding with the substrate is so strong. It is thus well documented that the MF in film does not exist as a trimer or dimer in the form ($Y@C_{82}$)-C/O-($Y@C_{82}$), contrary to the conclusion of the EXAFS study of $Y@C_{82}$ [19]. The EXAFS result needs to be explained in another way, based on the attractive interaction exerted between the unpaired electrons of two neighbouring $Y@C_{82}$ molecules (see Sect. 9.4.4) [66,67,70]. MF images can be obtained at small bias voltages as low as ±0.7 eV, which is the limit for imaging the normal Si(100)-2×1 clean surface. This implies that MFs on the Si surface have small band gap or may even be metallic, consistent with the calculation for $La@C_{82}$ [99].

In order to determine the shape of the $Sc@C_{74}/Sc_2@C_{74}$ and $Sc_2@C_{84}$ molecules, their images were compared with that of the C_{60} coadsorbed on the

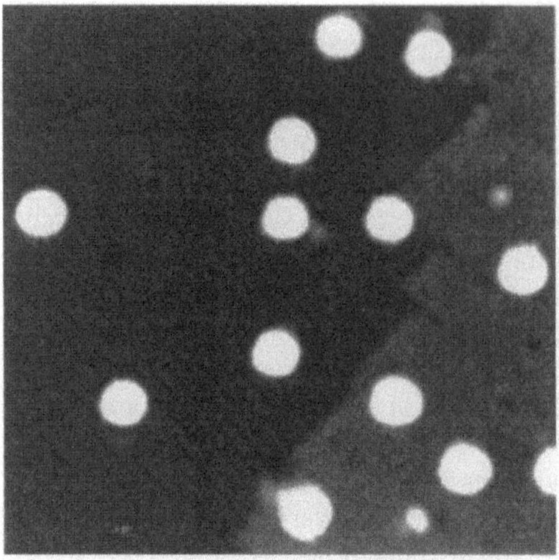

Fig. 9.18. A large scale STM image showing a random distribution of $Sc_2@C_{84}$ molecules adsorbed on the Si(100)-2×1 surface (300 Å×300 Å, V_s = +3.0 V). Coverage is approximately 0.01 ML

same surface using a separate C_{60} doser. Figure 9.19 shows the image of the surface coadsorbed with the $Sc@C_{74}/Sc_2@C_{74}$ and C_{60} molecules. It clearly shows that the MF molecules are imaged larger than C_{60}. The STM images for $Sc_2@C_{84}$ are essentially the same as Fig. 9.19 except for the difference in size. The fullerene molecules are imaged as large as 20 to 30 Å in diameter when they adsorb individually on the clean Si surface. This is because the STM imaging process of an isolated large molecule involves convolution between the tip and the object (Fig. 9.3b). Therefore, their images can be distorted and special care must be taken in analyzing molecular shapes.

Nevertheless, these STM images provide us with important information on the shape and size of molecules by comparing the data with those of the well known C_{60} molecule. Based on the observation that all the MF molecules are imaged to be spherical, similar to the case of C_{60}, we claim that the metal atoms are encapsulated in the carbon cage. If the metal atom is located outside the cage or part of the cage, the corresponding STM image should exhibit some asymmetry due to the presence of the metal.

The size of MF molecules can be deduced by the comparison of the MF and C_{60} images. Knowing that the C_{60} is ideally spherical and its diameter is 7.1 Å [99], and assuming that both C_{60} and MF molecules are imaged under the same tunneling process, the height of MF molecules can be determined from

$$H_{\mathrm{MF}} = h_{\mathrm{MF}} + (7.1 - h_c) , \quad (9.1)$$

where H_{MF} and h_{MF} are the real and measured heights of the MF, respectively, while h_c is the measured height of C_{60}, and 7.1 Å is the diameter of C_{60}. We have applied this method in the case of C_{84} [29,30] and C_{70} [36] and obtained good agreement with theoretical calculations [100]. The distributions of the diameter in the scanning direction and the height of the C_{60}

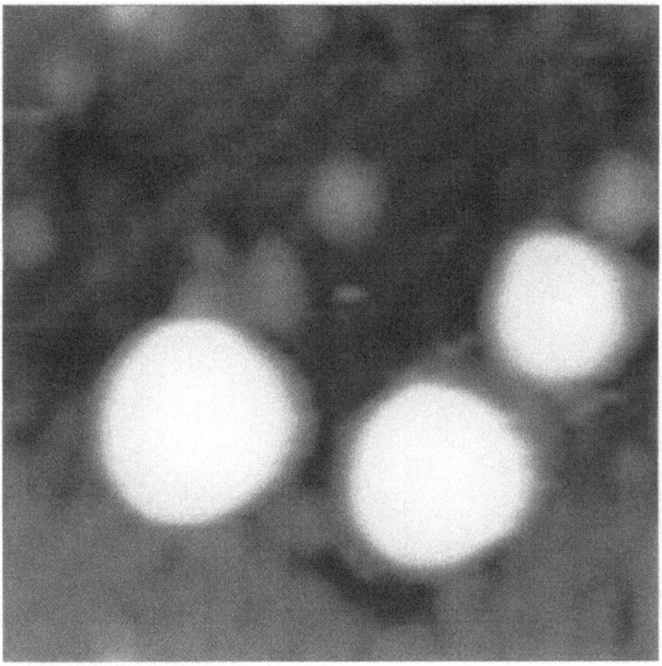

Fig. 9.19. An enlarged STM image of the Si(100)-2×1 surface coadsorbed with $Sc@C_{74}/Sc_2@C_{74}$ and C_{60} ($V_s = -2.5$ V)

and $Sc@C_{74}/Sc_2@C_{74}$ molecules coadsorbed on the same surface have been analyzed in detail and the result is shown in Fig. 9.20.

The data in this figure can easily be sorted into two groups A and B, and they are identified as C_{60} and $Sc@C_{74}/Sc_2@C_{74}$ molecules, respectively. Group B can be divided into two subgroups: B′ for the smaller ones and B″ for the larger ones. Using (9.1), the average heights for the two subgroups, B′ and B″ are estimated to be 8.6 and 9.6 Å, respectively. The distribution of $Sc@C_{74}/Sc_2@C_{74}$ heights is fairly wide, ranging from 8.0 to 10.5 Å, due most likely to the fact that they are not spherical and thus their height depends on the adsorption orientation and geometry. Figure 9.21 is the histogram of the height of the $Sc_2@C_{84}$ and $Sc_n@C_{74}$ ($Sc@C_{74}/Sc_2@C_{74}$) molecules. Similarly to Fig. 9.20, the peak located at 7.1 Å corresponds to C_{60}, while the broad distribution for $Sc@C_{74}/Sc_2@C_{74}$ has two peaks at 8.6 and 9.6 Å.

The diameter of $Sc_2@C_{84}$ is determined to be 8.9 Å, slightly larger than that of C_{84}, although the distribution of $Sc_2@C_{84}$ is wide, from 8.3 to 10.2 Å, compared with that of isotropical C_{60}. C_{74} is supposed to be smaller than C_{84}, whose diameter was calculated to be less than 9.4 Å by Saito et al. [100], and measured to be less that 8.6 Å [29,30]. Thus the measured height of 9.5 Å for $Sc@C_{74}/Sc_2@C_{74}$ may imply that the electronic states of the MF near the Fermi level are larger than those of C_{60} and C_{84}, consistent with several

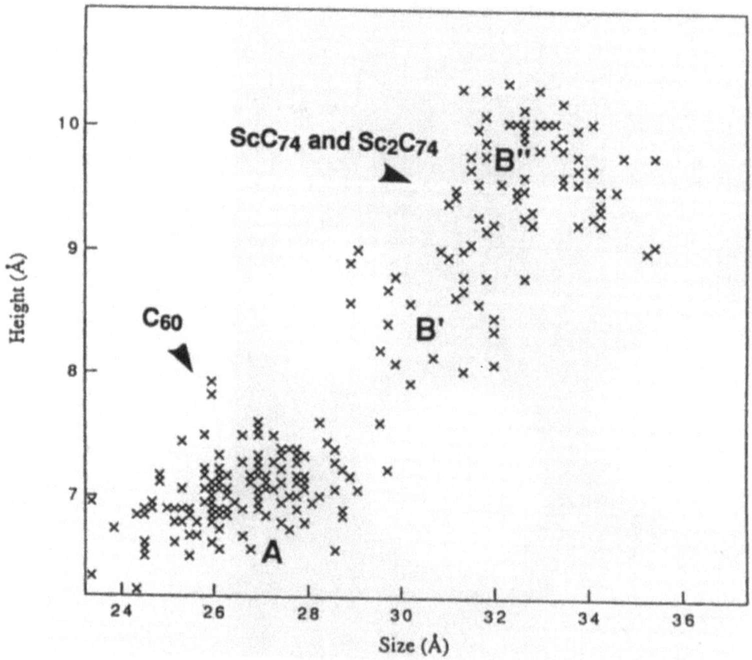

Fig. 9.20. Distribution of C_{60} and $Sc@C_{74}/Sc_2@C_{74}$ molecules according to their diameter and height

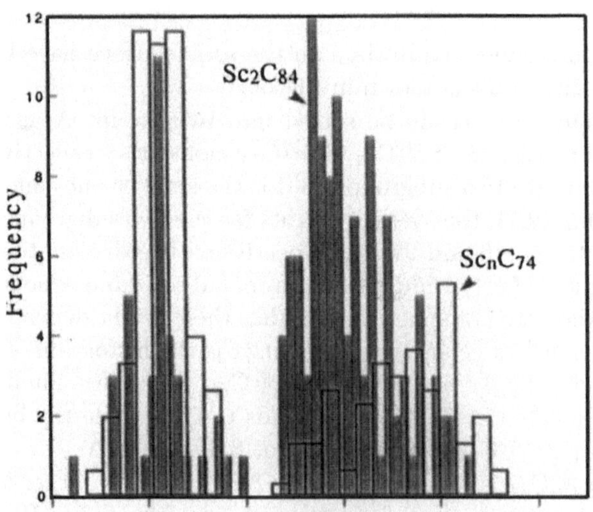

Fig. 9.21. Histogram of height distributions for STM images of $Sc_2@C_{84}$ and $Sc_n@C_{74}$ ($Sc@C_{74}/Sc_2@C_{74}$) molecules, calibrated by setting the peak for height of C_{60} at 7.1 Å. The *shaded histogram* represents the distribution of $Sc_2@C_{84}$ and the *bold line histogram* that for $Sc_n@C_{74}$ ($Sc@C_{74}/Sc_2@C_{74}$)

theoretical and experimental results. Shinohara et al. [18] and Yannoni et al. [19] suggested, according to their ESR results, that the Sc atom is in the +3 oxidation state and the C_{82} cage is in the −3 state in the case of Sc@C_{82}, and that the electrons of Sc are paired either in metal-dimer orbitals or in LUMOs in the fullerene cage for the case of Sc_2@C_n. Similar results were reported in the case of La@C_{82} [11,13] and Y@C_{82} [16]. A recent calculation applied to the La@C_{82} by Laasonen et al. [99] showed that the lanthanum atom is nearly in the La^{3+} state and the unpaired electron is somewhat delocalized on the cage, and that the HOMO–LUMO gap is approximately 0.4 eV. It is reasonable to assume that charge transfer takes place from Sc atoms to the fullerene cage, consistent with the general picture obtained by the present metallofullerene study. Therefore, the size difference between Sc_2@C_{84} and C_{84} molecules can be understood through charge transfer.

A puzzling result is that Sc@C_{74}/Sc_2@C_{74} is imaged larger than Sc_2@C_{84}. One possible explanation is as follows. Sc atoms in Sc@C_{74}/Sc_2@C_{74} transfer almost 3 electrons each to the fullerene cage and the C_{74} cage of Sc@C_{74} is in the −3 state while that of Sc_2@C_{74} is in the −6 state. This implies that the C_{74} cage of Sc@C_{74} may have higher density of states near the Fermi level than that of Sc_2@C_{74}. Therefore the subgroup B″ with estimated height 9.6 Å corresponds to Sc@C_{74}, instead of Sc_2@C_{74}. Theoretical calculations by Nagase and Kobayashi suggested that the electronic structure of Sc_2@C_{84} can be described by $(Sc^{2+})_2$@${C_{84}}^{4-}$, with 2 electrons transferred from each Sc atom to the cage [101]. It may be the case that the difference in electronic states between Sc@C_{74} and Sc_2@C_{84} overcomes the size difference and Sc@C_{74} is measured to be larger than Sc_2@C_{84} in the STM height measurement.

Further important information is obtained by analyzing multi-layer three-dimensional (3D) islands of Sc_2@C_{84} [32,33]. When Sc_2@C_{84} molecules are deposited on the Si(100)-2×1 substrate heated to approximately 150°C, crystalline 3D islands are formed. Figure 9.22a shows a typical crystalline island after depositing approximately 3 ML of Sc_2@C_{84}. A well-ordered hexagonal close-packing array of Sc_2@C_{84} is evident in the enlarged STM image (Fig. 9.22b). The nearest neighbour distance (nnd) is measured to be 11.7 Å. The result shows that the crystal structure of solid Sc_2@C_{84} is quite similar to that of pristine C_{84} and other fullerene crystals, i.e., the fcc or hcp structure. However, the measured nnd for Sc_2@C_{84} is slightly smaller than that of the pristine C_{84} (12.0 Å). A substantial permanent dipole moment in Sc_2@C_{84} must be playing a significant role in shrinking the close packing geometry in the 3D island.

The STM image of each molecule does not show evidence of intramolecular structures, while molecular sizes are not imaged uniformly even though they are located at the lattice positions of the hexagonal grid. The difference in size implies that several kinds of isomers coexist. The pristine C_{84} have mainly two kinds of isomers with symmetries D_2 and D_{2d}. Thus the difference in images of molecules shown in Fig. 9.22b may also reflect differences

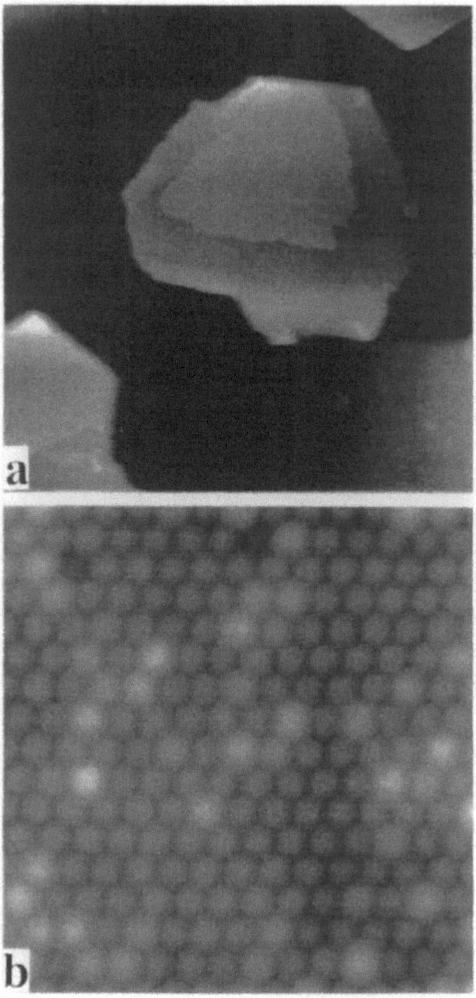

Fig. 9.22. (**a**) Large scale STM image of the Si(100)-2×1 surface after depositing approximately 3 ML of $Sc_2@C_{84}$, while keeping the substrate at approximately 150°C (2 300 Å×2 300 Å, V_s = −2.5 V). (**b**) An enlarged STM image of the crystalline 3D island showing a well-ordered hexagonal close-packed structure

in molecular orientations and/or different isomers. It is interesting to realize that the well-ordered close-packed film can still be formed even though the orientation of the $Sc_2@C_{84}$ in crystals is rather disordered and several isomers with different geometrical structure may coexist. The two isomers D_2 and D_{2d} (Iowa, round) [102] of C_{84}, are not perfectly spherical, with the maximum diameter difference being 0.7 Å. Even though the atomic structure of the $Sc_2@C_{84}$ molecule has not been determined yet because techniques are lacking, brightness differences of individual molecules in the $Sc_2@C_{84}$ film can be understood in the same way as for pristine C_{84} molecules, and we can exclude other possible isomers such as D6h, Td, and D2 (no. 21) [102] on the basis of STM analysis.

9.3.4 Interaction of C_{60} with Si(111)-7×7

Adsorption of C_{60} molecules on the Si(111)-7×7 surface is also characterized by interaction of the C_{60} molecule with Si dangling bonds, similar to the case of their adsorption on the Si(100)-2×1 surface [38]. Figure 9.23a shows a typical STM image of large areas of the Si(111)-7×7 surface covered with C_{60} molecules of 0.03 ML. The large spherical protrusions are images of C_{60} molecules adsorbed on the surface. The height of these protrusions is well defined and is either 6.0 Å or 3.6 Å with an error of 0.2 Å. However, their size in the lateral direction varies considerably in the range 10 to 20 Å, depending on the shape of the tip. When the C_{60} molecule is located at the corner vacancy (site B in Fig. 9.23b), a considerably smaller protrusion is observed with diameter 14 Å and height 3.6 Å, while adsorption on the faulted or unfaulted half of the 7×7 unit (site A or A′ in Fig. 9.23b) gives larger spherical protrusions, 20 Å across and 6.0 Å high.

The STM data show that the adsorption site A or A′ is preferred among the adsorption positions shown in Fig. 9.23b. There are four almost equivalent A or A′ sites for the C_{60} adsorption in each half, which are surrounded by three Si adatoms. In detail, among those four possible sites, C_{60} molecules choose most often (80%) the one with the highest symmetry (threefold symmetry) which does not have the rest-site Si atom at its center. These observations can be understood in terms of the highly symmetric characteristics of C_{60} molecules. Because of its large diameter (7.1 Å), there is only one C_{60} adsorbed per half unit, i.e., a maximum of two C_{60} molecules per 7×7 unit cell at this small coverage. The statistical analysis of our data indicates that C_{60} molecules favor the faulted half (56%) over the unfaulted half (30%). They occasionally take up the corner vacancy (10%). The C_{60} was rarely found on dimer lines (4%, site C in Fig. 9.23b).

The fact that C_{60} molecules prefer the faulted halves suggests that there is a significant amount of interaction between the C_{60} adsorbate and the Si(111)-7×7 substrate. It has been well accepted on the basis of nuclear magnetic resonance experiments that C_{60} molecules rotate rapidly (10^9 s^{-1}) at room temperature inside the bulk phase [94]. It is of great interest to C_{60} researchers to know whether C_{60} molecules still rotate on the adsorbate surface. Figure 9.24a shows a single C_{60} molecule positioned at site A and its details are shown in Fig. 9.24b. One can clearly see four slightly curved rows running from top to bottom in this figure. Some internal structure is also visible in each row. This result shows that Si dangling bonds interact with C_{60} molecules and the rotation of C_{60} molecules is suppressed or terminated, because of this strong bonding between Si and C_{60} molecules, as happens when it is adsorbed onto the Si(100)-2×1 surface. We can see three or four stripes in the individual C_{60} molecules. The intramolecular structures show twofold symmetry. In order to reveal the twofold symmetry of the C_{60} molecule, one of the carbon double bonds of a C_{60} molecule, interconnecting two hexagonal rings, should face down on the Si(111) surface, as in the case of the

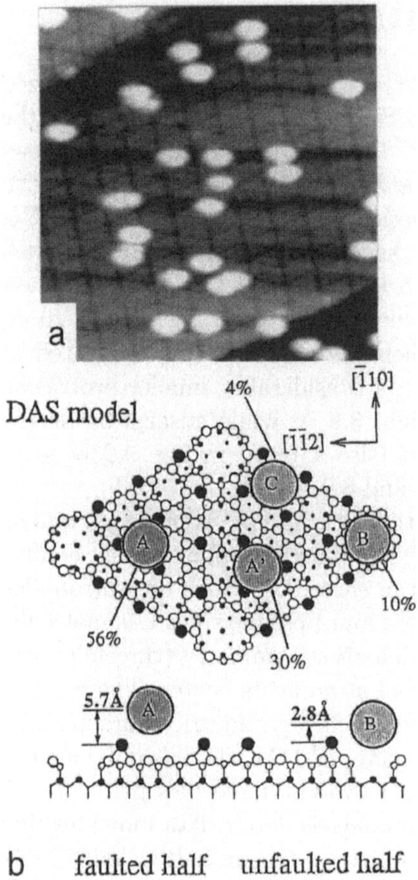

Fig. 9.23. (**a**) Large scale STM image of the Si(111)-7×7 surface with 0.03 ML C_{60} (240 Å×500 Å, $V_s = -2.0$ V) and (**b**) a schematic of possible adsorption sites A, A′, B and C for C_{60} molecules on the Si(111)-7×7 surface. The vertical positions estimated from the data for C_{60} on the Si(100)-2×1 surface are also given, assuming a rigid body model for both Si and C_{60}

Si(100)-2×1 surface. If one of the pentagonal rings in the molecule faced the Si(111) surface, the molecule would reveal a fivefold symmetry instead of the observed twofold symmetry. If one of the hexagonal rings in the molecule faced down on the surface, the molecule would reveal a threefold symmetry.

Unlike the case of C_{60} adsorption on the Si(100)-2×1 surface, the C_{60} molecule did not form ordered monolayer or multi-layers on the Si(111)-7×7 surface under the experimental conditions we used. However, Xu et al. have reported fcc-crystalline 3D-island formation of C_{60} on the Si(111)-7×7 surface [40]. Their analysis was based on the STM observation of two distinct orientations of the 3D islands, which deviated ±11° from the Si [211] direction. They claimed that first layer adsorption is dominated by the chemical interaction between C_{60} and Si, but the nature of the bonding between C_{60} first layer and Si substrate changed after 2nd layer formation, so that a disorder-to-order transition took place at the interface for higher coverages. They also concluded that the direction of the islands was determined by pinning of the C_{60} molecules at corner holes (site B in Fig. 9.23b) [40].

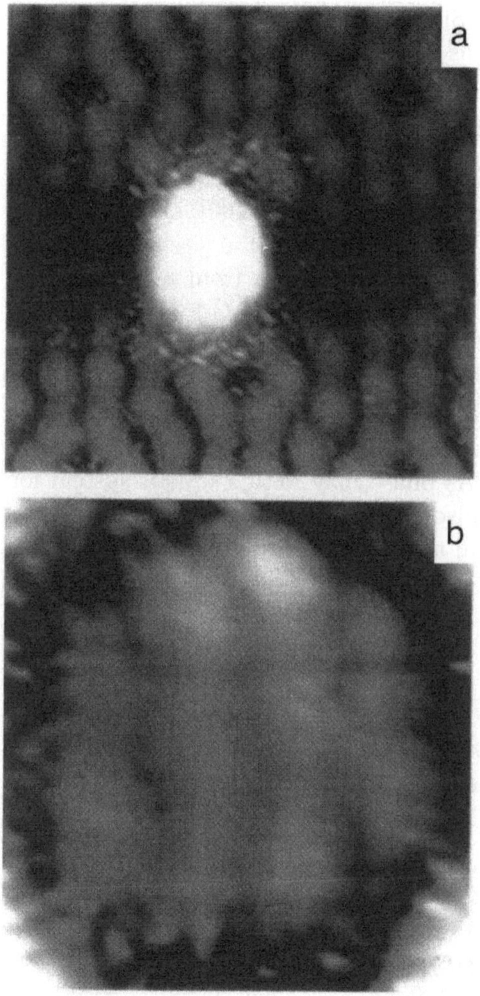

Fig. 9.24. A single C_{60} molecule sitting at site A exhibits some internal structure. The details are shown in the enlarged image (**b**). Each of the four bright lines, slightly curved, shows additional structure

This observation was later questioned by several groups who observed ordered first monolayer-film formation on the Si(111)-7×7 surface by adjusting experimental parameters such as substrate temperature and C_{60} deposition rate. Sato et al. observed the local ordering of the first C_{60} monolayer film with distinct double domains which had the same orientation angles as those observed by Xu et al. [43]. Kasuya et al. also observed that the quality of the local ordering could be changed by altering their experimental conditions [103]. Formation of local ordering on the strongly interacting interface can be understood as the result of small rearrangements of the C_{60} molecules caused by later impinging molecules. This was demonstrated by a molecular dynamics calculation for the case of C_{60} adsorption on the Si(100)-2×1 surface [104]. Thus film formation of C_{60} molecules on the highly corrugated Si(111)-7×7 surface can be well understood on the basis of the strong interaction between fullerene molecules and Si dangling bonds, without introducing a complicated interaction transition as a function of the coverage.

9.3.5 Strained C_{60} Overlayer on GaAs(001)-2×4-As

Adsorption of C_{60} molecules on the GaAs(001) As-rich 2×4-β phase (referred to as the 2×4-As phase in this section) shows a unique first-layer and strained fcc(110)-oriented multi-layer formation, as a result of the site-specified C_{60} interaction with the 2×4 substrate and charge transfer from substrate dangling bonds to the C_{60} adsorbate [50,51]. The GaAs(001)-2×4-As phase is the most stable and highly ordered with large domains among various 2×4 phases. The surface exhibits a strong one-dimensional corrugation which is as deep as 2.8 Å with a periodicity of 16.0 Å because of the two missing dimers and two missing Ga atoms per unit cell (Fig. 9.25) [80,81,105]. Therefore, the C_{60} molecules on the 2×4-As phase may form a unique system for studying adsorbate–adsorbate and adsorbate–substrate interactions and the 2×4-As surface may serve as a special template for epitaxial growth of these C_{60} molecules.

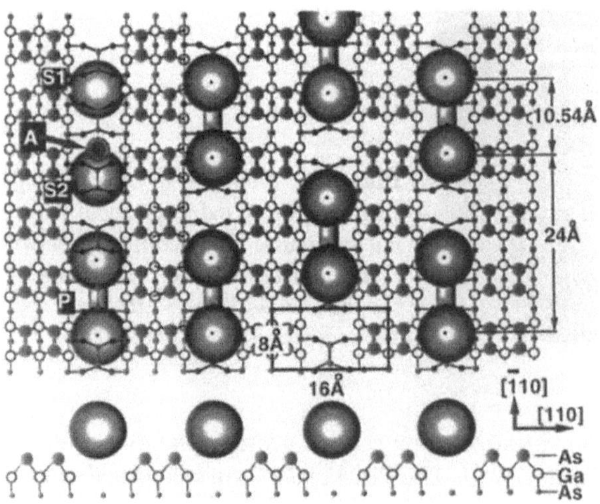

Fig. 9.25. Schematic diagram of the GaAs(001)-2×4-As surface structure and possible adsorption configurations of adsorbed C_{60} molecules for an isolated C_{60} (S1 and S2) and a paired C_{60} (P) adsorption

At small coverages, C_{60} molecules exclusively adsorb in the trough, as is expected from observation of C_{60} molecules on Si(100)-2×1 and Ge(100)-2×1 surfaces [28,29,44]. About 20–25% of the C_{60} molecules form a C_{60}–C_{60} pair with an intermolecular separation of 11.8 Å, even at low coverage (0.02 ML). Most pairs are located at the normal lattice site, while over 70% of individual C_{60} molecules are located at defects (mainly antiphase boundaries), indicating the mobile character of C_{60} molecules at room temperature. The C_{60} molecules reside 3.8 ± 0.6 Å higher than the first-layer As dimer. The relative position of individual molecules along the [$\bar{1}$11] direction is found to be shifted by 1.6 ± 0.4 Å from the center of the first-layer As dimer (marked S1 or S2 in Fig. 9.25). As for the pair, its center is shifted by 6.2 ± 0.4 Å along the $[\bar{1}11]$ direction with respect to the center of the first-layer As dimer (marked P in Fig. 9.25).

As C_{60} coverage increases, troughs are gradually filled with C_{60} molecules and the pairing of molecules becomes more pronounced. At saturation coverage, a unique structure forms, as shown in Fig. 9.26. The structure consists of an array of linear C_{60} chains running along the $[\bar{1}11]$ direction, separated by 16.0 Å. Each C_{60} chain is composed of C_{60} pairs whose intermolecular distance is 10.54±0.20 Å, evenly spaced by 24.0 Å, three times the substrate unit (Fig. 9.26). We note that this overlayer is remarkably stable and an annealing at 300°C, the desorption temperature of the multi-layer C_{60}, does not cause any change.

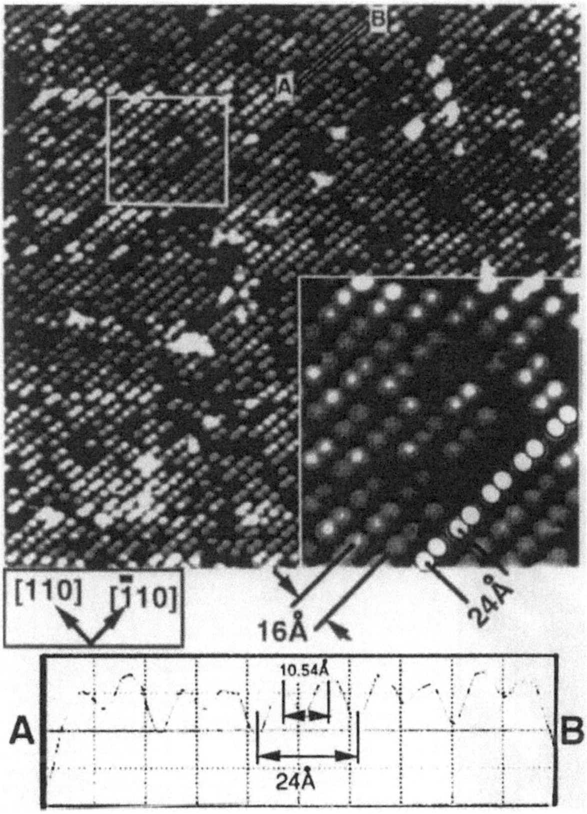

Fig. 9.26. STM image (650 Å×650 Å, $V_s = -2.4$ V) of the 2×4-As surface with approximately 1 ML C_{60} molecules. The *inset* image with higher magnification clearly shows that arrays of paired C_{60} molecules are building blocks for the first layer. The separation of paired C_{60} molecules is 10.54 Å

A well-ordered structure forms at approximately 2 ML, as can be seen in Fig. 9.27a, whose unit cell is a 16.0 Å ×11.3 Å rectangle with axis ratio 1.42 ± 0.02. Due to the uniform separation of C_{60} molecules along the $[\bar{1}11]$ direction, this overlayer is only uniaxially commensurate with the substrate. A surprising observation is that this configuration remains intact as film growth continues. In Fig. 9.27b, we show the STM image after approximately 10 ML of C_{60} deposition. As indicated by a white rectangle in the image, the unit cell of this ordered structure is the same as that shown in Fig. 9.27a. The long to short axis ratio of the rectangular unit cell is almost exactly $\sqrt{2}$, and hence the multi-layers shown in Fig. 9.27b correspond to the fcc(110)-oriented crystalline film of C_{60} molecules, although the lattice is highly 'strained' with as much as 13% expansion. The measured interplanar distance of the film

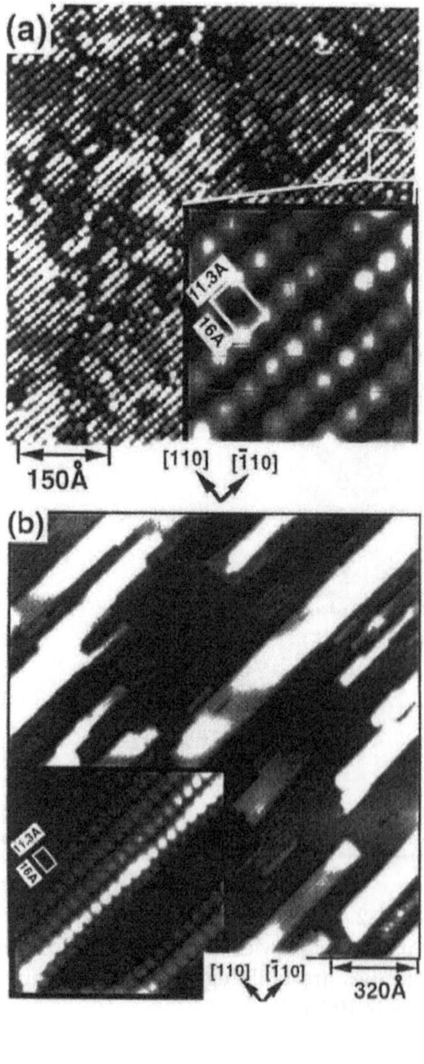

Fig. 9.27. (**a**) An ordered structure of the second layer (V_s = −3.8 V). The *inset* shows the details of the fcc(110) plane. (**b**) STM image of the C_{60} multi-layers

is 5.9 Å, again consistent with the fcc(110)-oriented film of C_{60} with 13% expansion.

These observations document the special role of the 2×4-As substrate in forming the fcc(110) overlayer, and raise two intriguing and important questions:

- why does the first monolayer form such a novel and stable paired-chain structure, instead of the rhombic or a uniaxially commensurate structure?
- what is the physical origin of the 'pseudo-morphic' growth of this strained (110) crystalline superlattice?

We first discuss possible bonding configurations of the C_{60} molecules to the substrate. According to the height analysis and relative registry of the C_{60} molecules with the 2×4-As structure at the initial stage of adsorption, the adsorption position of the isolated C_{60} molecule is determined to be site S1 or site S2, as shown in Fig. 9.25. Site P in Fig. 9.25 agrees well with the experimental result for the paired C_{60} position of the first layer. A molecular dynamics (MD) simulation, which has been successfully applied to C_{60} adsorption on the Si(100)-2×1 surface [104], was carried out in order to understand the current observations. The present MD simulation is based on the assumption that the interatomic interactions for Ga–As and C–C are described by a harmonic potential, the details of which are reported elsewhere [106]. For the C_{60} overlayer, the equilibrium pattern is determined by balancing the intermolecular and molecule–substrate interactions [44]:

$$\sum_i \frac{\partial V_{\mathrm{ms}}(\boldsymbol{R}_i)}{\partial R_i} + \frac{1}{2}\sum_{i \neq j} \frac{\partial V_{\mathrm{mm}}(|\boldsymbol{R}_i - \boldsymbol{R}_j|)}{\partial R_i}\,, \tag{9.2}$$

where $\boldsymbol{R}_i$, $\boldsymbol{R}_j$ are the position vectors of the i and jth molecules, respectively, V_{ms} is the molecule–substrate interaction and V_{mm} is the intermolecular van der Waals force. Since the 2×4-As reconstruction produces a steep periodic potential V_{ms} along the [110] direction and this one-dimensional configuration remains intact with adsorption of C_{60} molecules, the ordering of the molecules in this direction is dictated by the substrate periodicity, with all C_{60} molecules trapped in the trough.

The simulated result is shown in Fig. 9.28 for 20 C_{60} molecules. The result predicts that the most favorable adsorption sites for isolated and paired C_{60} molecules are S1 and P, respectively. Along the $[\bar{1}11]$ direction, V_{ms} has a discernible dependence on the substrate in the present system, unlike the cases of adsorption on the Si(100)-2×1 and Ge(100)-2×1 surfaces [28,29,44]. When another C_{60} approaches a stable pair marked by P in Fig. 9.25, the best bonding position for the molecule is site A shown in Fig. 9.25. This is because the molecule–substrate interaction overrides the weak attractive interaction among the C_{60} molecules and the site S2 cannot be taken. With further deposition, a subtle balance between the site-specific C_{60}–substrate interaction and the C_{60}–C_{60} van der Waals coupling results in a unique ordering of paired molecules in the $[\bar{1}11]$ direction (Fig. 9.28).

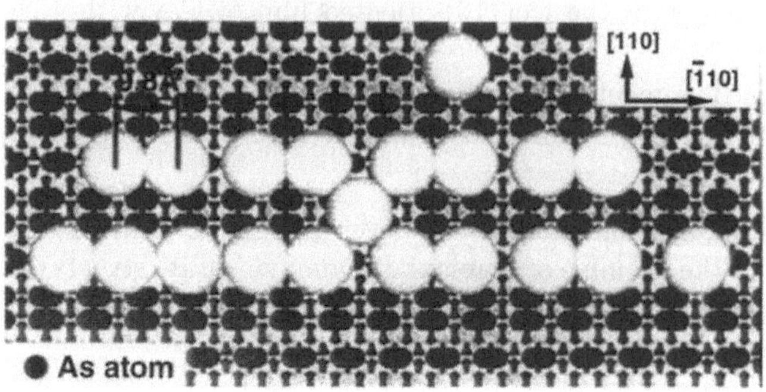

Fig. 9.28. The first layer formation of C_{60} molecules on the 2×4-As surface simulated by the molecular dynamics calculation, showing the formation of paired C_{60} molecules and their stable location

The intermolecular distance of the pair is 9.8 Å in the MD simulation, which is shorter by 2.0 Å than an isolated pair and shorter by 0.74 Å than the pair in the saturated first layer. The large differences suggest that an additional interaction must be involved. We now consider an interaction induced by a long range dipole–dipole repulsion since charge transfer between C_{60} molecules and substrate takes place in general for most of the fullerene adsorption on the semiconductor and metal substrate. The equilibrium separation can be quantitatively estimated by balancing the energy increase due to relaxation of the paired molecules and the energy decrease due to the charge-induced dipole–dipole coupling. The coupling gives rise to an interaction energy with the form

$$V_{\boldsymbol{PP}} = \frac{\boldsymbol{P}^2}{4\pi\varepsilon_0 R^3} , \tag{9.3}$$

where $\boldsymbol{P}$ is the induced dipole moment and R is the separation. The interaction energy between the two C_{60} molecules on the surface is approximated by the Girifalco potential $\Phi(R)$ [107,108]:

$$\Phi(R) = \frac{300\varepsilon_0\sigma}{r_0}\left[\frac{1}{R(R-r_0)^3} + \frac{1}{R(R+r_0)^3} - \frac{2}{R^4}\right] , \tag{9.4}$$

where $r_0 = 7.1$ Å. For an isolated pair, the intermolecular distance R_1 is given by

$$V_{\boldsymbol{PP}}(R_1) = \Phi(R_1) - \Phi(R_0) , \tag{9.5}$$

where R_0=9.8 Å is the equilibrium distance without charge transfer.

Starting with the simulated results, the interaction exerted on the individual molecule of a given pair for the saturated layer can be written,

$$V_{\boldsymbol{PP}}(R_2) - V_{\boldsymbol{PP}}(24 - R_2) + \sum_N [V_{\boldsymbol{PP}}(24N) - V_{\boldsymbol{P'P'}}(24N)]$$

$$+ \sum_N [V_{\boldsymbol{PP}}(24N + R_2) - V_{\boldsymbol{P'P'}}(24N + R_2)] \ , \tag{9.6}$$

where $N = 1, 2, 3, \ldots$ and R_2 is the C_{60}–C_{60} distance of the pair at saturation coverage. The term comes from the effect of different screenings for the two molecules that have the same separation as that of the molecule we are considering. The interaction with the equivalent molecules located in other troughs has a net zero effect in the $[\bar{1}11]$ direction and is not included in (9.6). The last two terms are very small and can be neglected. Combining (9.3) through (9.6), we obtain an analytic solution: $R_2/R_1 = 0.90$. We can now predict the distance of the paired C_{60} molecules in the paired-chain using the measured distance for a single pair (and vice versa). Taking $R_1 = 11.8$ Å, the observed value for the isolated pair, we obtain $R_2 = 10.62$ Å, which is in good agreement with the experimental value 10.54 Å. Adsorption of C_{60} molecules has been investigated on the GaAs(001) Ga-rich 4×2, InAs(001) As-rich 2×4 and In-rich 4×2 phases, as well as on GaAs(001) As-rich $c(4\times4)$ and 2×6 phases [49,109]. However, for most of the cases, only the fcc(111)-oriented multi-layers formed, except for the case of C_{60} adsorption on the InAs(001) As-rich 2×4 system [109]. These results further demonstrate the unique characteristics of the 2×4-As surface arising from charge transfer. As for the formation of the intriguing (110) multi-layer film, we believe that the first layer paired-chain structure serves as a perfect template for further overlayer growth. The uniform distribution of the second overlayer along the $[\bar{1}11]$ direction implies that the C_{60} monolayer adsorption screens and weakens the substrate template potential along the $[\bar{1}11]$ direction. On the other hand, the newly formed paired-chain structure still exerts the corrugated potential on the second layer C_{60} in the [110] direction with a periodicity that is basically the same as for the substrate.

9.4 Results and Discussions on Metal Substrates

9.4.1 Film Growth of C_{60} on Cu(111)-1×1

In contrast to the Si and GaAs surfaces, the fullerene molecules on the Cu(111)-1×1 surface show relatively weak bonding to the substrate and are mobile on the terrace at room temperature, segregating to the steps to form linear chains. However, when a highly ordered 4×4 commensurate phase forms in the case of C_{60}, C_{70} and $C_{60x}C_{70(1-x)}$ adsorption on the Cu(111) surface, fullerene molecules ratchet on the surface and intramolecular structures are observed [63–65]. As coverage increases, C_{60} molecules start to form two-dimensional (2D) islands and grow from the step edges in the direction of the ascending steps until they reach the next step edge, thereby producing

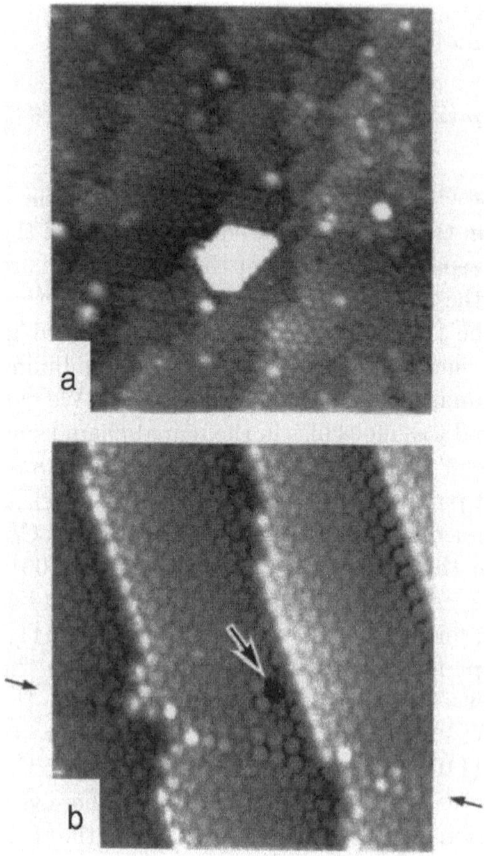

Fig. 9.29. STM images of a monolayer film of C_{60} molecules (**a**) deposited at room temperature (320 Å×320 Å, V_s = −2.0 V) and (**b**) prepared by annealing the surface (**a**) at 290°C (230 Å×230 Å, V_s = +2.0 V). The *central arrow* indicates a point defect and the *arrows on left* and *right* show domain boundaries

a monolayer film of C_{60} (Fig. 9.29a). In Fig. 9.29a, the steps replicate the substrate Cu(111) surface, except for the 2nd layer in the center of Fig. 9.29a.

The monolayer film grown at room temperature shows several domains. Figure 9.29b shows an STM image of the C_{60} monolayer film formed after annealing at 290°C. The surface is highly ordered, except for point defects (shown by an arrow) and domain boundaries (shown by arrows). The steps reflect the substrate Cu and are lined up along the $[0\bar{1}1]$ direction, which is the substrate atomic row direction. The steps are not straight for the case of room temperature adsorption (Fig. 9.29a). This suggests a significant mass transport of surface Cu atoms while the surface is being annealed at 290°C. The nearest neighbour distance (nnd) of the C_{60} molecules in the film is 10.2 Å, which is $4a_0$ (a_0 = 2.56 Å, the nnd of the Cu(111) surface). We observe the (4×4) LEED pattern for the monolayer film of C_{60} [26]. Individual C_{60} molecules are imaged at a bias voltage (V_s) close to zero. This suggests that gap states are induced by charge transfer from the substrate to the C_{60} overlayer, which enable us to image the C_{60} molecule for V_s within the

HOMO (Highest Occupied Molecular Orbital)–LUMO (Lowest Unoccupied Molecular Orbital) band gap [110,111].

We note that individual C_{60} molecules appear in a three-lobe clover shape at $V_s = +2.0$ V (Figs. 9.29b and 9.30c). At a bias voltage of $V_s = -2.0$ V (Fig. 9.30a), each C_{60} molecule appears in a doughnut shape. We observe triangular intramolecular structures for V_s close to zero (Fig. 9.30b). Similar to the case of C_{60} adsorption on the Si(100)-2×1 surface [28,29], the free rotation of the C_{60} molecule is terminated and intramolecular structures are observed.

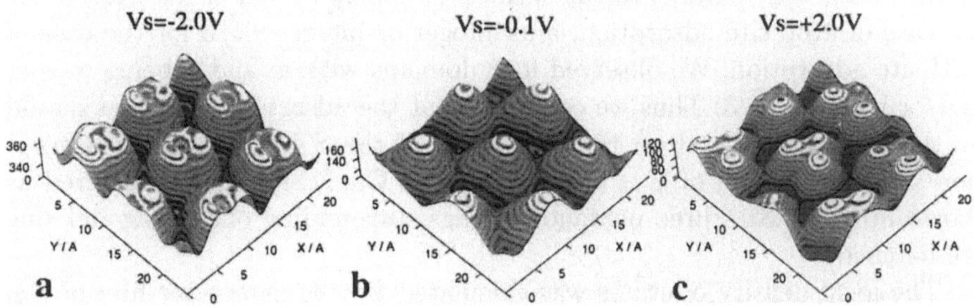

Fig. 9.30. Bias dependent intramolecular structures of C_{60} molecules (three-dimensional presentations). (**a**) Filled states image ($V_s = -2.0$ V) with doughnut shaped intramolecular structure, (**b**) near the Fermi level ($V_s = -0.1$ V) with triangular intramolecular structure and (**c**) empty states image ($V_s = +2.0$ V) with a three-lobe clover shaped intramolecular structure

We are able to discuss the symmetry of C_{60} molecules by examining the symmetry of intramolecular structures. The threefold symmetry of intramolecular structures suggests that the hexagonal rings of a C_{60} molecule should face down on the Cu(111) surface. Otherwise the intramolecular structures should reveal a fivefold or a twofold symmetry. Further information on the adsorption geometry can also be derived from the symmetry argument. The Cu(111) substrate has threefold symmetry and the hexagonal ring of a C_{60} molecule has apparent sixfold symmetry. Note that the size of the hexagonal ring facing the substrate almost perfectly matches the size of the triangle formed by substrate Cu atoms (the size mismatch is only 4%). Thus there are four possible adsorption orientations:

- three carbon atoms of the hexagonal ring take almost atop sites of the three Cu atoms,
- the centers of three bonds of the hexagonal ring take almost atop sites of the Cu atoms,
- the mirror symmetry of these two.

In Fig. 9.31, one of the lobes of the clover-shaped C_{60} molecules points in the $[0\bar{1}1]$ or $[01\bar{1}]$ direction, and none point in the $[\bar{2}11]$ direction. This observation limits the adsorption orientation of the C_{60} molecule to two directions, i.e., the first case above and its mirror symmetric orientation. Both the atop site and threefold hollow (3FH) site are possible for the adsorption position.

Two adsorption positions are possible for the 3FH site: one has the second layer substrate Cu atom (the so-called hcp site) and the other does not (the fcc site) (Fig. 9.31). For each adsorption position, the C_{60} molecule has two orientational freedoms, resulting in the four types of domains. The adsorption position is unique for the atop site except for the 60° rotational freedom, resulting in the two types of domains. We examine the shift of the C_{60} molecules in the two different adjacent domains and express it by a two-dimensional vector (ma_0, na_0) parallel to the surface. Numbers m and n are integers for the case of atop site adsorption, and integer or integer $\pm 1/3$ for the case of 3FH site adsorption. We observed four domains with m and n being integer and/or integer $\pm 1/3$. Thus we conclude that the adsorption position should be the 3FH site. Based on these analyses of the STM images, an adsorption structure model of C_{60} molecule on the Cu(111) surface is derived as shown in Fig. 9.31 (three pentagonal rings surrounding one hexagonal ring are hatched).

The local density of states was calculated for the monolayer film of C_{60} molecules [64,112,113]. We estimate a significant amount of charge transfer from Cu to C_{60}, at least more than for Si surfaces [28,29]. We were able to calculate the simulated images of intermolecular structures by evaluating the 3D mapping of the local electron density of states near the Fermi level. When we observe the empty states, the 181 (LUMO band for neutral C_{60} crystal), 182 and 183 levels should contribute to the image. When we observe the filled density of states, the 180, 179, 178, 177 and 176 levels should contribute [64,112,113]. Figures 9.32a, b and c show the calculated 3D map of

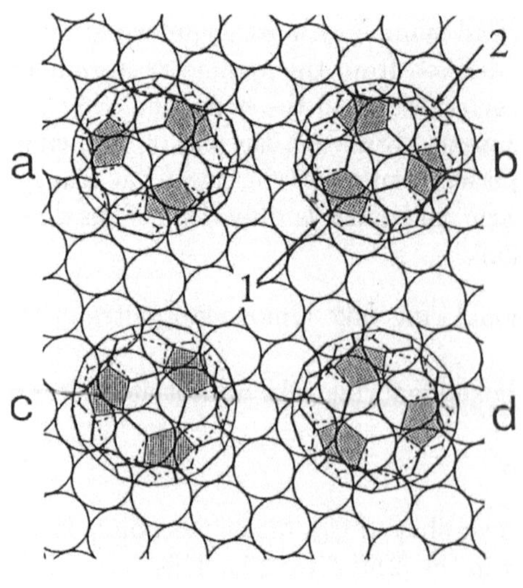

Fig. 9.31. Adsorption model for C_{60} molecules on the Cu(111)-1×1 surface. Three pentagonal rings of the C_{60} molecules are *shaded*. The adsorption position of (**a**) or (**b**) is different form that of (**c**) or (**d**) in terms of the existence of the Cu atom in the second layer of the substrate. *Arrows* 1 and 2 indicate the directions, which have a double bond and a pentagonal ring at the side of the molecule, respectively

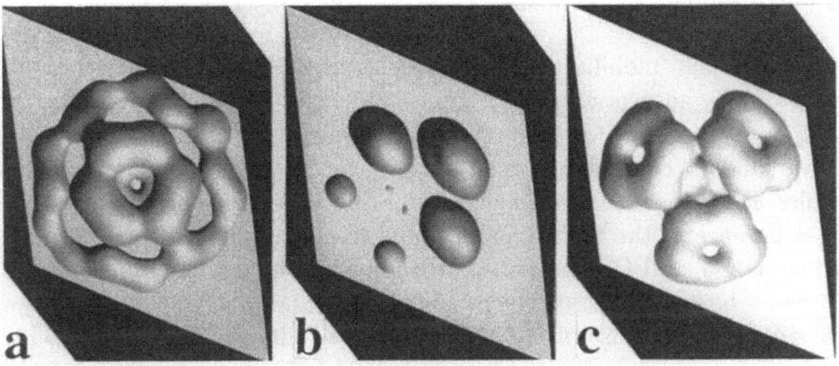

Fig. 9.32. Simulated STM images of a C_{60} molecule for V_s = (**a**) −2.0 eV, (**b**) −0.1 eV and (**c**) +2.0 eV, obtained by calculating the local density of states near the Fermi level

the local density of states that simulates the STM images in Figs. 9.30a, b and c, respectively. The clover lobes we observe at $V_s = +2.0$ V correspond to three pentagonal rings surrounding the hexagonal ring at the top of the molecule.

Individual C_{60} molecules in the fcc bulk phase are orientationally disordered at room temperature. The system undergoes the first order phase transition to the simple cubic (sc) phase below 258 K [114]. Four C_{60} molecules in an fcc unit cell become orientationally non-equivalent below the critical temperature. The orientational ordering has been explained by the electrostatic interactions between electron-rich double bonds, interconnecting two hexagonal rings, and electron-poor pentagonal rings [114]. In three terraces from the right of Fig. 9.29b, one of the three lobes representing the pentagonal rings is facing downward (in the $[01\bar{1}]$ direction), and in the leftmost terrace they point upward. We observe orientational ordering of C_{60} molecules in a highly ordered monolayer film on the Cu(111) surface. The domain boundary seen in the center of Fig. 9.29b is continuous beyond steps. If we examine the orientations of C_{60} molecules, we find a similarity with those in the sc bulk phase, namely, one of the double bonds almost neighbours one of the pentagonal rings, as shown in Fig. 9.31 by arrows 1 and 2. We also observe that some of the C_{60} molecules appear to rotate and are imaged round at domain boundaries or at sites next to point defects (Fig. 9.29b). The orientational freedom of C_{60} molecules adsorbed on the Cu(111) surface is limited by their being ratcheted to the surface. In the bulk sc phase, where they have more orientational freedom, they rotate about 4° and form better alignment with regard to the interaction between double bonds and pentagonal rings [114].

9.4.2 Monolayer Film of $C_{60x}C_{70(1-x)}$ on Cu(111)-1×1

We observe a high mobility of C_{70} molecules on the Cu(111) terrace, as in the case of C_{60} adsorption [65]. Figure 9.33 shows the monolayer films of $C_{60x}C_{70(1-x)}$ ($x = 0.06$), obtained by annealing at 320°C. We mainly observe two kinds of contrasts (dim and bright) for the individual C_{60} and C_{70} molecules. Based on the number density of dim molecules and the apparent height difference in the STM images, the dim and bright molecules are assigned to be C_{60} and C_{70} molecules, respectively. The C_{70} molecules form a well-ordered hexagonal arrangement (nnd = $4a_0$), i.e., the Cu(111)-4×4-C_{70} phase, similar to the case of C_{60}. The monolayer film shown in Fig. 9.33 is divided into several domains separated by domain boundaries, where we observe domain boundary segregation of the C_{60} molecules. We do not observe intramolecular structures either in C_{60} or C_{70} molecules, and attribute this fact to rotation of the fullerene molecules. The long axis of the C_{70} molecule measures 7.9 Å and the nnd of the bulk van der Waals-like crystal of C_{70} is 10.6 Å. It is probable that C_{70} molecules in the 4×4 phase are taking upright orientation, the long axis being perpendicular to the Cu(111) surface, since the bulk-like structure (free rotation) or lying-down orientation, with the long axis parallel to the Cu(111) surface, both form monolayer films with too much stress. Consequently, rotation of C_{70} molecules should be along the long axis, similar to the bulk crystal in the moderate temperature range between 70°C and 10°C [115,116]. The apparent height difference between the C_{60} and C_{70} molecules observed in the STM images is consistent with this argument. When a C_{70} molecule takes upright orientation, one of the pentagonal rings faces the Cu(111) surface and the symmetry of the C_{70} molecule no longer fits the substrate.

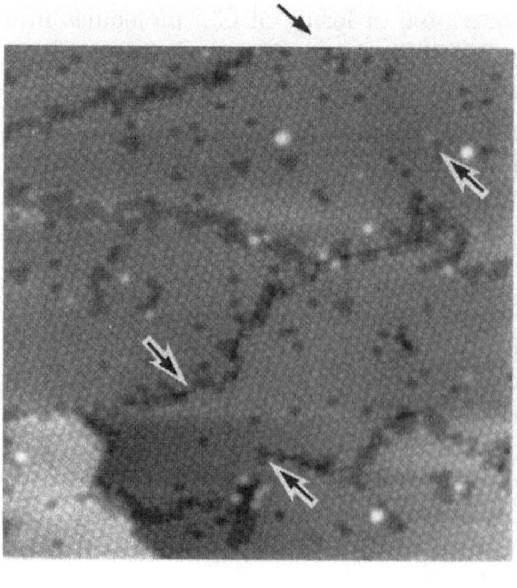

Fig. 9.33. STM image of the $C_{60x}C_{70(1-x)}$ ($x = 0.06$) monolayer film, showing the domain boundary segregation of C_{60} molecules (500 Å×500 Å, $V_s = +2.0$ eV). Note that particular domain boundaries do not exhibit any segregation (examples shown by *arrows*)

Figure 9.34a shows a detailed model of an example of the domain boundary. Two 4×4 phases form two kinds of domain boundary, shown as the lines AB and BC. At the domain boundary AB, the nnd is $4.6a_0$ and at the domain boundary BC, the nnd is $3.6a_0$. Figure 9.34b shows an nnd dependence of the C_{60} density at the domain boundary. No segregation is seen when the nnd is less than approximately $4a_0$ (examples are seen in Fig. 9.33, indicated by arrows), and the degree increases as the nnd becomes larger than $4a_0$. If we assume that the difference in substrate–adsorbate interactions for C_{60}

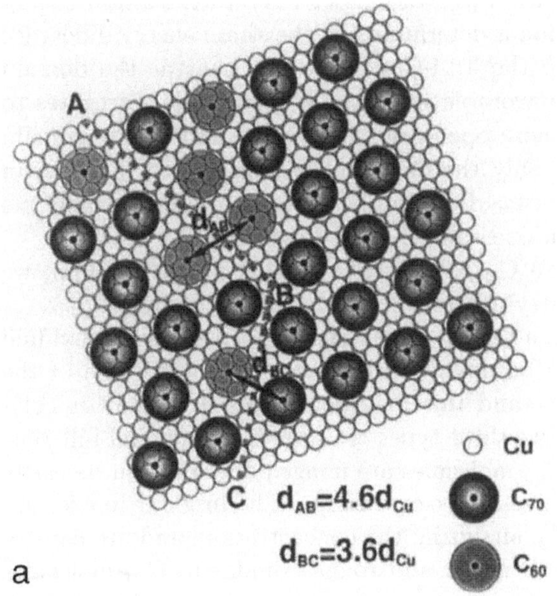

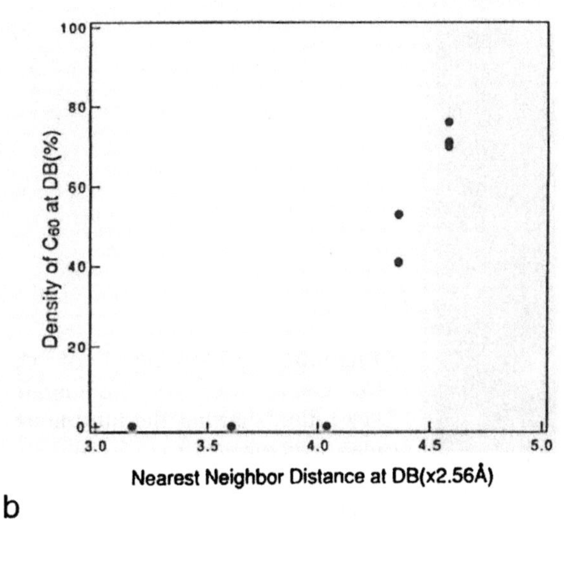

Fig. 9.34. (**a**) Structure model of domain boundary and (**b**) nnd dependence of the degree of domain-boundary segregation

and C_{70} molecules is negligible, this phenomenon may be treated as two-dimensional domain boundary segregation based on adsorbate–adsorbate interactions. Although we do not fully understand the mechanism for domain boundary segregation and its nnd dependence at this stage, we would like to point out two possible explanations.

We apply the strain energy model for the first possible explanation. We assume that the C_{60} impurity in the 4×4-C_{70} phase induces compressive stress. This is understandable when we recall that the C_{70} molecules are taking upright orientation in the 4×4-C_{70} phase and that the size of the C_{70} molecules is 7.0 Å along the short axis (along the 2D film in this case), slightly smaller than the diameter of the C_{60} molecule (7.1 Å). In the strain energy model, the degree of segregation is determined by the strain energy difference between the C_{60} impurity in the 4×4-C_{70} phase and that at the domain boundary. It is energetically favorable when the C_{60} molecule segregates to the domain boundary with more open area, i.e., larger nnd at the domain boundary. This explains not only the degree of segregation to the domain boundary with nnd larger than $4a_0$, but also the depletion of the C_{60} molecule at the domain boundary with nnd smaller than $4a_0$ in Fig. 9.34b.

In the case of the mixture $C_{60x}C_{70(1-x)}$ ($x = 0.3$) monolayer film, we observe an adsorption geometry different from the case of the $C_{60x}C_{70(1-x)}$ ($x = 0.06$). Figure 9.35 shows a typical empty states STM image. Individual fullerene molecules are again occupying lattice sites, as in the case of the 4×4-C_{60} and 4×4-C_{70} phases, and this may be expressed as the Cu(111)-4×4-C_{60}/C_{70} phase. We observe three types of image for individual fullerene molecules in Fig. 9.35. The C_{60} molecules are imaged dim and can be easily recognized by their unique three-lobe clover shape. The brighter images are judged to be C_{70} molecules by analyzing the concentration, and are categorized into two groups: one is brighter and round, similar to C_{70} molecules

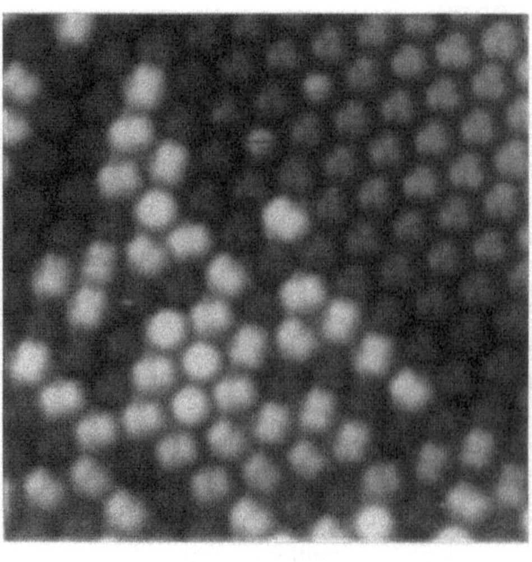

Fig. 9.35. STM image of the $C_{60x}C_{70(1-x)}$ ($x = 0.3$) monolayer film, showing the intramolecular structures of the C_{60} and C_{70} molecules (110 Å×110 Å, V_s = +2.0 eV)

in the 4×4-C_{70} phase, the other is less bright and exhibits twofold symmetry. If we inspect the latter case more closely, we find that the image of a molecule comprises two elongated protrusions, each of which consists of two spots. From a symmetry argument similar to the case of C_{60} molecules in the 4×4-C_{60} phase, we attribute the latter to C_{70} molecules with lying-down orientation. This assignment is supported by the stress argument as follows. When the concentration of the C_{60} molecule is low, the most favorable arrangement for C_{70} molecules is the upright orientation since the long axis of the C_{70} molecule is 7.6 Å and much larger than for the C_{60}. Given that the nnd in the bulk fcc phase of C_{60} is 10.0 Å, slightly smaller than the nnd in the Cu(111)-4×4 phase (10.2 Å), the co-adsorption of C_{70} molecules with C_{60} molecules reduces the stress surrounding the C_{70} molecules, and C_{70} molecules are able to take lying-down orientation in the 4×4-C_{60}/C_{70} phase. Even in the 4×4-C_{60}/C_{70} phase, a C_{70} molecule, when surrounded by C_{70} molecules, may feel too much stress and hence assume an upright orientation. These are observed as the brightest round molecules in Fig. 9.35.

By analogy with the intramolecular structure observed for C_{60} molecules, we claim that the adsorption orientation shown in the Fig. 9.36 is probable, assuming that pentagonal rings in the C_{70} molecules are observed bright in empty states STM images. Although we do not have an exact calculation for the 4×4-C_{60}/C_{70} phase, recent results by Maruyama et al. [113] for the hexagonal monolayer film of the lying-down C_{70} molecules (the same orientation as in Fig. 9.36), show that the LUMO of the C_{70} molecule localizes at the four pentagon rings facing up to the surface, thereby supporting the argument discussed above.

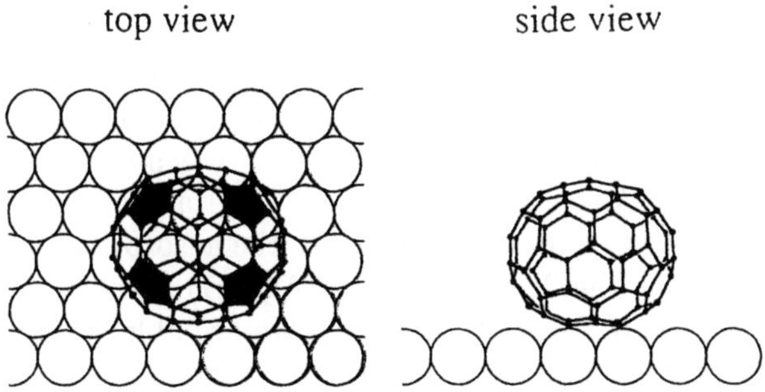

Fig. 9.36. Structure model of the lying-down C_{70} molecule in the Cu(111)-4×4-C_{60}/C_{70} phase

9.4.3 Rotational Epitaxy of C_{60} and $Sc_2@C_{84}$ on Ag(111)-1×1

When monolayer films of fullerenes form on the Ag(111) surface we observe several kinds of close packed hexagonal phases, which have almost equal nnd but different rotational angles with respect to the substrate Ag atomic row direction [72]. Figure 9.37 shows an example for the case of the C_{60} monolayer film on the Ag(111) surface after annealing at 300°C. We observe three phases in this image, which are consistent with STM observations reported by Altman and Colton [58]. The nnd in all these phases is measured to be 10.0 ± 0.1 Å, essentially the same as $2\sqrt{3}a_0$ (10.0 Å, since $a_0 = 2.89$ Å for Ag). The phase at top right in Fig. 9.37 has a rotation angle of 30° and is determined to be the Ag(111)-$2\sqrt{3} \times 2\sqrt{3}\,R30°$-$C_{60}$ phase. The phase in the lower part is rotated by 12° and the phase at upper left is rotated by 46°. We have examined many phases by STM and have found that the rotation angle can be grouped into two types, 11.5–13.5° and 45–47°.

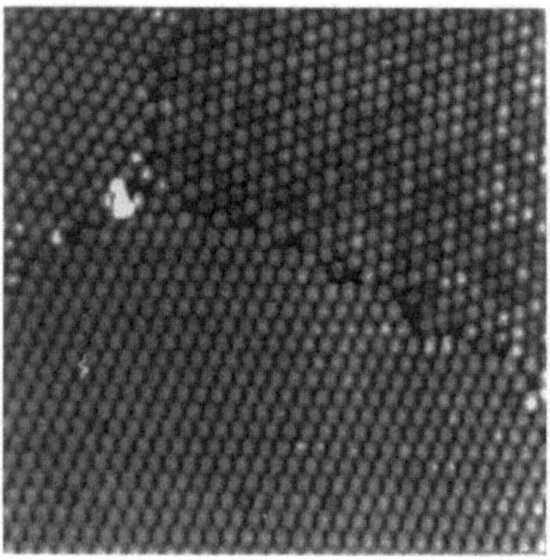

Fig. 9.37. STM image of the C_{60} monolayer film on the Ag(111) surface, showing three phases with almost equal nnd (300 Å×300 Å, $V_s = +2.0$ eV)

We have also found that prolonged annealing at 300°C for up to 36 hours results only in the $2\sqrt{3} \times 2\sqrt{3}\ R30°$ phase and disappearance of the other phases. This indicates that the $2\sqrt{3} \times 2\sqrt{3}\ R30°$ phase is the most energetically stable. As proposed by Altman and Colton [58] and analogous to the structure of the Cu(111)-4×4-C_{60} phase, we estimate the adsorption site of the $2\sqrt{3} \times 2\sqrt{3}\ R30°$ phase as the 3FH site on the Ag(111) surface. By analyzing the nnd and rotation angles of other phases, we are able to propose two structure models for one of the two groups (with rotation angle of 11.5–13.5°). These are shown in Fig. 9.38. We can obtain the other group of phases with rotation angle of 45–47° by considering the mirror symmetry of these two models. To make these models, we have tried to draw as many C_{60} molecules as possible at the 3FH sites, because of stability considerations. For the case of $Sc_2@C_{84}$ monolayer film formation on the Ag(111) surface, we can expect the Ag(111)-4×4-$Sc_2@C_{84}$ phase, since the $4a_0$ of the Ag(111) surface is 11.6 Å and is very close to the nnd (11.7 Å) measured from STM images of multi-layer $Sc_2@C_{84}$ islands [72]. We have examined many STM images and observed eight phases with nnd $= 11.6 \pm 0.2$ Å and rotation angles of 0°, 8°

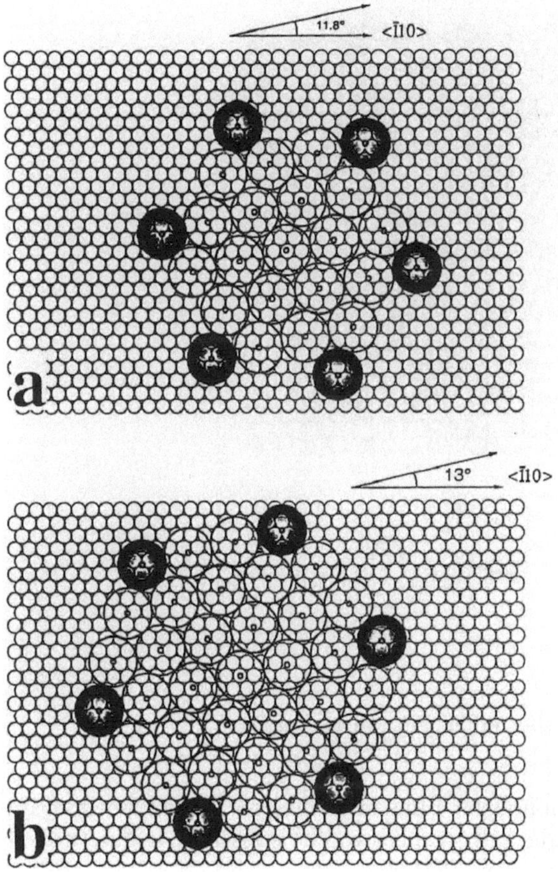

Fig. 9.38a, b. Structure models of the Ag(111)-C_{60} phases

(52°), 14° (46°), 19° (41°) and 30°. The inaccuracy in angle measurements is estimated to be ±1°.

Figure 9.39 shows structure models for the (a) 4×4 (nnd = 11.6 Å, R = 0°), (b) 7×7 R8.2° (nnd = 11.7 Å), (c) 7×7 R30° (nnd = 11.7 Å), (d) $3\sqrt{7}\times3\sqrt{7}$ R19.1° (nnd = 11.5 Å) and (e) $\sqrt{111}\times\sqrt{111}$ R45.6° (nnd = 11.5 Å) phases. In drawing these structure models, assumptions and considerations taken into account were similar to the case of the C_{60} monolayer film on the Ag(111) surface. As can be seen directly from Fig. 9.39, all the $Sc_2@C_{84}$ molecules in the (a) 4×4 phase occupy the 3FH sites, while two out of three $Sc_2@C_{84}$ molecules in the (b) 7×7 R8.2° and (c) 7×7 R30° phases occupy the 3FH sites. In the case of the (d) $3\sqrt{7}\times3\sqrt{7}$ R19.1° and (e) $\sqrt{111}\times\sqrt{111}$ R45.6° phases, only 1/4 and 1/7 molecules, respectively, occupy 3FH sites.

The monolayer formation of fullerenes on the Ag(111) surface can be understood by the following mechanism. In comparison with fullerene adsorption on the Cu(111) surface, the interaction between fullerene molecules and the Ag(111) surface is weak. Hence the interaction among fullerene molecules becomes more important and van der Waals-type close-packed hexagonal

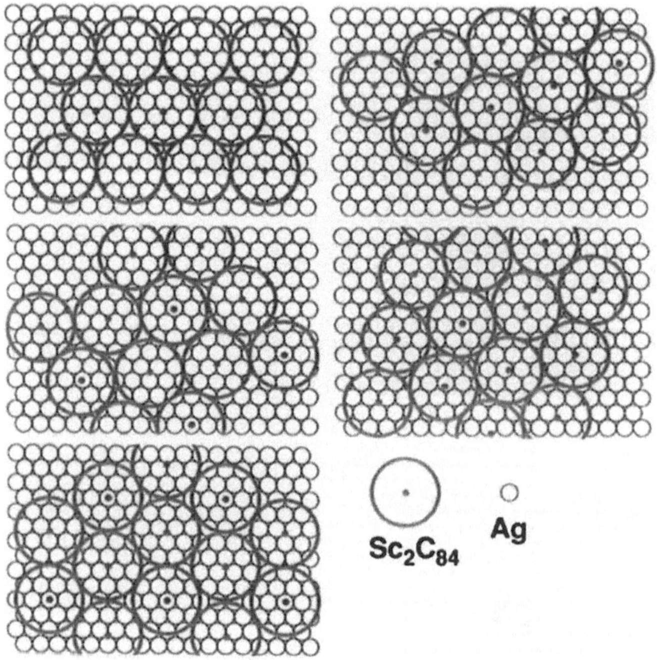

Fig. 9.39. Structure models of the Ag(111)-Sc_2@C_{84} phases

monolayer films form. The monolayer films still have rotational freedom on the surface. The rotation angle with respect to the substrate is determined by fullerene molecules which occupy 3FH sites, while the nnd is kept almost constant. Stability of the phases is closely related to the number density of molecules occupying 3FH sites. In the case of the Cu(111) surface, the interaction between fullerene molecules and substrate is more dominant and commensurate phases are preferred.

9.4.4 Pairwise Interaction of Y@C_{82} on Cu(111)-1×1

At the initial stage of Y@C_{82} molecule adsorption on the Cu(111)-1×1 surface, Y@C_{82} molecules segregate to step edges and form one-dimensional (1D) clusters, similar to the case for other fullerene molecules on the Cu(111)-1×1 and Ag(111)-1×1 surfaces. However, Y@C_{82} molecules predominantly form dimers, in contrast with the cases of Gd@C_{82} and C_{60} molecules [66,67,70]. Figure 9.40 shows the initial stage adsorption of Y@C_{82} on the Cu(111)-1×1 surface at room temperature. The coverage of Y@C_{82} molecules is 0.006 ML and corresponds to 3.4 Y@C_{82} molecules per 1 000 Å step edge for this particular miscut of the Cu(111) sample. We observe seven Y@C_{82} molecules adsorbing at the upper edge of the monoatomic step running from the upper right to lower left corners of the image. There are three dimers and one trimer

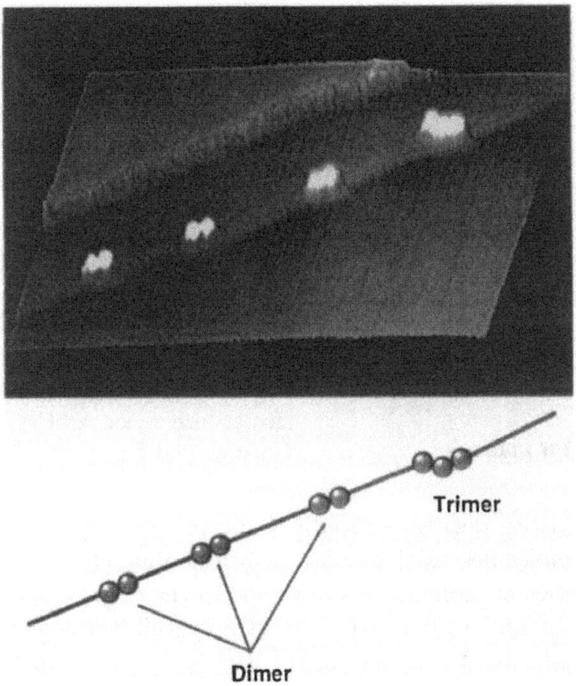

Fig. 9.40. Typical STM image (three-dimensional presentation) of the initial stage adsorption of $Y@C_{82}$ on the Cu(111)-1×1 surface (400 Å×400 Å, $V_s = -1.2$ V). We observe three dimers and one trimer at the upper step edge running from *upper right* to *lower left* corners of the image

of $Y@C_{82}$ molecules in this image. One of the most important findings is that $Y@C_{82}$ molecules predominantly form clusters, $(Y@C_{82})_n$, where $n = 2$ to 4 for the main part, with a small proportion of $n = 5$ and 6 clusters. Since molecules segregate to step edges, most clusters have one-dimensional (1D) shape at this coverage. This kind of cluster formation is in clear contrast with the case of C_{60} adsorption on the Cu(111) surface, where monomer adsorption dominates at low coverages. The observed distance between $Y@C_{82}$ molecules in the cluster is 11.2 Å, 0.6 Å smaller than the nearest neighbour distance (nnd) in the fcc bulk crystal (11.8 Å).

In order to analyze cluster formation quantitatively, we have taken more than 300 STM images of the initial stage adsorption of each of $Y@C_{82}$, $Gd@C_{82}$ and C_{60} molecules on the Cu(111) surface and counted the number of clusters as a function of cluster size (n = number of molecules in a cluster). More than 1 000 clusters are taken into the statistics for each molecule. The coverages for $Y@C_{82}$, $Gd@C_{82}$ and C_{60} molecules are 3.4, 3.9 and 4.1 molecules, respectively, per 1 000 Å step edge length. The small difference in the coverages of the three fullerenes do not affect the basic mechanisms of adsorption and cluster formation and the low adsorption rate at these small coverages. The cluster size distribution is summarized in Fig. 9.41. Since we observe only a small number of clusters with more than four molecules, we show the abundance of monomer ($n = 1$), dimer ($n = 2$), trimer ($n = 3$) and clusters with $n = 4$. In the C_{60} case, the monomer is dominant in the his-

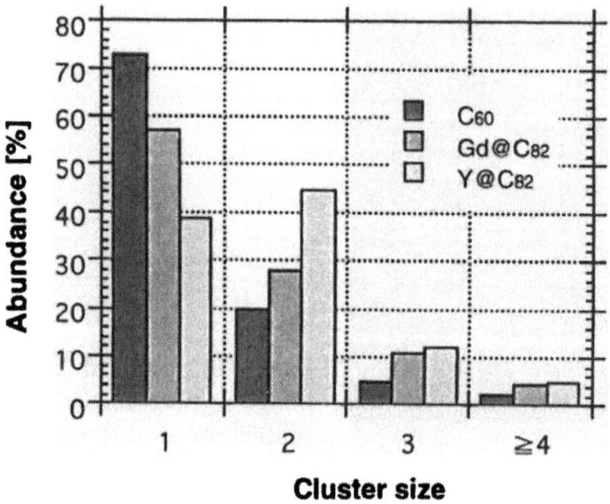

Fig. 9.41. Distribution of cluster size n for $Y@C_{82}$, $Gd@C_{82}$ and C_{60}

togram and the number of clusters decreases rapidly as cluster size increases. More than half of the C_{60} molecules exist on the surface as monomers. In the case of $Gd@C_{82}$, abundance of monomers is less than in the case of C_{60}, and cluster formation becomes more evident. However, the overall features of cluster formation are still similar to C_{60}. In the case of $Y@C_{82}$, it is clear that dimer formation dominates. The abundance of monomer formation is reduced to almost half of the value for C_{60}. Almost half of the $Y@C_{82}$ molecules exist as dimers, in clear contrast to the C_{60} case. We notice that abundances of trimer and larger clusters are similar to the $Y@C_{82}$ case.

Dimer formation of $Y@C_{82}$ molecules can be understood if we assume that $Y@C_{82}$ molecules sublimate from the doser as dimers and adsorb on the surface without breaking up to monomers. In order to check this point, we have observed the initial stage adsorption of $Y@C_{82}$, $Gd@C_{82}$ and C_{60} molecules on the Si(111)-7×7 surface. Fullerene molecules react with dangling bonds of Si surface atoms and adsorb randomly on the surface. The chemisorption energy is large enough to prevent surface migration of fullerene molecules on Si surfaces. We expect $Y@C_{82}$ molecules to adsorb as dimers or pairs at a nearby position on the surface if they sublimate as dimers. However, the observed results do not bear this out, and the $Y@C_{82}$, as well as $Gd@C_{82}$ molecules adsorb randomly on the Si(111)-7×7 surface, similar to the case of C_{60}. We thus conclude that dimer formation takes place after the fullerene molecules adsorb on the surface.

The basic process of the initial stage adsorption of $Y@C_{82}$, $Gd@C_{82}$ and C_{60} molecules can be modeled as follows:

I) The barrier height of each molecule for surface migration on the Cu(111) terrace (terrace corrugation energy $= E_{tc}$) is small and the individual molecule is mobile on the terrace at room temperature.
II) Since we observe monomer adsorption in a stable manner by STM, the barrier height of each molecule for surface migration along the step edge

(edge corrugation energy = E_{ec}) is large enough to ensure that individual molecules are not mobile at room temperature after energy relaxation. Thus the critical nucleation size (minimum size for nucleation) can be as small as a monomer for 1D islands on the step edge.

III) When fullerene molecules adsorb on the surface, each molecule obtains adsorption energy, which is in the order of 1 eV or several tenths of 1 eV depending on the adsorption system. The molecule gradually loses this energy during migration on the surface. The energy left on the molecule is defined as E_m here. When the molecule diffuses to the step edge, the molecule (a) stays at the same position if $E_m < E_{ec}$ and $E_m < E_{st}$, where E_{st} is the barrier height for migration from step edge to terrace, (b) migrates along the step edge if $E_m > E_{ec}$, or (c) diffuses back to the terrace if $E_m > E_{st}$. The molecule is eventually relaxed and adsorbs on the step edge.

IV) Since very low adsorption rates were used, the relaxation of the first molecule occurs before the second molecule adsorbs on the surface. The second molecule can adopt all the processes (a), (b) and (c) described in (III). When the second molecule comes close enough to the first molecule to interact with it, we can expect similar processes to (a), (b) and (c) in (III). The criteria can be obtained by replacing E_{ec} and E_{st} with $E_{ec}+E_2$ and $E_{st}+E_2$, respectively, where E_2 is the bonding energy between two molecules. When the process takes place, the molecules form a cluster.

Based on the model mentioned above, we have tried to reproduce the cluster size distribution for the case of Gd@C_{82} and C_{60}, using two methods. In the first method, we approximate the adsorption process by 1D random adsorption on the step edge. Because the coverage is very low, the cluster size distribution is well described by a Poisson distribution. If the molecules do not migrate on the step edge, the unique parameter is the averaged site occupation probability (λ), which is calculated from the coverage to be 0.047 and 0.041 for Gd@C_{82} and C_{60}, respectively. Here the nnd of possible adsorption sites is assumed to be 12 Å for Gd@C_{82} and taken as 10.0 Å for C_{60}. However, these parameters λ for Gd@C_{82} and C_{60} give enhanced distribution of monomer formation. The migration process mentioned in (IV) above enhances cluster formation. We introduce this effect by multiplying λ by a fitting parameter (m = migration parameter), calculated from the abundance of the monomer in the histogram (Fig. 9.41) to be 12 and 7.7 for Gd@C_{82} and C_{60}, respectively. This parameter includes the effects of both terrace migration and step edge migration of the molecule. In the second method, we estimate the cluster size distribution by a simple simulation. The adsorption positions of fullerene molecules on a 1D step edge are determined from a set of random numbers, and cluster formation is judged to have occurred if the distance between adjacent adsorbed molecules is smaller than the averaged migration length D. The values of D for Gd@C_{82} and C_{60} are taken to be

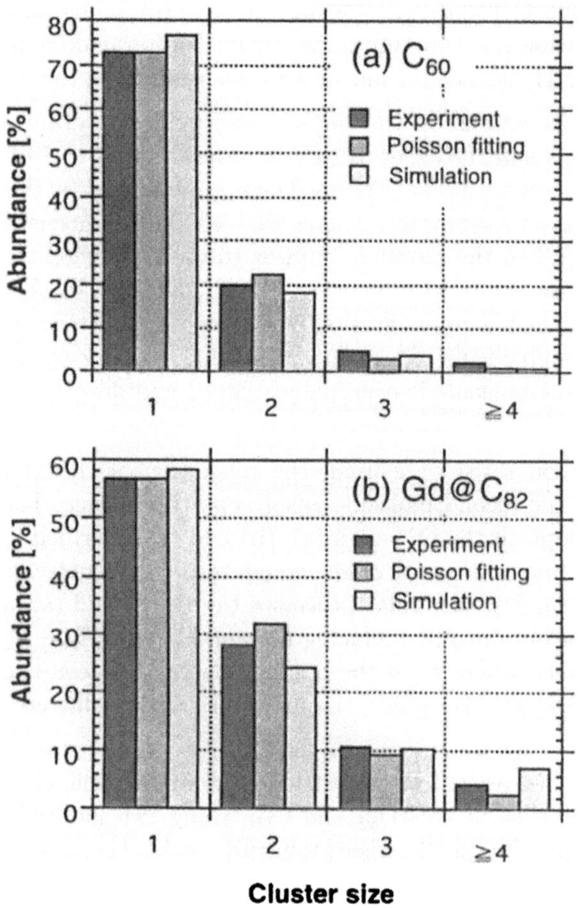

Fig. 9.42. Size distributions produced using (1) Poisson distribution and (2) computer simulation described in the text for (**a**) Gd@C_{82} and (**b**) C_{60}

the same as in the first method, viz., 12×12 Å and 7.7×10.0 Å, respectively. The results are shown in Fig. 9.42a and b for Gd@C_{82} and C_{60}, respectively.

The agreement between experiment and reproduced size distributions is very good for both methods. We thus conclude that the adsorption process can be modeled by a random adsorption with migration for the case of Gd@C_{82} and C_{60}. The difference in the cluster size distribution for Gd@C_{82} and C_{60} can be explained by the difference in the edge corrugation energy E_{ec}, the barrier height for migration from step edge to terrace E_{st}, the bonding energy E_2 and/or adsorption energy, although the detailed values are not known. In the case of Y@C_{82}, the cluster size distribution cannot be reproduced from the model based on random adsorption and migration mentioned above. By changing the value of the migration parameter m, we find that there is no case where the dimer abundance becomes dominant in the histogram (Fig. 9.43). We anticipate a special interaction for dimer formation in the case of Y@C_{82}. Walton treated cluster formation on the surface using an

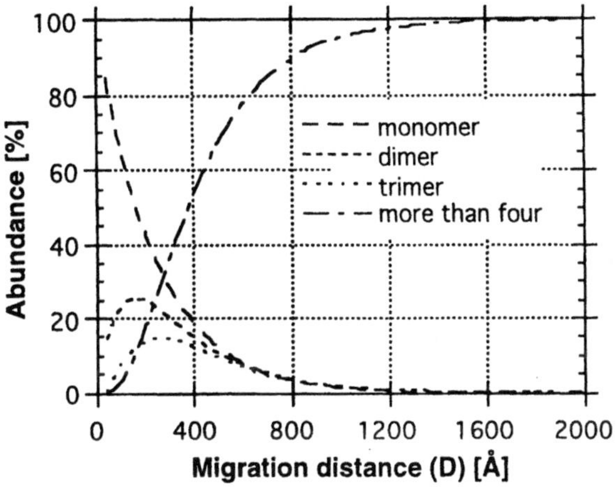

Fig. 9.43. Calculated size distribution as a function of migration length D

atomic model [117]. He assumed thermal equilibrium of the N atoms on the surface (2D) with n_0 lattice sites. The summation over all states is expressed as

$$\Im = \prod_i \frac{n_0 n_i}{n_i!} \exp \frac{n_i \varepsilon_i}{kT} , \tag{9.7}$$

where n_i is the number of clusters consisting of i atoms, ε_i is the potential energy of a cluster consisting of i atoms and kT is the thermal energy. By maximizing $\Im$ under the condition of

$$\sum_i i n_i = N(\text{constant}) , \tag{9.8}$$

the number of clusters n_i at the equilibrium condition is calculated as

$$n_i = n_0 \left(\frac{n_i}{n_0} \right)^i \exp \frac{\varepsilon_i - i\varepsilon_1}{kT} . \tag{9.9}$$

In (9.9), the energy term

$$\Delta\varepsilon_i = \varepsilon_i - i\varepsilon_1 , \tag{9.10}$$

is the cluster formation energy from i atoms adsorbing on the surface. It therefore corresponds to the sum of bonding energies of all nearest neighbours in the cluster.

We apply Walton's model to the adsorption of Y@C_{82}, Gd@C_{82} and C_{60} molecules on the Cu(111) surface. An assumption we need to make is that the cluster size distribution results from the thermal equilibrium. In the present cases, 1D shaped clusters consisting of fullerene molecules adsorb on step edges. We have analyzed the cluster size distribution shown in Fig. 9.41 and obtained the cluster formation energy ($\Delta\varepsilon_i$) as a function of cluster size (Fig. 9.44a). The cluster formation energy is a linear function of cluster size

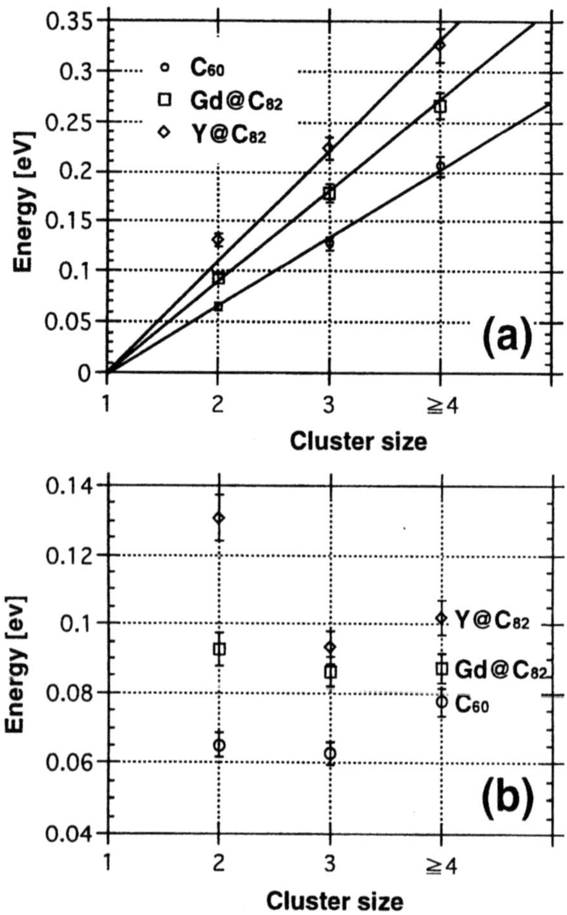

Fig. 9.44. (**a**) Cluster formation energy $\Delta\varepsilon_i$ as a function of cluster size for $Y@C_{82}$, $Gd@C_{82}$ and C_{60}. (**b**) Bonding energy $\Delta\varepsilon_{i+1} - \Delta\varepsilon_i$ as a function of cluster size for $Y@C_{82}$, $Gd@C_{82}$ and C_{60}

for the cases of $Gd@C_{82}$ and C_{60}, which means that the bonding energy is independent of cluster size. However, this is not the case for $Y@C_{82}$, where the cluster formation energy for dimers is larger than that for other clusters. In Fig. 9.44b, we show the increase in cluster formation energy ($\Delta\varepsilon_{i+1} - \Delta\varepsilon_i$) as a function of cluster size for $Y@C_{82}$, $Gd@C_{82}$ and C_{60}. This corresponds to the bonding energy of the ith molecule when forming the size i cluster. Each bonding energy is 90 ± 10 meV and 70 ± 10 meV for $Gd@C_{82}$ and C_{60}, respectively. It is 100 ± 10 meV for trimer and four-molecule clusters of $Y@C_{82}$, whereas it is 130 ± 10 meV for $Y@C_{82}$ dimers, approximately 30 meV larger than for trimer and four-molecule clusters. These results clearly indicate that the $Y@C_{82}$ dimer is formed by the special bonding energy between two $Y@C_{82}$ molecules.

From the bonding energy estimated here for C_{60}, we estimate the condensation energy of C_{60} to be approximately 840 meV. This is roughly 50% of the value in the bulk fcc C_{60} crystal. One possible explanation for the low estimate of the bonding energy is that this adsorption system is not operating in perfect thermal equilibrium conditions, but rather adsorption positions are partially limited by the adsorption kinetics, i.e., migration of molecules is not sufficient to realize the ideal equilibrium. Another possibility is that the interaction between C_{60} molecules is weakened because of charge transfer from the Cu(111) substrate to adsorbates. The difference in bonding energy between Gd@C_{82} and C_{60} can be ascribed to charge transfer from the encapsulated metal atom to the carbon cage. A similar increase in condensation energy has been discussed on the basis of the sublimation temperature difference for the case of Sc_2@C_{84}. However, if the system is not in thermal equilibrium conditions, the difference between Gd@C_{82} and C_{60} can be ascribed to the difference in corrugation energy. This in turn produces the difference in migration distance which enhances cluster formation.

On the basis of the adsorption model for Y@C_{82}, Gd@C_{82} and C_{60} on the Cu(111) surface, we have observed that the Y@C_{82} fullerene has a very special tendency for dimer formation. Although we expect some charge transfer from the Cu(111) substrate to the Y@C_{82} molecule, this dimer formation should be closely related to the structure and electronic properties of the Y@C_{82} molecule. The endohedral nature of the Y@C_{82} molecule was recently confirmed by X-ray diffraction studies [118]. The main two-line spectrum of ESR was observed and hyperfine splitting (hfs) analysis indicated that the three valence electrons of the encaged Y atom are transferred to the C_{82} cage to form an endohedral metallofullerene of the type Y^{3+}@${C_{82}}^{3-}$ [14,16,17]. For the case of La@C_{82}, similar analyses have been made and a La^{3+}@${C_{82}}^{3-}$ picture has been proposed, based on the eight-line ESR spectra [13]. The electronic structure of Y@C_{82}, with the C2 isomer of the C_{82} cage, has been confirmed by recent theoretical calculations [101]. A high-level ab initio calculation has suggested a $Y^{2.59+}$@${C_{82}}^{2.59-}$ picture, essentially of the Y^{3+}@${C_{82}}^{3-}$ type [101]. In addition, it was found that the charge on the encaged Y atom is little changed even when Y@C_{82} loses or gains an electron [101]. Since we expect some charge to be transferred from the Cu(111) substrate to the Y@C_{82} molecule, the C_{82} may have slightly more electrons than ${C_{82}}^{2.59-}$.

The special tendency for dimer formation in Y@C_{82} cannot simply be interpreted in terms of the enriched electron density of states on the carbon cage. For the case of C_{60} adsorption on the Cu(111) surface, charge transfer from the substrate to the fullerene molecule has been documented on the basis of low bias (10mV) observations by STM. However, the cluster size distribution is explained by van der Waals-type bonding among the C_{60} molecules. For the case of Gd@C_{82}, we expect charge transfer from the Gd atom to the carbon cage. However, it is reported to be ESR silent and so the electron on the C_{82} cage does not have a half-spin. One possible way of understanding this is to assume that the molecule can be pictured as Gd^{2+}@${C_{82}}^{2-}$, although the electronic structure is not yet known. We have observed the increase in condensation energy, similar to the Sc_2@C_{84} case, but the cluster size distribution shows that the bonding energy of Gd@C_{82} can also be described by the van der Waals-type interaction. We have previously observed the initial stage adsorption of $Sc_2@C_{84}$ on the Ag(111) surface. The system did not show any special tendency for dimer formation [31]. The $Sc_2@C_{84}$ molecule is reported to be ESR silent and the electronic structure has been reported as $(Sc^{3+})_2@{C_{84}}^{6-}$.

We thus anticipate that the half-spin on the carbon cage enhances dimer formation. In the $Y@C_{82}$ molecule, almost three electrons are donated to the C_{82} cage and it has a structure analogous to the Li atom, with a triply charged nucleus surrounded by a triply charged sphere, $Y^{3+}@{C_{82}}^{3-}$. The idea of a superatom has been discussed theoretically in relation to the endohedral metallofullerenes [119,120]. We claim that the dimer formation observed by STM clearly shows an example of the superatom. The increase in bonding energy for the case of the $Y@C_{82}$ dimer may be explained by the spin–spin interaction of the half-spin left on the C_{82} cage of the molecules. The observed intermolecular distance of $Y@C_{82}$ dimers (11.2 Å), 0.6 Å smaller than the nnd in the fcc bulk crystal (11.8 Å), can also be explained by the increased bonding energy in the $Y@C_{82}$ dimer. In this sense, we may consider the $Y@C_{82}$ dimer as a supermolecule, by analogy with the Li–Li molecule.

9.5 Conclusions

We have performed a systematic investigation of the adsorption and film growth of fullerenes C_{60}, C_{70}, $C_{60x}C_{70(1-x)}$, C_{84}, $Sc@C_{74}/Sc_2@C_{74}$, $Sc_2@C_{84}$, $Y@C_{82}$ and $Gd@C_{82}$ on the Si(100)-2×1, Si(111)-7×7, GaAs(001)-2×4-As, Cu(111)-1×1 and Ag(111)-1×1 surfaces.

Fullerene molecules form strong bonding with dangling bonds on the Si(100)-2×1 and Si(111)-7×7 surfaces and charge transfer from Si atoms to the fullerene molecules was observed. At the initial stage of adsorption, they adsorbed in a random and stable manner. At higher coverage, fullerene molecules form multi-layer islands with Stranski–Krastanov mode growth. Bulk-like fcc(111) single crystals are formed above the third layer. Because of the strong chemical interaction of C_{60} with the Si, intramolecular structures with twofold symmetry are observed. Reaction of fullerene molecules with the Si(100) substrate takes place only when the fullerene covered surface is heated above 1 000°C and formation of silicon-carbide islands is observed. STM observations of Sc-metallofullerenes ($Sc@C_{74}/Sc_2@C_{74}$, $Sc_2@C_{84}$) adsorbed on the Si(100)-2×1 surface show individual MF molecules imaged spherically, similar to the cases of C_{60}, C_{70} and C_{84}, and are consistent with the model whereby metal atoms are encapsulated within the carbon cage and charge is transferred to the sphere of the cage.

The first monolayer of C_{60} on the GaAs(001)-2×4-As β phase consists of dimer-like C_{60} pairs. Molecular dynamic simulations and phenomenological analysis concerning site-specific C_{60}–GaAs interactions and charge transfer from substrate dangling bonds to C_{60} adsorbates produce excellent agreement with experiment. A unique fcc(110)-oriented strained C_{60} multi-layer growth has been observed for the first time, while other systems of fullerene adsorption show the well known fcc(111)-oriented multi-layer growth.

At the initial stage of adsorption on the Cu(111) and Ag(111) surfaces, fullerene molecules are mobile on terraces and segregate to steps. After filling adsorption positions at the step, 2D islands form from the steps and then close-packed hexagonal phases form with monolayer coverage. Because of the strong interaction between the Cu surface and fullerene molecules, and because of the interaction among the orientationally ordered C_{60} molecules, the free rotation of C_{60} molecules ceases in the Cu(111)-4×4-C_{60} phase. Bias-dependent intramolecular structures of C_{60} molecules with threefold symmetry are observed and interpreted as 3D mappings of the local electron density of states corresponding to the bias voltage condition. Some of the C_{70} molecules in the Cu(111)-4×4-$C_{60x}C_{70(1-x)}$ ($x = 0.3$) phase also stop rotation and show intramolecular structures. The interaction between fullerene molecules and the Ag(111) substrate is relatively weak, in contrast to the Cu(111) surface, and close-packed fcc(111)-oriented monolayer films form with multi-domains. Rotation angles of the domains with respect to the substrate are determined by adsorbate–substrate interactions.

Preferential dimer formation of Y@C_{82} molecules is observed at the step edges of the Cu (111)-1×1 surface, in contrast to the case of C_{60} and Gd@C_{82}. This observation is interpreted in terms of the attractive interaction exerted between the unpaired electrons of two neighbouring Y@C_{82} molecules. Although its interaction energy is small (< 100 meV), the Y@C_{82} molecule on the terrace edge behaves essentially like a Li atom, which also possesses an unpaired electron and forms Li_2 molecules. The Y@C_{82} molecule is one of the first experimentally-documented cases of a so-called 'superatom', originally proposed for semiconductor heterostructure.

Acknowledgements

We would like to thank Dr. Xiang-Dong Wang, Dr. Kumi Motai, Dr. Yong Ling, Dr. Qi-Kun Xue, Dr. H. Kamiyama, Professor Akira Sakai, Professor Young Kuk, Professor Yukio Hasegawa, Professor Kaoru Ohno, Professor Yoshiyuki Kawazoe, Professor Hisanori Shinohara, Professor Yahachi Saito, Professor Yuichiro Nishina and Professor Howard W. Pickering for fruitful collaboration throughout this FI-STM investigation. We acknowledge financial support from the Department of Education of Japan (Nishina Project Contract: No. 07N0501).

References

1. H.R. Kroto, J.R. Heath, S.C. O'Brien, R.F. Curl and R.E. Smalley, Nature **318** (1985) 162; R.F. Curl and R.E. Smalley, Sci. American **265** (1991) 32.

2. W. Krätschmer, K. Fostiropoulos and D.R. Huffman, Chem. Phys. Lett. **170** (1990) 167.
3. W. Krätschmer, L.D. Lamb, K. Fostiropoulos and D.R. Huffman, Nature **347** (1990) 354.
4. A.F. Hebard, M.J. Rossinsky, R.C. Haddon, D.W. Murphy, S.H. Glarum, T.T.M. Palstra, A.P. Ramirez and A.R. Kortan, Nature **350** (1991) 600.
5. F. Diederich, R. Ettl, Y. Rubin, R.L. Whetten, R. Beck, M. Alvarez, S. Anz, D. Sensharma, F. Wudl, K.C. Khemani and A. Koch, Science **252** (1991) 548.
6. K. Kikuchi, N. Nakahara, T. Wakabayashi, M. Honda, H. Matsumiya, T. Moriwaki, S. Suzuki, H. Shiromaru, K. Saito, K. Yamauchi, I. Ikemoto and Y. Achiba, Chem. Phys. Lett. **188** (1992) 177.
7. F. Diederich, R.L. Wand and M.M. Alvarez, Sicence **254** (1991) 1768.
8. Kikuchi, N. Nakahara, T. Wakabayashi, S. Suzuki, H. Shiromaru, Y. Miyake, K. Saito, I. Ikemoto, M. Kainosho and Y. Achiba, Nature **357** (1992) 142.
9. H. Shinohara, H. Sato, Y. Saito, M. Takayama, A. Izuoka and T. Sugawara, J. Phys. Chem. **95** (1991) 8449.
10. J.R. Heath, S.C. O'Brien, Q. Zhang, Y. Liu, R.F. Curl, H.W. Kroto, F.K. Tittel and R.E. Smalley, J. Am. Chem. Soc. **107** (1985) 7779.
11. Y. Chai, T. Guo, C. Jin, R.E. Haufler, L.P.F. Chibante, J. Fure, L. Wang, J.M. Alford and R.S. Smalley, J. Phys. Chem. **95** (1991) 7564.
12. M.M. Alvarez, E.G. Gillan, K. Holczer, R.B. Kaner, K.S. Min and R.L. Whetten, J. Phys. Chem. **95** (1991) 10561.
13. R.D. Johnson, M.S. de Vries, J. Salem, D.S. Bethune and C.S. Yannoni, Nature **355** (1992) 239; M. Hoinkis, C.S. Yannoni, D.S. Bethune, J.R. Salem, R.D. Johnson, M.S. Crowder and M.S. de Vries, Chem. Phys. Lett., **198** (1992) 461.
14. S. Suzuki, S. Kawata, H. Shiromura, K. Yamaguchi, K. Kikuchi, T. Kato and Y. Achiba, J. Phys. Chem. **96** (1992) 7159.
15. K. Kikuchi, S. Suzuki, Y. Nakao, N. Nakamura, T. Wakabayashi, H. Shiromaru, I. Saito, I. Ikemoto and Y. Achiba, Chem. Phys. Lett. **216** (1993) 67.
16. J.H. Weaver, Y. Chai, G.H. Kroll, C. Jin., T.R. Ohno, R.E. Haufler, T. Guo, J.M. Alford, J. Conceicao, L.P.F. Chibante, A. Jain, G. Palmer and R.E. Smalley, Chem. Phys. Lett. **190** (1992) 460.
17. H. Shinohara, H. Sato, Y. Saito, M. Ohkohchi and Y. Saito, J. Phys. Chem. 96 (1992) 3571.
18. H. Shinohara, H. Sato, M. Ohkohchi, Y. Ando, T. Kodama, T. Shida, T. Kato and Y. Saito, Nature **357** (1992) 52.
19. C.S. Yannoni, M. Hoinkis, M.S. de Vries, D.S. Bethune, J.R. Salem and M.S. Crowder and R.D. Johnson, Science **256** (1992) 1191.
20. H. Shinohara, H. Yamaguchi, N. Hayashi, H. Sato, M. Inagaki, Y. Saito, S. Bandow, H. Kitagawa, T. Mitani and H. Inokuchi, Mater. Sci. Eng. **B19** (1993) 25.
21. T. Guo, M.D. Diener, Y. Chai, M.J. Alford, R.E. Hanfler, S.M. McClure, T. Ohno, J.H. Weaver and R.E. Smalley, Science **257** (1992) 1661.
22. T. Pradeep, G. Kulkarni, K. Kannan, T.G. Row and C. Rao, J. Am. Chem. Soc. **114** (1992) 2272.
23. D.S. Bethune, R.D. Johnson, J.R. Salem, M.S. de Vries and C.S. Yannoni, Nature **363** (1993) 605.

24. H. Shinohara, M. Kishida, T. Nakane, T. Kato, S. Bandow, Y. Saito, X.D. Wang, T. Hashizume and T. Sakurai, in Recent Advances in the Chemistry and Physics of Fullerenes and Related Materials, K.M. Kadish and R.S. Ruoff (Eds.) PV 94-24, The Electrochemical Society Proceedings Series, Pennington, NJ (1994) p.1361.
25. E. Gillan, C. Yeretzain, K.S. Min, M.M. Alvarez, R.L. Whetten and R.B. Kaner, J. Phys. Chem. **96** (1992) 6869.
26. L. Moro, R.S. Ruoff, C.H. Becker, D.C. Lorents and R. Malhotra, J. Phys. Chem. **97** (1993) 6801.
27. G. Binnig, H. Rohrer, C. Gerber and E. Weibel, Phys. Rev. Lett. **50** (1983) 120.
28. T. Hashizume, X.D. Wang, H. Shinohara, Y. Saito, Y. Nishina and T. Sakurai, Jpn. J. Appl. Phys. **31** (1992) L880.
29. X.D. Wang, T. Hashizume, H. Shinohara, Y. Saito, Y. Nishina and T. Sakurai, Phys. Rev. **B47** (1993) 15923.
30. T. Hashizume, X.D. Wang, Y. Nishina, H. Shinohara, Y. Saito and T. Sakurai, Jpn. J. Appl. Phys. Lett. **32** (1993) L132.
31. X.D. Wang, T. Hashizume, Q. Xue, H. Shinohara, Y. Saito, Y. Nishina and T. Sakurai, Jpn. J. Appl. Phys. **32** (1993) L866.
32. X.D. Wang, Q.K. Xue, T. Hashizume, H. Shinohara, Y. Nishina and T. Sakurai, Phys. Rev. **B48** (1993) 15492.
33. H. Shinohara, N. Hayashi, H. Sato, Y. Saito, X.D. Wang, T. Hashizume and T. Sakurai, J. Phys. Chem. **97** (1993) 13438.
34. X.D. Wang, T. Hashizume, Q. Xue, H. Shinohara, Y. Saito, Y. Nishina and T. Sakurai, Chem. Phys. Lett. **216** (1993) 409.
35. T. Sakurai, X.D. Wang, T. Hashizume, Y. Nishina, H. Shinohara and Y. Saito, Appl. Surf. Sci. **67** (1993) 281.
36. X.D. Wang, Q. Xue, T. Hashizume, H. Shinohara, Y. Nishina and T. Sakurai, Phys. Rev. **B49** (1994) 7754.
37. D. Chen and D. Sarid, Surf. Sci. **329** (1996) 206.
38. X. D. Wang, T. Hashizume, H. Shinohara, Y. Saito, Y. Nishina and T. Sakurai, Jpn. J. Appl. Phys. **31** (1992) L983.
39. Y.Z. Li, M. Chander, J.C. Patrin, J.H. Weaver, L.P.F. Chibante and R.E. Smalley, Phys. Rev. **B45** (1992) 13837.
40. H. Xu, D.M. Chen and W.N. Creager, Phys. Rev. Lett. **70** (1993) 1850.
41. D. Chen, J. Chen and D. Sarid, Phys. Rev. **B49** (1994) 7612.
42. D. Chen, M.J. Gallagher and D. Sarid, J. Vac. Sci. and Tech. **B12** (1994) 1947.
43. T. Sato, T. Sueyoshi and M. Iwatsuki, Surf. Sci. **321** (1994) L137.
44. D. Klyachko and D.M. Chen, Phys. Rev. Lett. **75** (1995) 3693.
45. H. Xu, D.M. Chen and W.N. Creager, Phys. Rev. **B50** (1994) 8454.
46. Y.Z. Li, J.C. Patrin, M. Chander, J.H. Weaver, L.P.F. Chibante and R.E. Smalley, Science, **252** (1991) 547.
47. Y.Z. Li, M. Chander, J.C. Patrin, J.H. Weaver, L.P.F. Chibante and R.E. Smalley, Science, **253** (1991) 429.
48. Y.Z. Li, J. C. Patrin, M. Chander, J.H. Weaver, L.P.F. Chibante and R.E. Smalley, Phys. Rev. **B46** (1992) 12914; Y.Z. Li, J.C. Patrin, M. Chander, J.H. Weaver, K. Kikuchi and Y. Achiba, Phys. Rev. **B47** (1993) 10867.
49. Q.K. Xue, T. Ogino, Y. Hasegawa, H. Shinohara and T. Sakurai, Phys. Rev. **B53** (1996) 1985.

50. Q.K. Xue, Y. Ling, T. Ogino, T. Sakata, Y. Hasegawa, T. Hashizume, H. Shinohara and T. Sakurai, Thin Solid Films **281–282** (1996) 618.
51. Q.K. Xue, T. Hashizume, Y. Hasegawa, H. Kamiyama, K. Ohno, Y. Kawagoe, H. Shinohara and T. Sakurai: submitted to Phys. Rev. Lett.
52. R.J. Wilson, G. Meijer, D.S. Bethune, R.D. Johnson, D.D. Chambliss, M.S. de Vries, H.E. Hunziker and H.R. Wendt, Nature **348** (1990) 621.
53. T. Chen, S. Howells, M. Gallagher, L. Yi, D. Sarid, D.L. Lichtenberger, K.W. Nebesny and C.D. Ray, J. Vac. Sci. Technol. **B9** (1991) 2461; ibid. **B10** (1992) 170.
54. L.D. Lamb, D.R. Huffman, R.K. Workman, S. Howells, T. Chen, D. Sarid and R.F. Ziolo, Science, **255** (1992) 1413.
55. Y. Zhang, X. Gao and M.J. Weaver, J. Phys. Chem. **96** (1992) 510.
56. S. Howells, T. Chen, M. Gallagher, D. Sarid, D.L. Lichtenberger, L.L. Wright, C.D. Ray, D.R. Huffman and L. D. Lamb, Surf. Sci. **274** (1992) 171.
57. E.I. Altman and R.J. Colton, Surf. Sci. **279** (1992) 49.
58. E.I. Altman and R.J. Colton, Phys. Rev. **B48** (1993) 18244.
59. H.P. Lang, V. Thommen-Geiser, C. Bolm, M. Felder, J. Frommer, R. Wiesendanger, H. Werner, R. Schlogl, A. Zahab, P. Bernier, G. Gerth, D. Anselmetti and H.-J. Gtherodt, Appl. Phys. **A56** (1993) 197
60. E.I. Altman and R.J. Colton, J. Vac. Sci. Tech. **B12** (1994) 1906.
61. J.K. Gimzewski, S. Modesti and R.R. Schlittler, Phys. Rev. Lett. **72** (1994) 1036.
62. Y. Kuk, D.K. Kim, Y.D. Suh, K.H. Park, H.P. Noh, S.J. Oh and S.K. Kim, Phys. Rev. Lett. **70** (1993) 1948.
63. K. Motai, T. Hashizume, H. Shinohara, Y. Saito, H.W. Pickering, Y. Nishina and T. Sakurai, Jpn. J. Appl. Phys. **32** (1993) L450.
64. T. Hashizume, K. Motai, X.D. Wang, H. Shinohara, Y. Saito, Y. Maruyama, K. Ohno, Y. Kawazoe, Y. Nishina, H.W. Pickering, Y. Kuk and T. Sakurai, Phys. Rev. Lett. **71** (1993) 2959.
65. X.-D. Wang, V. Yu. Yurov, T. Hashizume, H. Shinohara and T. Sakurai, Phys. Rev. **B49** (1994) 14746.
66. T. Sakurai, S. Yamazaki, Y. Hasegawa, Y. Ling, T. Hashizume and H. Shinohara, in Recent Advances in the Chemistry and Physics of Fullerenes and Related Materials, K.M. Kadish and R.S. Ruoff (Eds.) PV 95-10, The Electrochemical Society Proceedings Series, Pennington, NJ (1995) p.709.
67. H. Shinohara, M. Inakuma, M. Kishida, S. Yamazaki, T. Hashizume and T. Sakurai, J. Phys. Chem. **99** (1995) 13769.
68. M.T. Cuberes, R.R. Schlittler and J.K. Gimzewski, Appl. Phys. Lett. **69** (1996) 3016.
69. T. Hashizume and T. Sakurai, Surf. Rev. Lett. **3** (1996) 905.
70. Y. Hasegawa, Y. Ling, S. Yamazaki, T. Hashizume, H. Shinohara, A. Sakai, H.W. Pickering and T. Sakurai, submitted to Phys. Rev. Lett.
71. E.I. Altman and R.J. Colton, Surf. Sci. **295** (1993) 13.
72. X.D. Wang, S. Yamazaki, T. Hashizume, H. Shinohara and T. Sakurai, Phys. Rev. **B52** (1995) 2098.
73. L. Wragg, J.E. Chamberlain, H.W. White, W. Krüschmer, D.R. Huffman, Nature, **348** (1990) 623.
74. S. Saito and A. Oshiyama, Phys. Rev. Lett. **66** (1991) 2637.
75. D.J. Chadi, J. Vac. Sci. Technol. **A5** (1987) 834.
76. T. Sakurai, X.D. Wang, Q.K. Xue, Y. Hasegawa, T. Hashizume and H. Shinohara, Prog. Surf. Sci. **51** (1996) 263.

77. X.D. Wang, T. Hashizume and T. Sakurai, Mod. Phys. Lett. **B8** (1994) 1597.
78. T. Sakurai, T. Hashizume, I. Kamiya, Y. Hasegawa, N. Sano, H.W. Pickering and A. Sakai, Prog. Surf. Sci. **33** (1990) 3.
79. T. Hashizume, Y. Hasegawa, I. Kamiya, T. Ide, I. Sumita, S. Hyodo, T. Sakurai, H. Tochihara, M. Kubota and Y. Murata, J. Vac. Sci. Technol. **A8** (1990) 233.
80. T. Hashizume, Q.K. Xue, J. Zhou, A. Ichimiya and T. Sakurai, Phys. Rev. Lett. **73** (1994) 2208.
81. T. Hashizume, Q.K. Xue, A. Ichimiya and T. Sakurai, Phys. Rev. **B51** (1995) 4200.
82. Q.K. Xue, T. Hashizume, J.M. Zhou, T. Sakata, T. Ohno and T. Sakurai, Phys. Rev. Lett. **74** (1995) 3177.
83. H. Shinohara, H. Sato, Y. Saito, A. Izuoka, T. Sugawara, H. Ito, T. Sakurai and T. Matsuo, Rapid Commun. Mass Spectrom. **6** (1992) 413.
84. Y. Saito, K. Kurosawa, H. Shinohara, S. Saito, A. Oshiyama, Y. Ando and T. Noda, J. Phys. Soc. Jpn. **60** (1991) 2518.
85. Y. Saito, N. Suzuki, H. Shinohara and Y. Ando, Jpn. J. Appl. Phys. **30** (1991) 2857.
86. H. Shinohara, J. Mass Spectr. Soc. Jpn. **40** (1992) 203.
87. S. Bandow, H. Kitazawa, T. Mitani, H. Inokuchi, Y. Saito, H. Yamaguchi, N. Hayashi, H. Sato and H. Shinohara, J. Phys. Chem. **96** (1993) 9609.
88. H. Shinohara, H. Yamaguchi, N. Hayashi, H. Sato, M. Ohkohchi, Y. Ando and Y. Saito, J. Phys. Chem. **97** (1993) 4259.
89. H. Shinohara, M. Inokuma, N. Hayashi, H. Sato, Y. Saito, T. Kato and S. Bandow, J. Phys. Chem. **98** (1994) 8597.
90. J.A. Appelbaum and D.R. Hamann, Surf. Sci. **74** (1978) 21; R.J. Hamers, R.M. Tromp and J.E. Demuth, Phys. Rev. Lett. **56** (1986) 1972.
91. T. Hashizume, I. Sumita, Y. Murata, S. Hyodo and T. Sakurai, J. Vac. Sci. Technol. **B9** (1991) 742.
92. A.D. Novaco and J.P. McTague, Phys. Rev. Lett. **38** (1977) 1286; J.P. McTague and A.D. Novaco, Phys. Rev. **B19** (1979) 5299; H. Shiba, J. Phys. Soc. Jpn. **46** (1979) 1852.
93. Y. Kawazoe, H. Kamiyama, Y. Maruyama and K. Ohno, Jpn. J. Appl. Phys. **32** (1993) 1433.
94. O. Zhou, J.E. Fischer, N. Coustel, S. Kycia, Q. Zhu, A.R. McGhie, W.J. Romanow, J.P. McCauley, A.B. Smith and D.E. Cox, Nature, **351** (1991) 462.
95. A. Yajima and M. Tsukada, Surf. Sci. **357–358** (1966) 355.
96. A. Yajima and M. Tsukada, Surf. Sci. **366** (1966) L715.
97. A. Yajima and M. Tsukada, private communication.
98. X.Q. Wang, C.Z. Wang, B.L. Zhang and K.M. Ho, Phys. Rev. Lett. **69** (1992) 69.
99. K. Laasonen, W. Andreoni, M. Parrinello, Science **258** (1992) 1916.
100. S. Saito, S. Sawada, N. Hamada and A. Oshiyama, Jpn. J. Appl. Phys. **32** (1993) 1438.
101. S. Nagase and K. Kobayashi, Chem. Phys. Lett. **214** (1993) 57; S. Nagase and K. Kobayashi, Chem. Phys. Lett. **231** (1994) 319.
102. S. Saito, S. Sawada, N. Hamada and A. Oshiyama, Mater. Sci. Eng. **B19** (1993) 105.

103. A. Kasuya, S. Suto and Y. Nishina, to be published; S. Suto, A. Kasuya, O. Ikeno, Chang-Wu Hu, A. Wawro, R. Nishitani, T. Goto and Y. Nishina, Jpn. J. Appl. Phys. **33** (1994) L1489.
104. H. Kamiyama, Y. Maruyama, K. Ohno and Y. Kawazoe, Sci. Rep. RITU **A39** (1993) 7.
105. D.J. Chadi, J. Vac. Sci. Technol. **A5** (1987) 834.
106. H. Kamiyama, Y. Maruyama, K. Ohno and Y. Kawazoe, Sci. Rep. RITU (1997) in press.
107. Ch. Girard et al., Phys. Rev. **B49** (1994) 11425.
108. L.A. Girifalco, J. Phys. Chem. **96** (1992) 858.
109. Q.K. Xue et al., to be published.
110. G.K. Wertheim, J.E. Rowe, D.N.E. Buchanan, E.E. Chaban, A.F. Hebard, A.R. Kortan, A.V. Makhija, R.C. Haddon, Science **252** (1991) 1419.
111. J. Rowe, P. Rudolf, L.H. Tjeng, R.A. Malic, G. Meifs and C.T. Chen, Int. J. Mod. Phys. **B6** (1992) 3909.
112. Y. Kawazoe, Y. Maruyama, H. Rafii-Tabar, M. Ikeda, H. Kamiyama and K. Ohno, Mater. Sce. and Eng. **B19** (1993) 165.
113. Y. Maruyama, K. Ohno and Y. Kawazoe, Phys. Rev. **B52** (1995) 2070.
114. P.A. Heiney, J. Phys. Chem. Solids, **53** (1992) 1333 and references therein.
115. G.B.M. Vaughan, P.A. Heiney, J.E. Fischer, D.E. Luzzi, D.A. Ricketts-Foot, A.R. McGhie, Y.W. Hui, A.L. Smith, D.E. Cox, W.J. Romanow, B.H. Allen, N. Coustel, J.P. McGauley, Jr. and A.B. Smith III, Science **254** (1991) 1351.
116. T.J.S. Dennis, K. Prassides, E. Roduner, L. Cristofolini and R. DeRenzi, J. Phys. Chem. **97** 8553 (1993).
117. D. Walton, J. Chem. Phys. **37** (1962) 2182.
118. M. Takata, B. Umeda, E. Nishibori, M. Sakata, Y. Saito, M. Ohno and H. Shinohara, Nature **377** (1995) 46.
119. A. Rosen and B. Waestberg, Z. Phys. **D12** (1989) 387.
120. S. Saito, in Clusters and Cluster-Assembled Materials, R.S. Averback, J. Bernholc and D.L. Nelson (Eds.) MRS Proceedings Vol. 206, MRS, Pittsburg (1990) p.115.

Index

MIX
Papier aus verantwortungsvollen Quellen
Paper from responsible sources
FSC® C105338

If you have any concerns about our products,
you can contact us on
ProductSafety@springernature.com

In case Publisher is established outside the EU,
the EU authorized representative is:
Springer Nature Customer Service Center GmbH
Europaplatz 3, 69115 Heidelberg, Germany

Printed by Libri Plureos GmbH
in Hamburg, Germany